Building Services Engineering
Sixth edition

David V. Chadderton

Routledge
Taylor & Francis Group

LONDON AND NEW YORK

First edition published 1991
by Spon

This sixth edition published 2013
by Routledge
2 Park Square, Milton Park, Abingdon, Oxon, OX14 4RN

Simultaneously published in the USA and Canada
by Routledge
711 Third Avenue, New York, NY 10017

*Routledge is an imprint of the Taylor & Francis Group, an informa
business*

British Library Cataloguing in Publication Data
A catalogue record for this book is available from the British Library

Library of Congress Cataloging-in-Publication Data
Chadderton, David V. (David Vincent), 1944–
Building services engineering / David V. Chadderton. – 6th ed.
    p. cm.
Includes bibliographical references and index.
1. Buildings–Mechanical equipment. 2. Buildings–Environmental
engineering. I. Title.
TH6010.C4867 2012
696–dc23   2012004661

ISBN13: 978-0-415-69931-0 (hbk)
ISBN13: 978-0-415-69932-7 (pbk)
ISBN13: 978-0-203-12132-0 (ebk)

Typeset in Frutiger Light by
Cenveo Publisher Services, Bengaluru, India

Printed and bound in Great Britain by
TJ International Ltd, Padstow, Cornwall

Building Services Engineering

Engineering services within buildings account for ongoing energy use, greenhouse gas contribution and life safety provisions. This fully updated sixth edition of Chadderton's leading textbook is the perfect preparation for those intending to enter this increasingly important field.

Chapters addressing heating, climate change, air conditioning, transportation systems, water, gas, electricity, drainage and room acoustics cover all the key responsibilities of the building services engineer. As well as introductory material and the underpinning theory, practical guidance is provided in the form of sample calculations and spreadsheets.

New material includes:

- Trends and recent applications in lowering the energy use by mechanical and electrical services systems, heating, cooling and lighting of buildings.
- Case studies modelled from post-occupancy reports to provide realistic discussion topics.
- Examples of the use of photovoltaic solar panels, chilled beams, underfloor air distribution, labyrinths, ground-sourced heat pumps, district heating and cooling, energy performance certificates, energy auditing and wind turbines.
- Outlines of the concepts of global warming, carbon trading and zero carbon buildings.
- Exercises in each chapter and online self-study questions.

A significantly expanded companion website offers over 1,000 self-test questions, PowerPoint slides for lecturers, and an instructor's manual, enabling the rapid generation of lectures, assignments, and tests. *Building Services Engineering* is the ideal textbook for students of building services engineering, as well as a comprehensive guide for those about to start work in this field.

David V. Chadderton was formerly a Principal Lecturer in building services engineering at Southampton Solent University, a Senior Lecturer at Oxford Brookes University, and lectured in Melbourne and Ballarat. He has also worked as a self-employed energy auditor, and an energy performance contracting engineer in Melbourne before retiring recently.

In this new edition, innovative technologies such as renewable energy systems have been included which are particularly important to the understanding of energy reduction in mechanical and electrical services. Outlining the concept of zero carbon buildings is another timely addition.

Runming Yao, University of Reading

...a well-structured, all-inclusive introduction of the essential materials for BSE study. On top of the classic competencies required of building services engineers (i.e. air conditioning, electrical installations, fire protection, and water services) the increasingly important contexts in which building services engineers have a vital role to play (i.e. climate change, energy economics and post occupancy), are also covered.

Joseph H.K. Lai, Hong Kong Polytechnic University

Contents

19 Answers to questions 365

Figures

Tables

Preface to sixth edition

Building Services Engineering, sixth edition, is an updated and expanded version to include website learning resources. Reference is made to the UK Government's *HM Government Carbon Plan, 2011* (DECC, 2011a) as this is a foundational text for the subject of this book.

A new chapter on climate change explores how building services engineering interacts with the global intention to save Earth from the identified global warming trend (Chapter 1). Published emissions data is used to identify and track UK, country and global emissions. Worked examples and discussion assignments aid understanding of the means of reducing energy emissions and how emissions trading functions.

Another new chapter focuses on post occupancy analysis of energy use, the manner in which installed systems function, and whether the designer's intentions were realized (Chapter 2). A case study is made of the Queens Building, Leicester. The calculation of *Emission Performance Certificates* is shown and a workbook is provided. Building shape, sustainability and on-site generation are discussed.

Each chapter has additional multiple choice questions for self-assessment. Readers have access to self-test questions on the publisher's website (http://www.routledge.com/cw/chadderton). Questions may require the reader to look up answers in additional resources or use the internet with a search engine. There is only one correct answer to each multiple choice question unless it is specified as having more than one. Incorrect answers may be partially true but are not considered by the author to be the entirely correct response for the purpose of this book; these may stimulate additional study, discussion, questioning with peers or the instructor.

Instructors can download the *Building Services Engineering Instructors Manual* of over 1500 multiple choice questions covering every chapter. All the multiple choice questions in this book are also in the *Manual* so that Instructors can cut and paste test material easily. Questions in the *Manual* correspond to the book chapter subjects to facilitate easy selection of questions for class quizzes, on-line tests, assignments and examinations. The author is well aware of the constant requirement for instructors to generate teaching resources, assignments and tests, having done so for many years in his academic career.

Spreadsheet software files can be downloaded from the website: http://www.efnspon.com/spon/featuresl chadderton.html. Users usually have to adjust the screen display to optimize the viewed pages. It is expected that the reader can use the spreadsheet software that is available on their own computer, or is provided on a network system for their use. If this is not the case, introductory training in spreadsheet software use is necessary. The reader is recommended to consult Chapter 1, 'Computer and Spreadsheet Use', in *Building Services Engineering Spreadsheets* (Chadderton, 1997b) where sufficient introductory training in computer and spreadsheet use is provided.

Building Services Engineering, sixth edition, is intended to be a broad introduction to the range of subjects involved in services engineering, with the aim of stimulating discussion of current topics and further study. The engineering content and calculation methods are sufficiently rigorous to match most of what is done in the industry in the design of many building services applications. The subjects covered and the depth with which they are analysed and calculated are more than sufficient to meet the syllabus requirements of higher technician, undergraduate and some postgraduate courses in building services engineering, heating, ventilating and air conditioning, energy management, architecture, building and quantity surveying, housing management, estate management and property facility management. Those preparing for clerk of works examinations will also find the book useful. The advanced user will need to progress to specialized textbooks and the standard references.

The reader is challenged to become actively engaged in the design calculations carried out by design engineers, through step-by-step introduction of each stage. A standard of numerical competence is expected that some lecturers may consider to be higher than is necessary for some courses. This was deemed appropriate in order to broaden the potential readership and provide an adequate basis for a deeper design study.

Readers are encouraged to make use of the internet as a learning resource. Graphics included in a printed book are there to explain a basic principle. Use a search engine to view real plant items such as steel panel heating radiators, pumps, air handling units, fire extinguishers or lifts, as needed. Happy surfing!

Acknowledgements

I am particularly grateful to the publishers for their investment in much of my life's work. Such a production only becomes possible through the efforts of a team of highly professional people. For me, working with Taylor & Francis has always been an enthusiastic, harmonious and efficient working relationship. All those involved are sincerely thanked for their efforts and the result.

My wife Maureen is thanked for her encouragement and understanding while I have been engrossed in keyboard work, on the drawing board and shuffling through piles of proofs. I would specifically like to thank those who have refereed this work. Their efforts to ensure that the book has comprehensive coverage, introductory work, adequate depth of study, valid examples of design, good structured worked examples and exercises are all appreciated. Users and recommenders of the book are all thanked for their support, without them, it would not exist.

Units and constants

Système International units are used in this book and Table 0.1 gives the basic and derived units employed, their symbols and some common equalities. Table 0.2 presents the multiples and submultiples and Table 0.3 presents the conventional symbols used in this field.

Table 0.1 Units and constants

Quantity	Unit	Symbol	Equality
Mass	kilogram	kg	
	tonne	tonne	1 tonne $= 10^3$ kg
Length	metre	m	
Time	second	s	
	hour	h	1 h $= 3600$ s
Energy, work, heat	joule	J	1 J $= 1$ Nm
Force	newton	N	1 N $= 1$ kg m/s^2
Power, heat flow	watt	W	1 W $= 1$ J/s
			1 W $= 1$ Nm/s
			1 W $= 1$ VA
Pressure	pascal	Pa	1 Pa $= 1$ N/m^2
	newton/m^2	N/m^2	1 b $= 10^5$ N/m^2
	bar	b	l b $= 10^3$ mb
Frequency	hertz	Hz	1 Hz $= 1$ cycle/s
Electrical resistance	ohm	R, Ω	
Electrical potential	volt	V	
Electrical current	ampere	I, A	$I = V/R$
Absolute temperature	kelvin	K	$K = (^\circ C + 273)$
Temperature	degree Celsius	$^\circ$C	
Luminous flux	lumen	lm	
Illuminance	lux	lx	1 lx $= 1$ lm/m^2
Area	square metre	m^2	
Volume	litre	l	
	cubic metre	m^3	1 m^3 $= 10^3$ l

Table 0.2 Multiples and submultiples.

Quantity	Name	Symbol
10^{12}	tera	T
10^9	giga	G
10^6	mega	M
10^3	kilo	k
10^{-3}	milli	m
10^{-6}	micro	μ

Table 0.3 Physical constants.

gravitational acceleration	g	9.807 m/s^2
specific heat capacity of air	SHC	1.012 kJ/kg K
specific heat capacity of water	SHC	4.186 kJ/kg K
Stefan-Boltzmann constant	σ	5.67×10^{-8} W/m^2 K^4
density of air at 20°C, 1013.25 mb	ρ	1.205 kg/m^3
density of water at 4°C	ρ	10^3 kg/m^3
exponential	e	2.718

Symbols

Symbol	Description	Units
A	area	m^2
	electrical current	A
A_f	physical constant	dB
α (alpha)	electrical temperature coefficient of resistance	$\Omega / \Omega\,°C$
	percentage depreciation and interest charge	%
	absorption coefficient	dimensionless
$\bar{\alpha}$	mean absorption coefficient	dimensionless
B	sound reduction index	dB
B_f	physical constant	dB
b	barometric pressure	bar, b
CO_2	carbon dioxide	%, ppm
DI	directivity index	dB
DU	demand or discharge unit	
d	pipe diameter	m or mm
	distance	m
Δp	difference of pressure	N/m^2
d.b.	dry-bulb air temperature	°C d.b.
EL	equivalent length	m
e	exponential	
η (eta)	efficiency	%
f	frequency	Hz
G	moisture mass flow rate	kg/m^2
GCV	gross calorific value	MJ/kg
GJ	energy	gigajoule
g	gravitational acceleration	m/s^2
	air moisture content	kg water/kg dry air
H	height	m
h	time	hour
Hz	frequency	cycle/s
I	electrical current	ampere
J	energy	joule
K	absolute temperature	kelvin
kg	mass	kilogram
kJ	energy	kilojoule

Symbol	Description	Units
kW	power	kilowatt
kWh	energy	kilowatt-hour
LDL	lighting design lumens	lumen
LH	latent heat of evaporation	kW
l	length	m
λ (lambda)	thermal conductivity	W/m K
LPG	liquefied petroleum gas	
m	length	millimetre
MF	maintenance factor	
MJ	energy	megajoules
mm	length	millimetre
MW	power	megawatt
μ (mu)	diffusion resistance factor	
N	air change rate	h^{-1}
	force	Newton
	number of occupants	
NR	noise rating	dimensionless
olf	concentration of odorous pollutants	
Ω (omega)	electrical resistance	ohm
P	pressure	Pascal
	permeance	kg/N s
P_a	pressure	Pascal
P_1, P_2	area fraction	
p_s	vapour pressure	Pascal
ϕ (phi)	angle	degree
Q	fluid flow rate	m^3/s or 1/s
	geometric directivity factor	dimensionless
Q_e	extract air flow rate	m^3/s
Q_{ex}	exhaust air flow rate	m^3/s
Q_f	fresh air flow rate	m^3/s
Q_L	leakage air flow rate	m^3/s
Q_r	recirculation air flow rate	m^3/s
q	water flow rate	kg/s
R	resistance, electrical	Ω
	resistance thermal	m^2 K/W
	room sound absorption constant	m^2
R_a	air space thermal resistance	m^2 K/W
R_{si}	internal surface thermal resistance	m^2 K/W
R_{se}	outside surface thermal resistance	m^2 K/W
R_v	vapour resistance	N s/kg
r	distance	m
r_v	vapour resistivity	GN s/kg m
		MN s/g m
ρ (rho)	density	kg/m^3
	specific electrical resistance	Ωm
	soil electrical resistivity	Ωm
S	spacing	m
	surface area	m^2
s	time	second
SE	specific enthalpy	kJ/kg
SH	sensible heat transfer	kW

(continued)

Continued

Symbol	Description	Units
SPL	sound pressure level	dB
SWL	sound power level	dB
SRI	sound reduction index	dB
SHC	specific heat capacity	kJ/kg K
Σ (sigma)	summation	
T	absolute temperature	Kelvin
	reverberation time	s
t_a	air temperature	°C
t_{ai}	inside air temperature	°C
t_{ao}	outside air temperature	°C
t_b	base temperature	°C
t_c	operative temperature	°C
t_{dp}	dew-point temperature	°C
t_e	environmental temperature	°C
t_{ei}	internal environmental temperature	°C
t_{eo}	outside environmental temperature	°C
t_f	water flow temperature	°C
t_g	globe temperature	°C
t_{HWS}	hot water storage temperature	°C
t_m	mean water temperature	°C
	area-weighted average room surface temperature	°C
t_{max}	maximum air temperature	°C
t_{min}	minimum air temperature	°C
t_r	mean radiant temperature	°C
	return water temperature	°C
t_{res}	resultant temperature	°C
t_s	supply air temperature	°C
	surface temperature	°C
θ (theta)	angle	degree
	time	h
U	thermal transmittance	W/m² K
U_n	new thermal transmittance	W/m² K
U_w	wall thermal transmittance	W/m² K
UF	utilization factor	
V	volume	m³
V	electrical potential	volt
v	velocity	m/s
v_s	specific volume	m³/kg
W	width	m or mm
W	power	watt
w.b.	wet-bulb air temperature	°C w.b.
Y	admittance factor	
	annual degree days	

1 Climate change

Learning objectives

Study of this chapter will enable the reader to:

1. explain weather and climate;
2. know the meaning of the greenhouse effect;
3. know basic facts about CO_2;
4. use DECC emission data;
5. produce emission data trend graphs;
6. follow emission trends of any country;
7. observe emission trends for the whole world with IEA data;
8. observe what local climate means;
9. realize what emission targets mean to different parts of the world's population;
10. know what the *HM Government Carbon Plan, 2011* (DECC, 2011a) is;
11. follow the *HM Government Carbon Plan, 2011* data trend into the future;
12. know how CO_2 emissions can be reduced;
13. identify alternatives to fossil fuel emissions;
14. use IEA country and world emission data to plot trends;
15. comment on global progress following the Kyoto Protocol agreement;
16. know how carbon capture and storage (CCS) functions;
17. know the various meanings for zero carbon buildings;
18. know about the *Passivhaus* principle;
19. know what is meant by green buildings;
20. know about BREEAM, LEED and NABERS;
21. know what star ratings for buildings signify;
22. use embodied energy data;
23. understand regulated demands;
24. calculate unregulated demands and compare with regulated ones;
25. know how the carbon tax works;
26. know how the EU ETS functions;
27. calculate the effect of the EU ETS on emission reduction projects.

Key terms and concepts

atmospheric CO_2 2; BREEAM 14; cap and trade system 19; carbon cap 19; carbon capture 4; carbon capture and storage 22; carbon emission allowances 20; carbon emission analysis tool 8; carbon plan 4; carbon policies 23; carbon tax 18; climate 2; climate change 2; CO_2 emission 3; conclusions 25; country emissions 7; DECC 5; DUKES 15; embodied energy 16; emission trends 7; emissions reduction 6; emissions trading 4; ETS 19; EU 19; global injustice 4; global warming 4; green buildings 14; green stars 15; greenhouse effect 3; International Energy Agency 9; LEED 15; lignite 19; NABERS 15; *Passivhaus* 14; petroleum consumption 9; regulated demands 17; regulation 19; target emissions 6; unregulated demands 17; weather 2; world emissions 7; zero carbon buildings 12.

Introduction

Climate change is presented in a discursive manner in order to promote our understanding. The carbon plan, the global emissions trend, zero carbon buildings, embodied energy, green buildings, unregulated demands, emission trading system and carbon capture are explained with worked examples and questions. The construction industry's response to scientific data and government legislation in creating solutions to reducing energy consumption in buildings is nothing new; designers have always worked within such criteria.

Weather

Weather is what is currently happening to the air and ground temperatures, atmospheric pressure, moisture level, cloud cover, solar radiation, precipitation, wind direction and strength at a location, affecting what we do there. As the Earth rotates, the air mass of the atmosphere is subject to aerodynamic shearing forces between it and the uneven ground, causing wind and weather movement. Some regularity takes place in such a chaotic system of fluid flow with wind patterns, such as the Gulf Stream of warm water northwards that moderates the climate in the UK.

Climate

What is meant by climate? It is the average of certain variables over a long period of time, tens, hundreds and thousands of years, depending upon when data was recorded. These variables are air temperature near to the ground, rain, hail, sleet and snow, wind strength and direction, cloud cover and solar radiation. Time of day or night creates greatly different climates; if we measured all of the variables an hour prior to dawn around the world, what sort of climate do you think would be found? One that was a lot colder than during a normal day in the same location. Ground cover affects the immediate climate for a location. Is there continuous snow cover; is it a sea, an arid desert, a lake, river or wetland such as mud flats, a sand estuary or marsh? Has the ground been built over with areas of tarmac, concrete, industry and buildings? Is the location a natural forest, barren heath, bare rock, sand dunes, or snow-capped mountains, or farmed? The area where climate is determined may be a small location such as village, a large area such as a state, a country or the whole of the planet Earth. Climate varies within a country or state depending upon distance from the coast, significant areas of water and altitude above sea level.

   There have been many changes in the Earth climate due to natural causes throughout history. The heat-trapping effect of CO_2, water vapour, methane and other gases in the atmosphere was understood by scientists 150 years ago. These greenhouse effect gases occur naturally due to evaporation, rainfall, rotting vegetation, animals, volcanoes and natural fires. Emissions from human activity have significantly increased since the early 1900s. The UK Met Office informs us that there is a discernible warming trend across the globe due to human activities. CO_2 emitted now will last for 100 years in the atmosphere, while methane, CH_4, breaks down in 10 years into CO_2 and H_2O. The Earth has natural sources and sinks for CO_2 but the increases of the industrial era cannot be explained by natural causes (http://www.metoffice.gov.uk/).

EXAMPLE 1.1

What is meant by the greenhouse effect of the atmosphere?

High temperature, short wave, high frequency radiation from the sun warms the Earth. Re-radiation from the cool surfaces of the Earth is at low frequency, long wave, and some of these are reflected in the particles of moisture and other gases within the atmosphere, thus trapping heat.

Climate change

The UK Department of Energy and Climate Change (DECC) states: 'Climate change refers to an identifiable change in the climate that persists for an extended period, typically decades or longer, and is often taken to mean man-made changes that have occurred since the onset of the industrial revolution.' Is the world's climate changing? Of course it is. It changes all the time. Wasn't it the Ice Age that formed the geography of northern countries? Human activity has added to these natural changes.

EXAMPLE 1.2

Write a description of the climate where you live, with the minimum of data, noting the principal geographic features of the locality.

I lived in Milford on Sea, Hampshire, England, 3 km inland from the south coast, in a quiet, tree-lined, residential road bordering a large market garden farm. Manor Road bordered the northern extremity of the suburbs and was near to the B3058 Lymington Road; look at it in satellite mapping software. There were no large industries in the area, just small retail shops and farms. All the roads were lined with deciduous trees. The land was slightly undulating but within around 25 m altitude. Ice and snow lasted for only a few days each winter, less than a week, while black ice on the roads happened occasionally after rain.

EXAMPLE 1.3

In your experience, have you noticed any change in the climate in the country where you live?

I grew up through junior school in north London and remember the sooty fog, i.e. the smog that occurred in most winters up to 1955. There was snow and ice every winter, summers were not noticeably hot and it rained a lot. After 1955, I lived mainly in the south of Hampshire where the sea air kept the atmosphere fresh and mild. The Clean Air Act of 1956 stopped the London smog but November usually brought wet fog to the south of England, making any travel hazardous. The normal tidal movements of sea level at Milford on Sea never seemed to vary over 20 years of my living around the south coast. I was not aware of any significant variation in the mild coastal climate during my time there.

How do we view climate change? How we understand the scientific data of climate change, with its dire warnings and calls for precipitate action, depends on the way in which we wish to discuss the topic. Professor Mike Hulme, of the University of East Anglia, in the Hartwell Lecture,(CIBSE Annual Lecture, 23 November 2010) identifies differing views ranging from: the threat is serious; it is depressing, I will forget about it; the debate is decided and closed; we will all call it climatic genocide; we would prefer a warmer climate; climate variability is always wide; observed increases in global average temperature are very likely to be man-made. Climate change as a topic to study grew from being insignificant up to 1990 to 8000 research papers during 2010.

Emissions trading in greenhouse gases commenced in 2006, making the price of a tonne of CO_2 look like any other stock market share, with a similar pattern of peak and trough prices between €10–40 per tonne. The term 'low carbon' has appeared thousands of times in major publications since 1985, books on global warming are numerous and the daily press publishes hundreds of articles whenever a major conference takes place. Fossil fuel CO_2 emissions are due to double from their global 1980 level of 5.0 GtC/yr and continue rising unabated. Global warming from atmospheric CO_2 has become the fashionable topic on a world scale. Whether nations will ever agree what to do about it is more difficult to believe and may take generations to resolve. Professor Hulme presents climate change due to mankind as six different frames of reference:

1. *failure of the commercial markets*: to do the best thing for humanity;
2. *technological hazard*: we simply did not understand enough about what we were doing in burning fossil fuels;
3. *global injustice*: rich countries raise their standards of living by using most of the world's fossil resources while many countries live in poverty;
4. *overconsumption*: the world's population is heading to grow from 3.0 billion in 1960 towards 10.0 billion in 2050 at the present rate of births and deaths;
5. *natural cyclic phenomenon*: planet Earth has undergone huge climatic changes throughout history over countless years;
6. *planetary tipping point*: 'We only have 10–15 years to avoid crossing catastrophic tipping points' (Tony Blair, UK Prime Minister, 1997–2007, October 2006).

The Carbon Plan

The UK government expressed the severest concern for the future.

This Carbon Plan sets out a vision of a changed Britain, powered by cleaner energy used more efficiently in our homes and businesses, with more secure energy supplies and more stable energy prices, and benefiting from the jobs and growth that a low carbon economy will bring. Without action to curb emissions, there is a very high risk of global warming reaching well beyond 2°C relative to pre-industrial times. Such unmitigated global warming would increase the risk of accelerated or irreversible changes in the climate system. Cutting emissions by at least 34% by 2020 and 80% by 2050 below the 1990 baseline. Demand for electricity is likely to double by 2050 compared to today. Changes in other sectors (of the economy) are likely to mean greater reliance on electricity for applications such as transport and heating, pushing up demand. At the same time, our existing power plants are coming to the end of their lives. Almost half of the UK's greenhouse gas emissions are from the energy used to generate heat, with the vast majority of our homes still relying on fossil fuel-powered gas boilers and with much of our building stock still poorly insulated and inefficient.

(DECC, 2011a)

Any carbon plan is a work in progress, is related to what other countries are doing, whether they have succeeded or not, so the construction industry watches for new rapid pace developments. Bear in mind that the carbon plan measures CO_2 emissions for the nation, so whatever the construction industry might achieve, it is only a part of the overall outcome. It is the emission of CO_2 that is counted and not which energy source is used or how much fossil fuel is consumed. Carbon capture and storage from flues might prove to be widely beneficial.

Growth in the number of UK households from 1970 to 2009 increased energy consumption by 18%, with an increasing population and internal air temperatures for comfort. Electrical appliance growth and the start of the consumer electronics boom in the 1970s caused the consumption of electricity to grow by 559% by 2009. This will likely continue as home-based working and entertainment develop. Where land and floor space allow, new homes have a large screen home theatre, further reducing the need for travel. Thermal efficiency improvements to the housing stock still have a long way to go as the current average SAP rating is 51.4 out of a possible 100 for a self-sufficient house. Rather than accept the benefits of using less energy, home thermal efficiency improvements may be made by a raised indoor air temperature (*Digest of UK Energy Statistics*, 2011).

Carbon dioxide basics

What is so bad about having CO_2 in the air we breathe? CO_2 is a colourless, invisible, non-flammable, odourless gas, around 360 ppm 0.036% in air but rises, denser than air, unhealthy at concentrations above 5,000 ppm, which is 0.5% in air. It is soluble in water where it can form carbonic acid and is used as a dry ice solid coolant. It is used as a modern refrigerant in some vapour compression systems. Dry ice sublimates directly into the gas phase, leaving no liquid residue, making it suitable for pressure cleaning. Solid dry ice pressure cleaning has many applications in industry, such as production equipment in the food industry, gas turbine engines, and printing presses and on moulds for tyre production. Such industrial uses of CO_2 may come from food and drink fermentation processes and may not add to overall emissions.

Humans exhale 1 kg of CO_2 daily, depending upon activity level; plants photosynthesize CO_2 from the air and emit O_2. CO_2 continuously exchanges between the oceans and the biosphere but large emissions reside in the atmosphere for hundreds of years. 1 kWh of electricity from fossil fuels causes an emission of around 1 kg CO_2. Water vapour is a very strong greenhouse gas and we cannot live without it (Carbon Dioxide Information Analysis Center, Oak Ridge National Laboratory, USA).

Is CO_2 bad for humans? We breathe it and eat plants grown with CO_2 taken from the air. It is not poisonous in low concentrations and is generally harmless and inert to many practical applications. Some call it a pollutant, maybe to justify putting a price on it. Anything can be dangerous to humans if we are exposed to too much of it, such as water, O_2 or bread.

DECC data

The UK Department of Energy and Climate Change (DECC) annual energy use data is available to download. Access the spreadsheet files, save them with a new file name of choice and then make use of the data for analysis and assignments. The Energy Consumption in the UK Domestic Data Tables relate entirely to energy used in UK housing.

> ### EXAMPLE 1.4
>
> Does it seem possible for the UK to achieve the stated emission reduction? Explain briefly what such a trend means.

Download the current UK spreadsheet data file 'Carbon dioxide emissions by fuel, 1990–present' and save the file. Plot the data and compare with the stated target reduction. Figure 1.1 shows the results up to 2010, your data will be current.

The 1990 baseline for total carbon dioxide emission is 589.7 Mt carbon dioxide emissions from the UK.

$$Target\ reduction\ in\ 2020 = 34\%\ of\ 1990\ emissions$$

$$Target\ for\ 2020 = (100-34)\,\%\times 589.7\ Mt\ CO_2$$

$$= 66\%\times 589.7\ Mt\ CO_2$$

$$= 389.2\ Mt\ CO_2$$

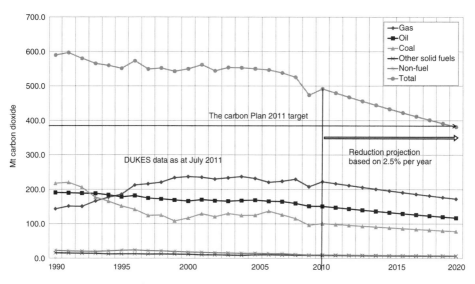

1.1 Target UK CO_2 emissions for 2020.

Emission in 2010 = 491.7 Mt CO₂

Steady rate of reduction needed = 2.5% p.a.

Extend the downloaded data list to 2020; multiply the last measured year emission by 0.975, extend that calculation across each cell until 2020.

Calculated emission in 2020 = 382 Mt CO₂

Figure 1.1 shows what your data and chart should look like, except that yours will have more published data. This indicates that the downward trend in all emissions may be sustainable if future usage remains consistent with recent past results. This means that despite rising standards of living, more homes, more commercial and industrial buildings being constructed, greater use of electricity for car and rail transport, developments such as greater use of nuclear and renewable energy power generation, efficiency improvements in all types of building and cleaner emissions from fossil fuel plants, the UK can emit less.

Now that you are experts in handling spreadsheet charts, copy your data table and chart onto additional worksheets and extend them to 2050 when the target emission is 80% below the 1990 total, i.e. 117.9 Mt CO_2. Experiment with the annual rate of reduction and you will find that it needs to be 4% of the previous year emission, each year, to achieve 112.0 Mt CO_2 in year 2050. Bear in mind that this is only 38 years ahead of the publication date of this book. If you were born in the baseline year 1990, expect to be alive and well, anticipating retirement to coastal or rural living while using one-fifth of the fossil fuel resources that you used when you were born, and are 60 years young in 2050, what sort of energy profile will you have in the UK? You will not be using natural gas, oil or any solid fuel to heat your house or workplace; there will be hardly any chimneys emitting flue gas. Carbon capture and storage have stopped the emission of large-scale flue gas into the atmosphere. You drive an electrically powered vehicle; aviation has found an alternative fuel to oil by using H_2; there is complete electrification of railways. H_2 has replaced diesel for trucks; renewable wind, solar, hydro and tidal energy generation is everywhere and nuclear power stations are the primary source for electricity as H_2 is produced from sea water.

EXAMPLE 1.5

Find the latest data for the world CO_2 emission from fuel combustion and the Kyoto Protocol targets from the International Energy Agency (IEA) site and comment on how progress is being made in the UK and globally. Repeat this assignment as new data is published.

Download the publication *CO₂ Emission from Fuel Combustion 2010 Highlights*, in portable document format and spreadsheets, the most recent edition, from the IEA site. Data can also be linked to smart phones. Plot the CO_2 emissions from the world, the UK, North America, OECD Europe, the Middle East, Asia, and China from 1971 to the present. Have a look at other countries to see what they are doing. Figures 1.2 and 1.3 show recent trends.

Published data is shown up to 2008; 2009–2020 is dummy data. The world and most countries are increasing their emissions. UK emissions are reducing. Smaller countries are also increasing emissions at high rates. If the UK closed down and ceased all emissions, would it make

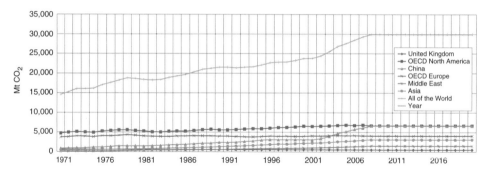

1.2 World CO$_2$ emissions.

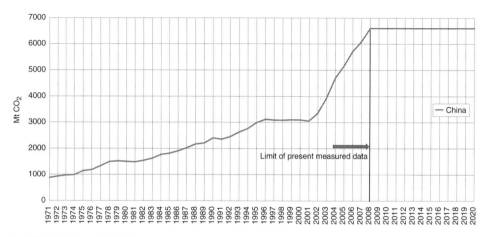

1.3 Country CO$_2$ emissions.

much difference? Hardly, as the increase in one year of China's emissions would replace that of the UK in total. If governments are serious about atmospheric CO$_2$ increases, every country in the world has to be involved.

'We need to ensure that the homes and buildings being built now and in the future are as energy efficient as possible' (DECC, 2011a). That is the reality check, meaning, we cannot achieve the impossible. We, the UK government, can only expect whatever our architects and engineers are able to do. That is you and me. It also means, what is possible to achieve with the money that is available for the job. There is no source of unlimited finance to reduce the nation's use of fossil fuels. Such an investment has to be on a reasonable payback time or the finance industry, the Bank of England, the Government Treasury, the International Monetary Fund, house-owners, landlords and building owners will quickly decide that too much money is needed to create more savings of primary energy and pose unjustifiable costs upon the nation. Investment in buildings has to stop somewhere between no thermal insulation, single glazing, openable windows with natural ventilation, and a building requiring no energy at all to run it; perhaps a technical impossibility for commercial buildings and most homes. No one suggests razing all existing houses and rebuilding with zero carbon new buildings, if that were possible, so the best we can ever do is improve what we have.

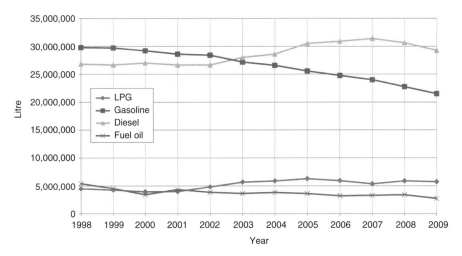

1.4 UK petroleum consumption trend.

For the purposes of this discussion, we will not include fossil fuels needed for transportation, agriculture and manufacturing as these are outside our remit. It was interesting to note that UK annual demand during 1998–2009 for gasoline (petrol) has declined steadily each year, as shown in Figure 1.4.

The Carbon Emission Analysis Tool (CEAT)

We are able to download energy usage and CO_2 emission data freely from online national sources, enabling a watch to be made on worldwide progress. UK data from the Department of Energy and Climate Change website (http://www.decc.gov.uk) is maintained quarterly while the International Energy Agency publishes annual world data. Some familiarity in using a spreadsheet application is needed for this assignment. The following Carbon Emission Analysis Tool (CEAT) provides a means of observing and commenting on what is happening around us, auditing the information and thinking about how the future could evolve:

1. Open the *International Energy Agency* website (http://www.iea.org).
2. Select *Home* and then the *Statistics & Balances* page.
3. Find the *Publications/Surveys Free for Download* area.
4. Click on the *CO_2 Emissions from Fuel Combustion 2010 – Highlights* link; the year available will be the most recent.
5. Download the PDF file if you wish to read the whole document.
6. Click on the *separate EXCEL file* link to download the current data file of CO_2 emissions.
7. Save the file to be downloaded in a directory of your choice on your computer with the file name *IEA original CO_2 data*.
8. Open your spreadsheet software and open the downloaded file, save it again, it is not important if macros in the original file are not available for use.
9. Click on the *CO_2 SA* tab, this *CO_2 Sectoral Approach* is the data we wish to use for analysis.
10. Have a look at the tabulated data; observe what countries are doing during the period of 1971 to the present; look for trends, countries that are significantly increasing or reducing their emissions.

11. Decide on a country of interest for data analysis, for example, China; this includes data from the People's Republic of China and Hong Kong, China; this data line really does change over time and is a major driver of industry to the world so it must be of significant interest.
12. Select the entire row of data from the title box of China all along to the end of the data row up to the latest data, 1971–2008 in the case of the 2010 file.
13. Copy this row of data and paste it onto a new blank workbook sheet.
14. Rename that tab as *world data*.
15. Select the row of year dates, 1971–2008, copy and paste the row onto your new worksheet above the CO_2 data row; type in *Year* as the row title.
16. Extend the row of year dates up to 2020 so that when new data is published, all you have to do is type in the value to keep the file up to date.
17. Save your new workbook in a directory of your choice as filename *World CO_2 Charts*.
18. Select the row of CO_2 data from the country title box to the latest year box of your own workbook; click *Copy*; this becomes the chart data.
19. Click *insert, line chart, chart layouts* that have axis titles and data labels; a chart appears on the same sheet as the data; cut and paste the chart onto the next blank sheet of the workbook; rename that tab as *China* or whatever country you chose.
20. Select the horizontal axis data on the chart, edit the axis to use the title *Year* and the year data row.
21. Format the chart to the size, labels, font and line types that you prefer.
22. Repeat the data collection and chart data sheets in your new workbook for as many countries as needed; only one chart per worksheet.
23. Save the file every time that a new sheet is added and formatted.
24. Are the country's emissions increasing or decreasing?
25. Did a significant change occur in the emissions at any particular time? Why was that? Has the trend since then remained consistent?
26. What was the trend of emissions since the baseline year of 1990?
27. Calculate the average annual rate of change in emissions for the country from 1971 to the most recent year.
28. Can you find any reason for that change, or rate of change, in emissions since 1971?
29. Does it seem that this country is working to reduce its CO_2 emissions from 1990 by 34% less by 2020?
30. Calculate the target emission for 2020 by reducing the 1990 value by 34%.
31. Does it seem that this country is working to reduce its CO_2 emissions from 1990 by 80% less by 2050?
32. Calculate the target emission for 2050 by reducing the 1990 value by 80%.
33. Do you imagine any possibility for this country achieving the 34% reduction from 1990 emissions by 2020?
34. Do you imagine any possibility for this country achieving the 80% reduction from 1990 emissions by 2050?
35. Compare this country with a nearby neighbour that has a similar geography and/or society by adding the new country's emission to the chart. What do you observe from the chart in relation to the earlier parts of this Carbon Emission Analysis Tool?
36. Apply this CEAT to countries that are in an OECD group such as Europe, North America and Asia. The Middle East is particularly interesting. Are the countries in that group changing their emissions in a similar manner? If not, which are going against the group trend? Are there reasons for this trend that you can observe?
37. Create a data chart for the *World* data.

38. Does it appear that the world is unified in reducing CO_2 emissions from this data and your chart?
39. Is there a particular country or OECD group that is continuing to increase its CO_2 emissions?
40. Is there a particular country or OECD group that is clearly reducing its CO_2 emissions?
41. Are you pessimistic or optimistic that the world will significantly reduce its CO_2 emissions by 2020, 2050 and beyond from the evidence of the IEA data published up to now? What is your analysis for the future?
42. What have you learned from government statements and policies, world news, technical innovations, investment decisions or published commitments in recent times that indicates that CO_2 emissions can be reduced and climate change reversed?
43. What are your conclusions about how climate change is being counteracted?

EXAMPLE 1.6

Apply the Carbon Emissions Analysis Tool to the International Energy Agency World emissions data. Figure 1.5 shows recent IEA World CO_2 emissions.

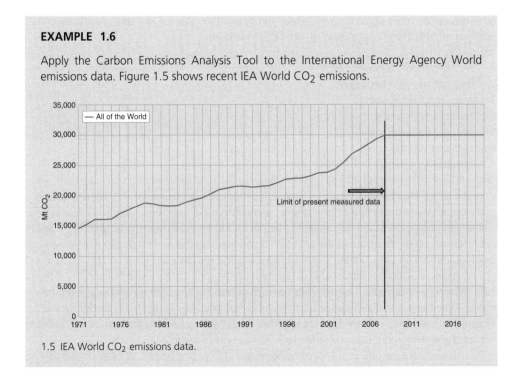

1.5 IEA World CO_2 emissions data.

World emissions continue to increase, unfortunately. A downward trend occurred during 1980–1983 during a US-led international recession with high unemployment, reduced manufacturing, and the Thatcher government in the UK battling against inflation. Emissions resumed increase from 1983 and have not slowed. Emissions have increased each year since 1990. Emissions in 1971 were 14,618 and 29,939 Mt CO_2 in 2008, an average annual rise of 2.8% pa. Significant increases in emissions came from China, the Middle East and North America. It does not seem as though the world is trying to reduce emissions yet; a target reduction of 34% from 1990, which would be 14,205 Mt CO_2 appears to be fanciful for 2020, while a reduction of 80% by 2050, down to 4,305 Mt CO_2, could be considered to be beyond imagination, even if every possible measure is implemented. UK emissions are reducing but Europe has a flat trend while all the other countries have increasing trends. The 2010 IEA data gives no indication that rising emissions may slow or reduce. Future use of technology is going to have to change if reductions of emissions and improvements to standards of living worldwide are to be achieved. Carbon capture and storage (CCS), technology to remove CO_2 from fossil fuel combustion products, renewable and nuclear electricity generation, all these are ways to achieve reductions. A positive view is that when governments set targets, the response is likely to be that far greater reductions are achieved, once industry finds the right technology. In conclusion, parts of the world have awoken to the problem, many countries may take a generation to respond seriously and we are all challenged to rethink how we use energy more efficiently.

Zero carbon buildings

The *HM Government Carbon Plan, 2011* (DECC 2011a) states the intention to deliver zero carbon new homes from 2016 and zero carbon non-domestic buildings by 2019. Does this mean that buildings can be constructed from materials, and by builders, that use no fossil energy resources? Surely an impossibility? Some ideas of what this means are:

1. zero net energy consumption;
2. zero net source energy use;
3. zero emissions building;
4. zero net energy cost;
5. zero off-site energy use;
6. zero grid supply;
7. allowable solutions.

Whether or not any building consumes energy depends upon its climate locality. There are regions of the world where the outdoor climate remains consistently within human comfort requirements where only protection from rain is needed, such as on the Mediterranean coast and on the South Pacific Ocean coast. What about the UK? Could you live at home and work in buildings without heating, artificial lighting, heated water for washing and refrigeration for food storage at least? No, of course not. Neither could anyone in the neighbouring countries of Europe. Try it if you wish by living in a caravan in winter with no form of heating and lighting and you might freeze due to hypothermia. No amount of fibreglass thermal insulation in the walls of the caravan or triple-glazed windows will ever be enough. Also, forget about washing in warm water, cooking hot food and drinking hot tea; got the idea? Do not bother trying it.

The idea of zero carbon is still developing in the minds of planners. It may be possible to design a new home that is so well insulated and with mechanically controlled heat reclaim ventilation that it requires no heating energy use in winter other than the warmth provided by the occupants, lighting and refrigerators. Whether this is possible with the existing stock of homes is questionable; expenditure on such a retrofit would seem to be beyond the occupants' willingness to meet the cost; it often takes seven years to repay the capital outlay for double glazing replacement windows. Another aspect of calling a building zero carbon regards the internal electrical appliances, artificial lighting, food refrigeration, water heating, home appliances, home entertainment system, computers, spa pump, swimming pool pump and power tools in the garage or home workshop. Who could live without most of those? Even if the solar and internal heat gains equal the winter heat losses through the fabric of the home, and the air conditioning system remains switched off, where does the rest of the energy use come from?

Initially, a zero carbon building was thought to mean net zero carbon dioxide emissions for the year, that is, heat losses of, say, 5,000 kWh in winter would be met by using fuel or electricity, while heat gains in summer, the same 5,000 kWh, would equal the heat losses, an ultra low energy building. As losses equal gains, the house would require no net energy to maintain a comfortable internal temperature. Energy bills for winter heating, summer cooling, domestic appliances and lighting still have to be paid. It may be comfortable in the British Isles to live without summer air conditioning in a suitably designed home, or work in a commercial or industrial building ventilated with outdoor air, although that is arguable. Especially so in the glass tower commercial office buildings that are air-sealed and rely on central air conditioning to maintain acceptable indoor air conditions and adequate mechanical ventilation, with power-hungry plant that cannot be turned off. How would you like to live and work in non-air-conditioned buildings when the outside air temperature remains in the 28°C–40°C range with blazing sunshine for many weeks? No amount of window shading, wall thermal insulation, reflective glazing and heat reclaim-assisted natural ventilation will satisfy liveable comfort conditions without risking hyperthermia and sun stroke.

Let's look at the different zero carbon building possibilities. A zero net source energy use building has on-site electricity generation that exceeds its consumption, including the energy used in transporting the energy to the building; this is a zero energy-plus building. This may include the gas and oil used in the building, making it an exporter of energy into the grid. This is the power station solution.

A net zero emissions building counterbalances the emissions used in supplying the building with on-site generation of energy through renewable zero emission means. Included in this definition is the energy embodied in the construction materials, the construction process and even the travel by its users. This is the solar panel and wind turbine on the roof solution.

The zero net energy cost approach is where the cost of purchasing energy used in the building is counterbalanced by the sale of energy to the grid from on-site generation. This is the accountant's solution, making the building a non-profit business.

The net zero off-site energy use method purchases energy from off-site, entirely renewable sources. This is the large-scale solar, wind, geothermal and wave power public utility solution that costs nothing on site.

A zero grid supply building is not connected to the supply grid for electricity, gas, district heating, possibly water and sewerage but may have a telephone line and computer network cable connection. All energy used is generated on site. This is the ship solution that floats free of ties to anywhere; however, an ocean liner consumes vast quantities of liquid fuel in its propulsion engines and essential plant to cope with widely varying climates, so it is not a zero emission site.

An allowable solution model for buildings recognizes the impracticality of achieving zero energy use with that embodied in the construction materials and process of building, and with the energy needed to run the building during its service period before demolition and recycling. All buildings are refurbished at least once, continually maintained, ultimately demolished and their materials recycled or dumped into landfill, so the whole picture needs to be considered. This model requires the building to have the highest practically attainable energy efficiency through fabric thermal insulation, solar and ventilation control, energy-efficient plant and systems with on-site renewable energy equipment appropriate to the building's location and architecture. On top of this, an allowance of energy use to comply with legislation is made, for example, to allow the ingress of outdoor air for gas appliances, to replace air exhausted from toilets, kitchens or chemical fume hoods. When a building complies with all the best practices at an acceptable cost, it may then be granted a 'zero carbon' rating. Whether you agree with this description is another matter that depends upon your perspective; we might see this as maintaining the present status of new building design. The EU defined the nearly zero energy building as having a significant amount of its net energy needs from renewable sources, on- or off-site. This is the engineer's solution.

The *Passiv Haus* puts these ideas into practice. Two of the qualifying standards for *Passiv Haus* design are that the specific heat load at the design temperature should not exceed $10.0 \, W/m^2$ floor area and that the maximum air leakage rate at a test air pressure of 50 Pa is not to exceed 0.60 air changes/h (http://www.passivhaus.org.uk/). Such houses are highly insulated with an exterior surface thickness around 500 mm, and airtight so that they consume only 25% or so of those energy levels set by the building regulations of most countries. Their mechanical and electrical services should be simple and reliable with ventilation air heat recovery.

EXAMPLE 1.7

Calculate whether the following design qualifies for the *Passiv Haus* standard. A single storey house has a floor plan of 20 m × 10 m and a room height of 2.5 m. Mechanical air change rate is 0.5 per hour through a heat exchanger that preheats incoming outdoor air from −1°C to 10°C in winter when the indoor air is maintained at 21°C. Eight windows are each 1.0 m × 1.5 m, two external doors are 1 m × 2 m. Thermal transmittances are $0.15 \, W/m^2 K$ for the floor and walls, flat plaster ceiling with a pitched tiled roof $0.1 \, W/m^2 K$ while the windows and doors are $0.8 \, W/m^2 K$. House construction is a concrete slab on the ground with 250 mm insulation, brick and block walls with 275 mm insulation, 400 mm insulation in the roof, triple-glazed windows and PVC insulated doors.

The data is entered into the simple workbook, *PassivHaus Calculation*, provided on the website, use other software if preferred or calculate the heat loss as shown in Chapter 6. Specific heat load is calculated as $18.5 \, W/m^2$ against an allowed $10 \, W/m^2$, so this house does not comply, despite having very high thermal insulation and mechanically controlled heat reclaim ventilation. Its designers need to work harder on the potential solution. Download the real *Passiv Haus PHPP Demo* workbook from the reference site to see how extensive the procedure is.

Green buildings

Green buildings are intended to reduce their impact on the environment, not necessarily to minimize energy use from any source. They may have similar means for passive and active solar energy use and natural day lighting as well as high standards of thermal insulation to maximize the efficient use of energy, ground space, locally sourced and natural materials but are not necessarily focussed on being zero carbon buildings. Turn back the clock to times when a timber-feller cleared a forested plot of land, used carpentry and local stone to build a house or medieval castle, burnt trees to keep warm, trapped rabbits, planted potatoes, drew water from a nearby stream, composted human and food waste, fished and sat back to watch the trees felled to grow up again. That was a green building; it was fully sustainable architecture, engineering, animal husbandry and agriculture. This was the recyclable timber solution. It is more difficult today as we expect to look through glazed aluminium windows, have insulated concrete foundations, park two cars in a home garage, illuminate every room with countless electrical lamps, have multiple appliances, communicate with the world-wide web every minute of the day or night, and travel 50 km to work and recreation locations.

The UK Building Research Establishment Environmental Assessment Method (BREEAM) is an environmental and sustainability standard methodology to rate buildings and on how buildings are used. It includes energy emissions, use of renewable energy, construction materials, and water use, recycling of waste and life cycle impacts of the building. Also it concerns management resources needed, land use issues, transportation methods in getting to the building, air and water pollution as well as the health and well-being of the users. It is not necessarily focussed on creating zero carbon buildings. Stars are awarded for the overall result. Some inputs may be open to interpretation but reports are overseen by trained auditors. It is the social responsibility index method.

Some countries have a Green Building Council and a Green Star rating tool that rates homes and non-domestic buildings. They are similar to BREEAM in that they include energy use, energy emissions, water use, waste recycling and the indoor comfort environment. They may be used as design tools to rate a new building prior to construction and can then audit the building in use. Existing buildings can be rated from their energy use per m^2 floor area to identify potential energy-saving measures and ongoing tracking. It is not necessarily focussed on creating zero carbon buildings but does help promote a building's energy efficiency, environmental factors and comfort. This is the marketing plan solution.

In North America, the Leadership in Energy and Environmental Design (LEED) green building rating system is used for residential and commercial buildings. It functions in the same way to BREEAM and Green Star schemes to enhance the value, energy saving and marketability of buildings and is not a zero carbon building tool.

The National Australian Built Environment Rating System (NABERS) rates public, commercial and domestic buildings on a 1–6 star scale with the aim of aiding existing building owners to reduce energy by 20–40% and assess performance of water, waste, comfort and air quality. A 5-star office building in Melbourne could be using 125 kWh/m^2 per year for all the electricity and gas consumed, so it cannot be a zero energy approach.

EXAMPLE 1.8

What are the trends in UK natural gas use for housing, public and commercial buildings? Do the trends appear to support the carbon plan? Download and plot the data available and add a trend line. Do different types of trend line show differing indications? Explain briefly what appears to be happening. Repeat this assignment as new data becomes available.

Download the current spreadsheet data file from DECC (http://www.decc.gov.uk) 'Natural gas and colliery methane production and consumption DUKES 4.1.1, 1970–present', and save the file. Plot the data and compare with the stated target reduction. Figure 1.6 shows the results up to 2008, your data will be current; upper line is domestic.

The consumption of natural gas for housing, public and commercial buildings appears to have peaked in 2004, and then started a downward movement that may be continued as the result of energy-saving measures. However, falls in earlier years, 1989, 1995 and 1997, followed increases, and that may happen again. Adding a logarithmic trend line shows an upward future consumption might be expected, while a polynomial gives a downward forward trend; such mathematical calculations do not take into account what is really happening to create the data, i.e. how gas is used, and should not be taken seriously. Gas use has increased rapidly since the 1970s and earlier, due to its competitive price, availability and convenience while burning coal and oil use decreased. Perhaps a different source of energy has become popular since 2004? Gas use is related to the weather, if its price is affordable, improvements in the energy efficiency

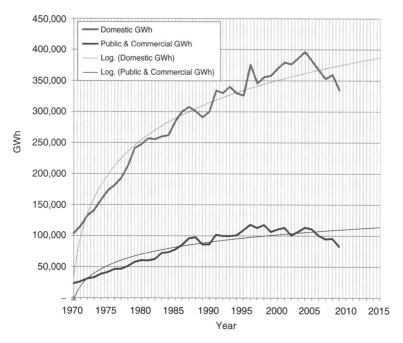

1.6 Gas data trends.

of buildings and plant, as well as gross national product, i.e. how many sites are able to use it; further investigation into these factors would be needed to draw valid conclusions.

Embodied energy

Embodied energy and CO_2 are the cost to the environment of using those materials in the construction of a building. It is a lot less than the materials pass in terms of heat transferred during their constructed period of service but needs to be considered at the design stage. Refer to the Inventory of Carbon and Energy (University of Bath, http://www.bath.ac.uk/mech-eng/sert/) for current data. Wide variations exist in the data. Where the energy comes from is important. Recycled material requires the lowest demands on the environment. How the material is won from the ground, processed, transported, fabricated into components such as bricks, how it is built into the structure, maintained over 100 years or so, then demolished and removed from the site, are all connected information. Sample data from the reference is aluminium 155 MJ/kg, heavy carpet tiles 378 MJ/kg, cement 4.5 MJ/kg, clay brick 3 MJ/kg, concrete 0.75 MJ/kg, copper tube 42 MJ/kg, glass 15 MJ/kg, fibreglass insulation 28 MJ/kg, a ream of A4 paper 70.5 MJ/kg, PVC pipe 67.5 MJ/kg, steel 20.1 MJ/kg, stainless steel 56.7 MJ/kg, timber 10 MJ/kg, and plywood 15 MJ/kg.

EXAMPLE 1.9

Compare the embodied energy used in the construction of single-glazed timber and aluminium-framed windows of 1.2 m × 1.2 m, with the energy caused by the winter heat flow during the first 30 years of use. This should show up the worst case as glass and aluminium windows have high embodied energy while their heat loss is greater than for all other parts of the building. Use the data: aluminium-framed window U value 6 W/m^2 K, timber-framed window 5.3 W/m^2 K, embodied energy per window 5,470 MJ for aluminium and 286 MJ for timber, indoor and outdoor design air temperatures 24°C and 0°C, annual heating for 2,250 hours and load factor 0.7.

Heating energy used by aluminium-framed window

$$= (1.2 \times 1.2)m^2 \times 6\frac{W}{m^2 K} \times (24-0)K \times 2,250\frac{h}{yr} \times 0.7 \times \frac{3,600s}{1h} \times \frac{1J}{1Ws} \times \frac{1MJ}{10^6 J}$$

$$= 1,176\frac{MJ}{yr}$$

$$5 \text{ years use} = 1,176\frac{MJ}{yr} \times 5yr$$

$$= 5,880MJ \text{ in 5 years}$$

$$30 \text{ years use} = 1,176\frac{MJ}{yr} \times 30yr$$

$$= 35,272 \text{ MJ in 30 years}$$

Embodied energy in an aluminium-framed single-glazed window is 5,470 MJ (Inventory of Carbon and Energy) so five years of use has a greater energy demand, while it is expected to last at least 30 years.

Heating energy used by timber-framed window

$$= (1.2 \times 1.2)m^2 \times 5.3\frac{W}{m^2 K} \times (24-0)K \times 2,250h \times 0.7 \times \frac{3,600}{10^6} \frac{MJ}{yr}$$

$$= 1,039\frac{MJ}{yr}$$

Embodied energy in a timber-framed single-glazed window is 286 MJ (Inventory of Carbon and Energy). That is far less than its annual heat loss but it is not expected to last 30 years and they are often replaced by PVC double glazing as a retrofit when they rot. PVC framed windows have half the embodied energy of aluminium framed ones.

Regulated and unregulated demands

The Building Regulations have a controlling influence over new buildings and those that are planned; these regulations create regulated energy demands and therefore CO_2 emissions for the future. Designers are free to exceed the legislated standards and produce a building that uses no energy or is able to export surplus to the grid or sell it down the road as district heating and cooling; this is a zero energy plus building, but these are very rare. Mostly, designs comply with regulations and with the optimum minimum cost, with the emphasis on the word minimum, from competitive tendering. It may be practical and economically justifiable to build new regulated zero energy homes, but once the occupiers move in, the energy demand becomes unregulated.

EXAMPLE 1.10

A newly constructed four-storey office building complies with the Building Regulations Part L2A. It is to be used as a 24-hour/7-day-a-week call centre, has a floor plan of 30 m × 20 m, a room height of 3 m. Mechanical air change rate is 1 per hour, indoor air is maintained at 21°C at an external temperature of −2°C. Each floor has eight windows of 2 m^2. There are four external doors of 2.5 m^2 each. Thermal transmittances are 0.25 W/m^{2}K for the ground floor and top floor roof, walls are 0.35 W/m^{2}K while the windows are 2.2 W/m^{2}K and doors are 3.5 W/m^{2}K. Calculate the regulated peak heat demand.

The regulated peak cooling demand is to be taken as equal to that for heating. Take the heating seasonal weather load factor as 0.7 for 4,000 h due to cold overnight operation, and 0.4 for 1,200 h as the cooling season, as nights provide free cooling to the building. Heating system overall efficiency is 80% and the coefficient of performance of the cooling system is 2. Estimate the annual regulated metered energy input using the workbook supplied, other software or manually as shown in Chapter 6. Lighting runs continuously at 10 W/m^2. Each floor has 75 continuously used computer work stations, using 250 W each. A computer server room on the top floor has a continuous electrical load of 10 kW. Refrigerators, water heaters and catering items add 3 kW loads to each floor with a load factor of 0.25. Sensible heat emission of each employee is 110 W. Estimate the annual unregulated demands and compare with those that are regulated. What do you observe from the calculated data?

The data is entered into the simple workbook, 'Commercial building BR Part L2A' provided on the website, use other software if preferred or calculate the heat loss as shown in Chapter 6. Note that the workbook provided only applies to examples within this book. This may be an extreme example but several types of buildings function continuously such as some manufacturing facilities, hospitals, hotels and police stations.

Regulated demands for heating and cooling amount to 48.16 kW each at peak load with an annual regulated load of 180,117 kWh for heating and cooling energy consumption.

Energy demands falling outside of those specified by the Building Regulations are found from this type of calculation:

$$Lighting\ load = 10\frac{W}{m^2} \times 2,400\,m^2 \times \frac{1\,kW}{1,000\,W}$$

$$= 24\,kW$$

$$Lighting\ energy = 24\,kW \times \frac{24\,h}{day} \times \frac{365\,day}{year}$$

$$= 210,240\,kWh/yr$$

In summer, a cooling system needs to remove heat gains through the building fabric and outdoor air ventilation, plus the internal electrical loads and the heat emitted from the occupants, to maintain the desired comfort conditions; this amounts to:

$$Summer\ cooling\ load = (48 + 121 + 33)\,kW$$
$$= 202\,kW$$

Other results are given in the workbook. Peak electrical load from internal sources of 121 kW are far greater than the maximum expected winter heat loss from the building of 48 kW, and then there is the additional heat emission of 33 kW from body heat. The building requires continuous cooling to remove 106 kW during winter. This example shows that however energy efficient the fabric of a building may be due to regulation of its design features, what users put into the building can overcome attempts to reduce CO_2 emissions.

The EU Emissions Trading System (EU ETS) 2012

Following the Kyoto Protocol 1997, most countries committed to capping emissions on a reducing scale. Three mechanisms are used:

1. *Regulation*: such as the building codes, efficiency of appliances and selective investment by governments from competitive tender for new projects.
2. *Carbon tax*: power stations and the largest industries have to pay for annual permits to emit CO_2 for a fixed term of years before a system of permit trading is allowed.
3. *The EU Emissions Trading System (EU ETS)*: regulated users are given a cap on their emissions, provided with paper certificates that are returned at the end of the year, sell surplus certificates or purchase more to make up for excess emissions.

EXAMPLE 1.11

A power station burning lignite has a capacity of 1,600 MWe, from eight equal-sized generator sets it annually supplies 11.4 million MWh of electricity (http://en.wikipedia.org/wiki/Hazelwood_Power_Station). It burns 16.7 million tonnes of lignite, open-cut brown coal, with a GCV of 26 MJ/kg when dried, and emits 1.5 tCO2/MWh. Calculate the overall efficiency of the power station, how many tonnes of CO_2 it emits in a year, how much a carbon tax rate of €20/ tCO2 would cost the plant operator in a year and how much would be added to the price of a kWh supplied to the grid. What incentive does the plant operator have to continue using lignite?

$$Energy\ input = 16.7 \times 10^6\,\frac{t}{yr} \times \frac{10^3\,kg}{1\,t} \times \frac{26\,MJ}{kg} \times \frac{1\,MWs}{1\,MJ} \times \frac{1\,h}{3600\,s}$$
$$= 120.6 \times 10^6\,\frac{MWh}{yr}$$

$$Overall\ power\ station\ efficiency = \frac{output}{input} \times 100\%$$

$$= \frac{11.4 \times 10^6\, MWh}{120.6 \times 10^6\, MWh} \times 100\%$$

$$= 9.5\%$$

This is very low, as power station efficiency should be around 28%. Usually, lignite is locally excavated at minimum cost to the power station operator.

$$Flue\ gas\ emission = 1.5 \frac{tCO_2}{MWh} \times 11.4 \times 10^6\, MWh$$

$$= 17 \times 10^6\, tCO_2$$

$$Annual\ carbon\ tax = 17 \times 10^6\, tCO_2 \times \frac{\text{\euro}20}{t\,CO_2}$$

$$= \text{\euro}340 \times 10^6$$

$$Production\ cost\ added = \frac{\text{\euro}340 \times 10^6}{11.4 \times 10^3 \times 10^6\, kWh}$$

$$= \frac{\text{\euro}0.03}{kWh}$$

Government imposition of a direct tax on CO_2 emissions from the power station adds 3 cents/kWh to the cost of production, and that will be passed on to distributors and final customers. A household that consumes 10,000 kWh per year will pay an additional €300.

If the power station operator converted one of its steam boiler flues to carbon capture and storage (CCS), it would save paying carbon tax of:

$$Annual\ tax\ saved = \frac{\text{\euro}340 \times 10^6}{8}$$

$$= \text{\euro}42.5 \times 10^6$$

The tax saving might apply for several years depending upon government allowances and may be sufficient to stimulate investment in emission reduction technology.

The European Union Emissions Trading System (EU ETS) commenced in 2005 to cover sites over 20 MW input, power stations, oil refineries and major industrial operators such as cement, steel, food and vehicle manufacturing. Each country has a national registry to administer the ETS. To minimize the effect of short-term weather and other variations in energy use, the EU set three trading period of 3, 5 and 8 years between 2005 and 2020. Each trading period commences on 1 January with period 3 from 2013–2020. Each EU member state sets an allowance of emissions for each year of the trading period; one allowance, one credit, is $1\,tCO_2$, i.e. 1 tonne of carbon dioxide emission. The allowance is calculated from verified emissions in the stated industries from annual audits and an assessment of what is likely to happen in the subsequent trading period. The UK had verified and estimated industrial emission of 272.4 Mt CO_2 in 2005 and was allowed a cap of 246.2 Mt CO_2, 9.6% less, for each year during 2008–2012.

Each industrial operator is given, free of charge, annual allowances at the start of the trading period. After the end of the first year, emissions are validated through an audit according to the assigned allowance. The operator gives back the allowances used in that year. More allowances are purchased from the open market, and handed over, to make up for excess emissions above the capped total. If the operator emitted less than the allowances given, those surplus credits can be rolled over to the next year, passed to another operator within the same company, moved across national borders or sold on the stock market for cash. At the end of each trading period, unused allowances may be retired or retained, depending upon the type of allowance.

Large industrial operators have buildings using energy for heating, ventilation, cooling, lighting, data centres, catering, distributed IT, transport and general power. Building services engineers and energy auditors have a professional involvement in the overall energy usage of these regulated sites, both for the buildings, staff accommodation and industrial energy use. For example, every major car manufacturing site has a continuously run air-conditioned paint spray vehicle tunnel and gas-fired ovens along a production line, using vast amounts of heating, ventilation and refrigeration energy and equipment.

Repeats of recent years' financial crises reduce energy demands, leading to oversupply of carbon emission allowances and their falling in value. EU energy efficiency initiatives also reduce emissions, which is their aim, leaving more allowances in the market than are needed. The aviation sector is due to join the EU ETS in 2012 but the largest emitters, the USA and China, may not wish to join, upsetting the aim of the scheme. If Germany closes its nuclear power generation by 2022, as indicated recently, the country may increase emissions by as much as the whole of those from the UK, 500 Mt CO_2/yr. Look at the current prices of allowances (EUA), and certified emission reductions (CER), on the website and notice the € volatility (Trading Carbon, at http://www.pointcarbon.com/, accessed August 2011).

Allowances or credits are a financial instrument that can be traded on the open market worldwide. Trade in any stock is based on people's confidence to invest, availability of the stock and fear of losing its value; all are human emotions. EU allowances opened with a cash value of €30/tCO_2 and oscillated wildly, going as low as €10, and even dropping to €0.03 in December 2007 due to oversupply of allowances in the market and the industrial recession. Large reductions in CO_2 emissions from energy savings and greater use of nuclear power will create a surplus of allowances, and their traded price can fall as there may be fewer buyers, but nothing is certain. A carbon allowance price of €20 to €30 is anticipated for trading periods up till 2020 but with significant volatility. This is the financial services stock market solution by trading shares.

EXAMPLE 1.12

Explain the aim of implementing the EU ETS.

Without government regulation, energy use and emissions are only penalized by their price. Energy prices for the largest consumers, power stations, oil refineries and major manufacturers are very low due to the high quantities consumed with a very high degree of security of demand from energy users. Industrial and commercial energy demands are highly predictable, unlike the weather and time-dependent peaks of demand from homes. There was no penalty for emitting CO_2 and other so-called pollutants high into the atmosphere through chimneys as long as smoke was controlled. With the EU ETS, there is a cost to emitters, a direct encouragement to invest in countermeasures and a benefit in selling surplus allowances afterwards. Commencing in 2013, member states will auction allowances, that is sell and not issue them free of charge, with the aim of raising funds for investment in emission reduction projects.

Carbon Capture and Storage (CCS)

CCS claims to have the potential to capture 90% of the flue gas emissions from fossil fuel-fired plant. This will almost certainly be developed for power stations and large-scale sites, maybe hospitals and industrial facilities. Captured CO_2 is compressed with gas or steam-driven turbines, pumped as high pressure liquid, in the same way as LPG, out to bulk transport ships or sea production platforms where it is pumped down into the same geological strata where the oil or methane gas came from (The Carbon Trust Ltd, http://www.carbontrust.co.uk).

   Three methods are used:

1. *Pre-combustion separation* by producing syngas from pulverized coal, removing CO_2 from syngas leaving H_2 that is fired into steam boilers or passed on to the transport industry. H_2 combusts to H_2O vapour only. This is the hydrogen economy solution.
2. *Post-combustion* by washing the flue gas with ammonia and water to separate out CO_2 gas. The remaining flue gas is mainly N_2, some O_2 and sometimes SO_2 if the fuel contained S plus possibly some NO_2.
3. *Oxy-fuel combustion* by removing N_2 from combustion air and burning O_2 plus re-circulated flue gas to reduce the flame temperature with the fuel. CO_2 is washed from the flue gas with ammonia and water.

EXAMPLE 1.13

A 2000 MW oil-fired power station constructed in 1975 converts one of its 500 MW_e generator steam boilers to CCS technology. Overall efficiency of the generator set is 28%. That generator set emits no more CO_2 into the atmosphere from 1 January 2013; this is the commencement of the 3rd EU ETS trading phase of 2013–2020. Design, installation and commissioning take all of 2012. The generator only provides peak lopping power due to the high cost of oil and ran at full load for 1,000 hours during 2011 and around the same in 2012. CO_2 emission intensity for the use of oil is agreed to be 1 kg CO_2/kWh_e. Predict the outcome for this investment decision.

Calculate the verified emission of the generator set in 2011, the last full year of registered data:

$$Emission = 500\,MW_e \times \frac{100}{28} \times \frac{10^3\,kW}{1\,MW} \times \frac{1,000\,h}{yr} \times \frac{1\,kg\,CO_2}{kWh_e} \times \frac{1\,t}{10^3\,kg} \times \frac{1\,Mt}{10^6\,t}$$

$$= 1.8\,Mt\,CO_2$$

A capped allowance of 1.8 Mt CO_2 was granted for this generator, set for the duration of the trading period until 2020 by the EU. Allowances traded at €20/tCO_2 during early 2013 with the market expectation of price volatility to continue with no guarantee of a rising value, due to many other carbon emission projects coming into use and there being a surplus of allowances for sale. The EU awarded a capital grant towards the cost of pumping the recovered CO_2 high pressure liquid into containers for export. The sale of liquid CO_2 did not produce any revenue for the power station but future streams might as inventors found uses for large amounts of it. The power station sold its surplus allowances each year to raise capital for further investments in

CCS and keep the power station using the low cost oil from the nearby oil refinery.

$$Revenue\ from\ allowances = \frac{€20}{tCO_2} \times \frac{10^6\,t}{1\,Mt} \times \frac{1.8\,Mt}{yr} \times 8\,years \times \frac{€M}{€10^6}$$

$$= €288\,M$$

It may be possible for the power station operator to sell allowances for something like €288 million, but he has no guarantee that amount can be realized from sales on the free market; more might be gained, or a lot less. The investment decision to install CCS may be made for several reasons, one of which is the financial return from sales of unused allowances. There is no indication at the present time what might happen to EU policy or the ETS market after 2020. The power station operator has to decide whether to invest in CCS under highly variable financial scenarios.

Taxing carbon

Extensive discussions around the world have focussed on how, or if, taxation on the consumption of hydrocarbon natural resources can reduce usage. The problem was reported to be that combustion of hydrocarbons, petroleum products such as petrol, diesel fuel and heating oil, plus natural gas, produced carbon dioxide emission into the atmosphere. CO_2 trapped warmth in the atmosphere and that had to be reduced. Buildings are major consumers of energy, are largely of existing stock plus new developments, and will continue to use energy into the foreseeable future. That is, just as long as humans need them. Transportation, manufacturing, construction and agriculture make up the overall picture. A renewable means of providing energy is being actively pursued but winning hydrocarbons from the earth is and will remain of vital importance for generations.

More than one thousand carbon policy measures were identified in the nine major countries (Productivity Commission 2011) that covered energy-saving engineering projects, government policies, taxation of hydrocarbon use and financial trading methods for carbon-use credits. The advice to government was that there could not be one single price to charge for greenhouse gas emissions. It was too difficult to compare hydrocarbon use in transportation with those used in buildings, for example.

The problem with taxation was that someone had to pay the tax; this was always the final consumer of products and services. Businesses pass on their direct costs by raising the prices of their products. So the consumer paid more for electricity, petrol, diesel, oil, natural gas and liquefied petroleum gas. The aim of the tax was to encourage the use of less hydrocarbon primary fuel sources, reduce national emissions and make the country look good as a world leader in helping to save the planet from extinction.

Ask yourself a question here. Has a tax on petrol ever made you drive less distance, downsize your vehicle and fly less? Does taxation on fuel reduce its use? How about the tax on tobacco, alcohol and luxury cars? The answer may be, not greatly, if at all. Car design evolves to be more fuel efficient but did the national consumption of petrol and diesel fuels reduce over your lifetime as a result? Would a tax on heating oil and natural gas lower their consumption? Answers are found in the DECC statistics used earlier and that is where we can track the progress of the carbon plan and other initiatives. Yes, is the answer as petroleum use is reducing in the UK, but was tax the reason?

Houses and other buildings evolve into being more efficient users of primary energy and use less. The old stock of buildings is gradually being replaced by new more energy-efficient designs, but more houses, offices, commercial and industrial buildings will always be built, increasing the number using energy resources. So, will energy use reduce on a national scale? Maybe it could in the UK as building land becomes full, but the world is a large place and there are vast reserves of land for future building development around the globe. Just drive between the major cities and towns in Scotland, Wales and Victoria in Australia, also probably any country that could be mentioned, and you will see unused or under-used land, not all flat land either, that, one day, future generations will see covered in houses, schools, hotels, hospitals, commercial premises and shopping centres. So, return to the question. Does taxation on fuel reduce its use? Maybe not. Other life factors determine how much is consumed of everything.

How could a price be decided for any type of carbon dioxide emission tax? Answer: find out what it would cost to achieve abatement of primary fuel usage from extra investment in energy-saving measures such as greater thermal insulation for buildings, higher efficiency heating and cooling systems, reducing lighting power, controlling energy systems better or by accepting that we have to live with widened standards of thermal comfort. And that only considers what the building services engineer could directly influence. Governments need to have a 'price of abatement', that is, how much it costs to achieve reductions in the use of primary energy, or more precisely any CO_2 or other emissions from the use of primary energy, as necessary to meet international commitments.

EXAMPLE 1.14

A London commercial building is to be retrofitted for letting at higher rental to maintain occupancy in competition with nearby new buildings. A complete upgrade to the windows, air conditioning, lighting, and installing photovoltaic power cells to the roof are estimated to cost £500,000 to potentially reduce electricity consumption by 500 MWh and natural gas by 100 MWh per year. Take the CO_2 emissions intensity as 1 tCO_2/MWh for electricity used and 0.5 tCO_2/MWh of gas used at the site. Electricity costs 10 p/kWh and gas costs 5 p/kWh at the site. Calculate the price of abatement for this project in £/tCO_2 and the annual cost saving. Is the building owner likely to make this investment? Could the building owner gain by selling carbon allowances to increase the financial return?

$$Emission\ reduction = 500\,MWh \times \frac{1\,tCO_2}{MWh} + 100\,MWh \times \frac{0.5\,tCO_2}{MWh}$$

$$= 550\,tCO_2$$

$$Cost\ saving = 500\,MWh \times \frac{10^3\,kWh}{1\,MWh} \times \frac{10}{100}\frac{£}{kWh} + 100\,MWh \times \frac{10^3\,kWh}{1\,MWh}$$

$$\times \frac{5}{100}\frac{£}{kWh}$$

$$= £55,000\ per\ year$$

$$Price\ of\ abatement = \frac{£500,000}{550\,tCO_2}$$

$$= \frac{£909}{tCO_2}$$

$$Simple\ payback\ period = \frac{£500,000}{£55,000}\ years$$

$$= 9\ years$$

The building owner is unlikely to make a commercial investment of over three years' simple payback unless there is a realistic opportunity of selling the building for a much higher price and recouping the investment early. Tenants may demand a rent reduction as the building costs less to operate after the refit. Commercial buildings are not given carbon allowances, only certain industries and power stations, so there is no gain to be made from the EU ETS.

Conclusion

Are we able to make any conclusions about the world's progress towards reducing CO_2 emissions in response to the threat of damaging climate change?

It is apparent from the International Energy Agency data that whatever individual countries may do in response, however many building are retrofitted in the UK and Europe, however clever we become at attempting to have zero carbon buildings, converting to electric cars, building nuclear power stations and wind generators, other countries are likely to continue increasing their emissions. Worldwide emissions are destined to increase for the foreseeable future, perhaps for decades or centuries. But the worldwide reduction of emissions has to start somewhere, or find another solution. The future is in our hands.

Questions

Descriptive answers must be made in your own words based on a thorough understanding of the subject. Copied work displays a noticeable discontinuity of style between your own production and the reference material. Also, the work has not been fully comprehended if it can only be answered by copying. The ability to pass on reliable and concise information is a vitally important part of business and government work, and you should realize that as much practice and experience as possible is needed to become an effective communicator.

1. How could open market trading in the EU CO_2 allowances make investment decisions in emission reduction plant and systems difficult?
2. Why should building services engineers, architects and construction companies be concerned about the trading or value of EU CO_2 allowances? These are not registered sites or users who are awarded or having to purchase allowances.
3. Which of these is correct about building services engineering and the EU ETS?

    1. There is no connection.
    2. All building services systems have CO_2 emissions.
    3. Without services in buildings, there would be no registered emission sites.
    4. Large commercial buildings need to trade in allowances.
    5. Services need to minimize CO_2 emissions on EU registered sites.

4. Which of these is correct?

    1. Every large energy-using site is registered with the EU ETS.
    2. The EU ETS charges fees for allowance traders.
    3. Only financial institutions can hold CO_2 allowances.
    4. Power stations and large industrial sites can register with the EU ETS.
    5. No organization profits from trading in allowances.

5. Which is correct about the carbon plan?

    1.   The most important atmospheric pollutant to be reduced is methane.
    2.   Carbon monoxide from vehicle exhausts is polluting the air and must be eliminated.
    3.   Carbon dioxide emissions from burning and combusting hydrocarbon fuels must be reduced.
    4.   All hydrocarbon emissions to the atmosphere are to be phased out.
    5.   Nitrous oxides will always be capped at zero.

6. What is the level of CO_2 emissions from the UK?

    1.   500 tonnes per year.
    2.   500 giga tonnes per year.
    3.   500 mega tonnes per year.
    4.   There aren't any.
    5.   500 million kilograms per year.

7. What are the problems caused by a market price for carbon credits?

    1.   Site owners do not have a guarantee for the value of emission reductions.
    2.   A fluctuating market price for allowances is attractive for registered sites.
    3.   Trading in allowances becomes an industry of its own.
    4.   Allowance price will always increase in line with inflation.
    5.   Carbon credit price is tied to the strength of the Euro in world markets.

8. What chemical compounds are formed during combustion of a fossil fuel and what are their effects on the atmosphere?

9. Which of these is an EU ETS registered site?

    1.   District heating systems.
    2.   Large air-conditioned buildings.
    3.   A cement-making plant.
    4.   A hospital.
    5.   An apartment building.

10. Which best describes the EU ETS?

    1.   Tax on energy use.
    2.   A disincentive to invest in energy efficiency.
    3.   An unavoidable cost for industry.
    4.   An outcome of the Kyoto Protocol.
    5.   Unnecessary administrative burden from the EU.

11. How does nature maintain a balance of O_2, N_2 and CO_2 in the atmosphere?

    1.   By manufacturing oxygen from the earth.
    2.   Oceans and soil absorb all CO_2 from the atmosphere.
    3.   Chemical reaction between H_2O, CO_2 and N_2 in the clouds returns C to the soil and oceans, leaving O_2 in the atmosphere that is necessary for life on Earth.
    4.   Oceans absorb CO_2 causing water acidification.
    5.   It doesn't.

12. How long does human-generated CO_2 remain in the atmosphere?

    1.   Minutes.
    2.   Hundreds of years.
    3.   12 months.

4.   It never decreases.
5.   Plants consume it quickly.

13. How has the world created what is now said to be an atmospheric catastrophe in the making?
14. Explain how the post-combustion capture of CO_2 works.
15. Explain what happens to CO_2 that is captured from fossil fuel combustion.
16. Which is approximately the concentration of CO_2 in the atmosphere at a few metres above ground level?

   1.   360 ppm.
   2.   1,000 ppm.
   3.   5%.
   4.   Negligible.
   5.   500 mg/m^3.

17. A two-storey inner terrace house has a floor plan of 8 m x 5 m and a room height of 2.5 m. The adjoining houses are of the same standard and are along the 8 m dimension. Mechanical air change rate is 0.5 per hour through a heat exchanger that preheats incoming outdoor air from $-3°C$ to $11°C$ in winter, when the indoor air is maintained at $22°C$. Four windows are each 1 m^2; two external doors are 2 m^2. Thermal transmittances are 0.14 W/m^{2}K for the floor and walls, flat plaster ceiling with a pitched tiled roof 0.1 W/m^{2}K while the windows and doors are 0.8 W/m^{2}K. House construction is a concrete slab on the ground with 250 mm insulation, brick and block walls with 275 mm insulation, 400 mm insulation in the roof, triple-glazed windows and PVC insulated doors. Does the design qualify for the *Passivhaus* standard, if not, what simple measure is permitted by *Passivhaus* for compliance?
18. A new single-storey office building complies with the Building Regulations Part L2A. It is to be used as a 12-hour/5-day a week office, has a floor plan of 40 m $\times$ 30 m, a room height of 3 m. Mechanical air change rate is 1 per hour, indoor air is maintained at $22°C$ at an external temperature of $-3°C$. There are 10 windows of 3 m^2. There are two external doors of 2.5 m^2 each. Thermal transmittances are 0.2 W/m^{2}K for the ground floor and roof, walls are 0.3 W/m^{2}K while the windows 2 W/m^{2}K and doors are 3 W/m^{2}K. Calculate the regulated peak heat demand. The regulated peak cooling demand is to be taken as equal to that for heating. Take the heating seasonal weather load factor as 0.7 for 2,250 h and 0.4 for 1,000 h of the cooling season. Heating system overall efficiency is 80% and the coefficient of performance of the cooling system is 2. Estimate the annual regulated metered energy input using the workbook supplied, other software or manually, as shown in Chapter 3. Lighting runs continuously at 12 W/m^2. Each floor has 100 computer work stations using 250 W each. A computer server room on the top floor has a continuous electrical load of 8 kW. Refrigerators, water heaters and catering items add 4 kW loads to each floor with a load factor of 0.25. Sensible heat emission of each employee is 110 W. Estimate the annual unregulated demands and compare with those that are regulated. What do you observe from the calculated data?
19. A 500 MW$_e$ gas-fired power station has an overall efficiency of 32% and runs at full load for 6,000 h/yr. Emission intensity is 0.5 tCO$_2$/MWh. Calculate how many tonnes of CO_2 it emits in a year, how much a payable carbon tax rate of €30/tCO$_2$ would cost the plant operator in a year and how much would be added to the price of a kWh supplied to the grid. What incentive does the plant operator have to start installing carbon capture and storage to reduce emissions?
20. A 500 MW$_e$ oil-fired power station with an overall efficiency of 25% converts to carbon capture and storage technology and saves 90% of its CO_2 emissions from 1 January 2013;

this is at the commencement of the 3rd EU ETS trading phase of 2013–2020. The generator only provides peak lopping power due to the high cost of oil and ran at full load for 1,000 hours during 2011 and around the same in 2012. CO_2 emission intensity for the use of oil is agreed to be 1.0 kg CO_2/kWh$_e$. Emissions for the 3rd EU ETS trading period are capped at 2011 levels. Traded allowances realize a maximum of €20.00/tCO_2. Predict the outcome for this investment decision.

21. Explain the relationship between the price of abatement, that is, the cost per tCO_2, for emission reduction retrofits from engineering and structural improvements, and the value of carbon allowances traded on the open market.

22. What is the correct current level of annual CO_2 emissions from the whole of the world?

    1.   10 million tonnes.
    2.   1,000 megatonnes.
    3.   500,000 Gt.
    4.   5 Gt.
    5.   30,000 Mt.

23. What is the approximate percentage increase in world CO_2 emissions from 1971 to 2008?

    1.   105%.
    2.   55%.
    3.   350%.
    4.   1,000%.
    5.   120%.

24. Which is correct about a zero carbon building?

    1.   Cannot consume any fossil fuel energy.
    2.   Must be provided only by renewable energy systems.
    3.   Consumes a minimum amount of energy for all uses.
    4.   Has a zero cost energy bill.
    5.   None of these.

25. How much energy does a green building use?

    1.   None.
    2.   A lot.
    3.   Only that used for lighting, computers and hot water.
    4.   Less than a 1960s design.
    5.   Any winter heat loss is replaced by summer heat gains.

26. The plant process for CCS is which of these?

    1.   Flue gas filtered through charcoal.
    2.   Suction pump removes CO_2 from flue exhaust.
    3.   Flue gas washed with alcohol.
    4.   Flue gas washed with ammonia.
    5.   Catalytic converter in exhaust gas absorbs CO_2.

27. State what is meant by these summary solutions of the apparent need to reduce global warming caused by human activities.

    1.   Renewable timber buildings.
    2.   Improvement of 1950s building designs.
    3.   Social responsibility index method.
    4.   Marketing plan aid.
    5.   The USA is ahead.

28. State what is meant by these summary solutions for buildings to reduce global warming caused by human activities.

 1. Down-under reduced energy scheme.
 2. Seems impossible.
 3. The no-television in the home solution.
 4. Dark in here at night.
 5. I keep my beer in the river.

29. State what is meant by these summary solutions for buildings to reduce global warming caused by human activities.

 1. A very tall all-glass walled building in a 40°C summer location, having natural ventilation and no refrigerated air conditioning.
 2. Natural daylighting with unshaded glazed perimeter to collect and store solar heat gains for winter.
 3. Accountants' solution.
 4. Engineers' solution.
 5. Very small energy use building.

30. State what is meant by these summary solutions for buildings to reduce global warming caused by human activities.

 1. Solar atrium office.
 2. Power station solution.
 3. Solar panel and wind turbine solution.
 4. Grid-connected renewable sources solution.
 5. Ship solution.

2 Post occupancy

Learning objectives

Study of this chapter will enable the reader to:

1. comprehend EPC, DEC, SER, BER, AR, iSBEM and TM22;
2. use building CO_2 emission calculations;
3. understand post occupancy evaluations;
4. recognize who is interested in post occupancy matters;
5. use CIBSE PROBE Reports;
6. identify differences between design and use of buildings;
7. have access to real data from occupied buildings;
8. realize that design does not reveal actual building energy usage;
9. compare real building energy use with benchmarks;
10. understand how buildings are air pressure tested and rated;
11. work on many case studies from the PROBE Reports;
12. calculate building Asset Ratings;
13. use Building Emission Rate, BER and SER;
14. use the best practice emission data;
15. understand the Queens Building case study;
16. relate actual building energy issues to *HM Government Carbon Plan 2011*;
17. have an understanding of the technical issues in reducing energy use;
18. relate modern buildings to lessons from history;
19. comprehend what is meant by sustainability;
20. relate building shape to energy use;
21. discuss the use of on-site generation.

Key terms and concepts

AR 31; BER 31; Best practice 35; CO_2 emissions 31; DEC 31; EPC 40; generation 33; iSBEM 00; PROBE 32; Queens Building 35; SER 34; shape 39; sustainability 38; TM22 33.

Introduction

How a building performs in reality is often different from what its designers intended. This chapter takes us behind the impressive façade of technical excellence into the reality of how many buildings perform for their everyday users. Once built, occupied and filled with computer workstations, long-term energy use becomes fixed. Retrofit measures may be able to reduce future carbon emissions but not by much; the pattern is set. Energy audits lead to an Emission Performance Certificate that rates a building from excellent performance down to expensive. Software use is introduced and a simple workbook provided for examples in this chapter. Extensive Post Occupancy Review of Building Engineering (PROBE) reports are summarized and data provided for assignments. A case study of the Queens Building, Leicester, is given to explore how such investigations are used. Sustainability, building shape affecting energy use and on-site generation are discussed.

The Building Emission Performance Certificate

This is about what we need to know about EPC, DEC, iSBEM and TM22 at this stage of our understanding. Existing buildings are required to have an energy audit to determine the appropriate level of emissions for an Emission Performance Certificate (EPC). Public buildings must display this rating on a Display Emission Certificate (DEC). The Asset Rating (AR) for the building is found from the total of the audited utility bills and calculated in kg CO_2/m^2 yr. Each building is given a Standard Emission Rating (SER), based on compliance with the Building Regulations. Although called emission ratings, these are only based on energy consumed by an existing building or designed to be consumed for a proposed building.

$$Asset\ Rating = 50 \times \frac{Building\ Emission\ Rating}{Standard\ Emission\ Rating}$$

$$AR = 50 \times \frac{BER}{SER}$$

The emission bands used to declare the EPC and DEC are shown in Table 2.1. An asset rating of zero, A+, means the building complies with the Building Regulations, Part L, Fuel and Power. This is said to have zero net energy annual emissions and has a Building Emission Rating (BER) of zero from regulated demands. However, the building does use energy for heating, cooling,

Table 2.1 Emission rating bands for EPC and DEC.

Emission band	Asset rating
$AR \leq 0$	A+
$AR \leq 25$	A
$AR \leq 50$	B
$AR \leq 75$	C
$AR \leq 100$	D
$AR \leq 125$	E
$AR \leq 150$	F
$AR > 150$	G

Source: *A User Guide to iSBEM* (2011).

lighting and from unregulated demands from electrical equipment and computers, as we saw in Chapter 1; it subsequently pays for energy use and causes a lot of CO_2 emissions.

Post occupancy

This is where we look back at what designers did and subsequently walked away from. Did the building that was thoughtfully designed and presented by the architect, computer modelled and calculated by the building services engineer, constructed, commissioned and handed over to the owner or end user, function as intended? The structural engineer made sure that it would not collapse under its own weight or blow over in a high wind, but what about the efficiency of the heating, ventilation, air conditioning, lighting, plumbing, lifts and daylight control? Did the energy consumption match predictions? Did the uncontrolled energy demands from computer systems and lighting cause unplanned energy consumption? Did users operate the building differently from that in the design brief? Did the users know how to make the energy efficiency measures work as intended? Was the building comfortable? Who would know? Probably, no one. Designers of the new building forgot about it the moment their final invoice had been paid, and have worked on countless other projects since then. Handover of the building to its owner by the building and services contractors, then a defects liability period, drew a line under professional interest. What then? An untrained owner received a pile of technical documentation that was hard to understand and not as user-friendly as the manual for a new car, and filed it all. A facility manager or contractor was hired to ensure the new building functioned, and he could resolve operational faults such as stuck lifts, door handles loose, air conditioning too hot, too draughty at a desk, and too noisy; multiple complaints came from tenanted floors. The building owner might be a property management company based in another country or simply not interested in anything but receiving rent and making a profit every year. Self-managed small owner-occupied buildings may have no budget for maintenance and no technical expertise to know what to do about faults. Who kept a check on the energy consumption? Did the building operate at its intended energy rating? Who checked? Again, some organizations have an Energy Manager or employ consultants to audit energy use, but there would be many users who knew nothing other than paying the bills that arrived.

Answers to these questions come from the Post Occupancy Review of Building Engineering (PROBE) investigations conducted 1995–2002 (*CIBSE Journal*). PROBE reports are available from https://www.cibse.org. Go to Freely download selected CIBSE publications, CIBSE Low Carbon Consultants Training Material, and download PROBE Reports, Guide F, and also, TM22 software.

Feedback from the investigations provided a link between the building, its mechanical and electrical services hardware, and the soft targets, the users, the non-experts in buildings, the ones who have to sit in them every day. Their reactions to the design and working conditions come from whether it is overpowering; if it is liked; does it give a feeling of security?; is it clean, tidy, too bright, too dark, too much glare, cramped, too spacious, overly warm, cold, draughty?; how much personal control can they have over lighting, heating and air movement, reflections on computer screens, background noise, cross-talk interference?; is there a view of outdoors?; does it feel claustrophobic?; is there traffic noise intrusion?: does the atrium feel like the edge of a cliff?; are circulation spaces comfortable to move around in?; are washrooms handy?; and most important of all, where is the catering facility? A long list, but everything is of importance to the user, to whom this becomes a second home.

It will be of little concern to most users whether the building is fully air conditioned, naturally ventilated, low or high technology, low energy or costs a fortune to run. All of the technical means of controlling the indoor air environment can and do function well. Complex ventilation mechanisms relying on high user interaction and understanding, as well as those that are fully automated requiring no user input, came out of the reviews as under-performers. Users appreciate being able to make adjustments to their immediate environment. But there are limits; users do not expect to have to act as a frequent control mechanism to maintain the air conditioning. Too much distraction and tasks become annoying to the point of being ignored; emergency measures, such as wedging a window or door open, become permanent. The PROBE team found that trying to relate productivity to comfort conditions was not quantifiable.

Calculations of Asset Ratings (AR) in this book are for illustration purposes and approximate, as they do not use the approved software and are not conducted by a registered consultant. Download *iSBEM* software (*A User Guide to iSBEM*, version 4.1c, March 2011) and TM22 or use commercial software. Download the workbook file *Asset RatingAR.xls* from the website to calculate the examples and questions in this book. Energy benchmarks are taken from *CIBSE* (*Guide F*) where good and typical practice data are listed. Good practice means compliance with Building Regulations Part L or better, while surveys of many existing buildings generated the typical practice benchmarks for comparison. Metered electricity and fossil fuel consumption for building in use include both the regulated (Part L) energy emissions and those from unregulated use of computer systems, lighting and all other sources. This makes the regulated energy consumption a small part of the total CO_2 emission due to the building in some cases.

Air leakage through a building is measured by removing a main door and connecting a ducted fan to the opening, as described in the PROBE reports. Outdoor air is blown in to pressurize the building to a target static air pressure of 50 Pa. The air flow through the fan and duct is measured and this is the air leakage rate for the building. Reversing the direction of flow through the axial fan allows negative pressurization so that inward air leaks can be found by holding smoke candles around their sources. Some buildings cannot maintain a 50 Pa pressure due to extensive opening, such as when the building has natural ventilation openings, openable windows, permanent ventilation louvres, swing doors or is excessively leaky. A lower static pressure is held for air flow measurement and the equivalent at 50 Pa is calculated. Air leakage flow is related to the floor area where a tight building passes around 8 m^3/m^2 h at 50 Pa while leaky buildings go up to 35 m^3/m^2 h or more.

EXAMPLE 2.1

The John Cabot Academy (*BSJ*, May 1994) was completed in Bristol in 1993, for 1,050 pupils, 140 staff and 230 computers on a gross floor area of 8,800 m^2. It has cross-flow manually controlled natural ventilation and gas-fired central heating. The computer server room and a seminar room have refrigerated cooling. Electricity consumption in 1997 was 509 MWh and gas averaged 869 MWh. The air leakage test revealed a rate of 35.4 m^3/m^2h at a pressure of 50 Pa; this was high due to the natural ventilation arrangements, toilet, kitchen and laboratory exhaust fan systems. Good practice energy benchmark for a secondary school (*CIBSE Guide F, 20.1*) is 25 kWh/m^2 yr for electricity and 108 kWh/m^2 yr for gas use. Take the electrical emission as 0.422 kg CO_2/kWh and gas emission 0.194 kg CO_2/kWh. Calculate the academy asset rating. Identify what energy saving measures could be applied to the academy to improve its asset rating. Comment on the likelihood of the measures you identify being implemented, with reasons.

Calculate the good practice energy benchmark emission for the building. This will be taken as the Standard Emission Rate (SER):

$$Electrical\ emission = 25\frac{kWh}{m^2yr} \times 0.422\frac{kgCO_2}{kWh}$$

$$= 10.55\frac{kgCO_2}{m^2yr}$$

$$Gas\ emission = 108\frac{kWh}{m^2yr} \times 0.194\frac{kgCO_2}{kWh}$$

$$= 20.95\frac{kgCO_2}{m^2yr}$$

$$SER = (10.55 + 20.95)\frac{kgCO_2}{m^2yr}$$

$$SER = 31.5\frac{kgCO_2}{m^2yr}$$

This is the standard benchmark that the academy will be graded against for its asset rating. Now find the metered emissions from the utility bills that include both regulated and unregulated demands, the Building Emission Rate (BER):

$$Electrical\ demand = 509,000\frac{kWh}{yr} \times \frac{1}{8,800m^2}$$

$$= 57.8\frac{kWh}{m^2yr}$$

$$Electrical\ emission = 57.8\frac{kWh}{m^2yr} \times 0.422\frac{kgCO_2}{kWh}$$

$$= 24.4\frac{kgCO_2}{m^2yr}$$

$$Gas\ demand = 869,000\frac{kWh}{yr} \times \frac{1}{8,800m^2}$$

$$= 98.75\frac{kWh}{m^2yr}$$

$$Gas\ emission = 98.75\frac{kWh}{m^2yr} \times 0.194\frac{kgCO_2}{kWh}$$

$$= 19.2\frac{kgCO_2}{m^2yr}$$

$$BER = (24.4 + 19.2)\frac{kgCO_2}{m^2yr}$$

$$BER = 43.6\frac{kgCO_2}{m^2yr}$$

$$Asset\ Rating = 50 \times \frac{Building\ Emission\ Rating,\ BER}{Standard\ Emission\ Rating,\ SER}$$

$$AR = 50 \times \frac{BER}{SER}$$

$$AR = 50 \times \frac{43.6}{31.5}$$

$$AR = 69$$

This calculated estimation of the asset rating for the John Cabot Academy corresponds to an AR of C grade. Potential energy-saving measures are: reduce air leakage around windows, doors, ventilators; turn off exhaust fans when not needed; install weighted dampers in exhaust ducts; double-glaze windows and doors; add thermal insulation to walls and roof; review automatic controls of the heating system; install movement sensors and light level sensors to control lighting. There may be other opportunities identified from an on-site audit as the building is nearly 20 years old.

To improve the asset rating from C to B, the AR must be reduced to 50 or less; the BER becomes:

$$BER = AR \times \frac{SER}{50}$$

$$BER = 50 \times \frac{SER}{50}$$

$$BER = SER$$

$$BER = 31.5 \frac{kgCO_2}{m^2 yr}$$

This means the building must equal the best practice benchmark. Use the workbook provided to find the electricity and gas consumptions that would be needed to reduce BER to grade B. Electricity usage needs to be reduced by 25% to 380,000 kWh/yr and gas reduced by 31% to 600,000 kWh/yr in order to achieve an AR of grade B. The Academy may not have a budget for such major investment in the building from either government or industry, so such a scale of project seems unlikely on a 20-year-old site.

Case study: Queens Building, De Montfort University, Leicester

The Queens Building, De Montfort University, Leicester (http://duall.iesd.dmu.ac.uk/1010buildings/), built in 1993, featured in The Government's Energy Efficiency Best Practice Programme (New Practice case study 102, June 1997). There was a full performance study (Probe 4, *BSJ*, April 1996). It won the HVCA Green Building of the Year award in 1995 (Heating and Ventilating Contractors Association). Day lighting and low energy design produced an academic facility that consumed half the energy required by a conventional university building. Offices, studios, classrooms, lecture theatres, mechanical and electrical laboratories occupied four storeys around a central light well. Planned usage was for 1500 students. Natural ventilation from manually controlled low level air inlets was heated with finned pipes. Ridge vents and 133 m air outlet stacks, or chimneys, gave the building a striking and almost a gothic look that reflected the appearance of nearby older buildings, suggestive of a prison or Victorian-era workhouse.

The School of Engineering and Manufacture had a landmark building of brick, L-shaped in plan with high thermal mass and very low thermal transmittances. Walls were brick/cavity/block; exposed brickwork and concrete internally provided thermal storage mass. Many small windows were openable and had internal and fixed external shading; some were triple glazed. The building is occupied by few staff and researchers during the hottest summer weeks; GFA 10,000 m^2; TFA 8,400 m^2. Northerly glazing admitted daylight, minimized solar heat gains and glare. The BEMS switched lighting on/off for normal working hours while manual switches and passive infra-red sensors provided local control. A LTHW gas-fired weather-compensated heating system of radiators, convectors and underfloor pipes used a condensing boiler to supplement a combined heat and power (CHP) system; air handling units with heating coils serve specialist technical rooms; radiators had TRVs. Low energy lamps were used throughout and had manual or occupancy sensing switching.

Metered electricity bills were 436.8 MWh/yr and gas 1,201.2 MWh/yr; these corresponded to an emission of 65 kg CO_2/m^2 yr. The Queens Building was compared to the Energy Efficiency Office low target for university academic buildings, 75 kWh/m^2 yr for electricity and 185 kWh/m^2 yr for gas; an emission of 90 kg CO_2/m^2 yr (no comparable data in *CIBSE Guide F, 2004*). Carbon dioxide atmospheric emission for electrical energy used was 0.52 kg CO_2/kWh and gas 0.2 kg CO_2/kWh (*A User Guide to iSBEM*).

Download the report to understand more about the operation of the building, its services systems, lighting, ventilation and any user issues found. Identify at last three technical issues found by the PROBE team. Calculate the annual CO_2 emission from the building with either the manual method shown in Example 2.1, the workbook provided (asset rating), TM22 or commercial software. Find the Emission Performance Certificate grade A–G that this building would have by using only the information provided for this question. Estimate the energy usage to improve the EPC grade; recommend three energy-saving measures to improve the building's EPC grade with reasons, and the likelihood of adoption, or rejection by the owners or tenants of the building. How worthwhile has the combined heat and power installation proven to be financially and in CO_2 emissions? How much electrical power is used at night and why? What can be observed about the energy used for artificial lighting? How useful was the occupancy survey? What issues did the occupancy survey highlight? What helpful conclusions can you make from this case study? What may happen to this building in the medium to long-term future?

Answers

1. *EPC Grade*: Calculated estimation of the asset rating for the Queens Building corresponds to 37, a B grade; this is a good result. Relating this to the *HM Government Carbon Plan 2011*, CO_2 emissions present now are going to remain in the long term.
2. Technical issues:

    (a) Monitoring and tuning the building energy management system (BEMS) with CO_2 sensing, for the complex modulating damper ventilation strategy took the first year of occupation.
    (b) Building defects stretched into the second year of use.
    (c) Cleaning the inlet air plenums proved difficult.
    (d) An air leakage test could not be conducted due to the intentionally leaky nature of the design.
    (e) Failure of LTHW flow control valves caused summer overheating.
    (f) Contractual disputes delayed implementation of corrections.
    (g) BEMS-controlled actuator arms sheared due to frozen roof lights.
    (h) Called a passive design, but it relied on computer control.
    (i) The ability to automatically control an innovative natural ventilation strategy took more than two years to finalize.
    (j) The combined heat and power system did not reach its design objective.

3.   Energy-saving measures:

(a)   Reduce lighting and PC operation when not needed.
(b)   Reduce electrical demand in unoccupied hours.
(c)   Ensure the automatic ventilation controls function to minimize heating energy.
(d)   Improvement to an A grade would mean reducing electricity use by 40% and gas by 25%; this seems insurmountable as major infrastructure changes to the building are highly unlikely.

4.   CHP:

(a)   Insufficient base thermal demand to maintain full operation.
(b)   Maintenance cost almost outweighed electrical energy-saving value.
(c)   Reduces electrical CO_2 emissions.

5.   Night-time power:

1.   Overnight electrical demand 40 kW.
2.   Lighting on all night for cleaning staff.
3.   Equipment running when building was totally unoccupied.

6.   Lighting:

(a)   Lighting proved to be the largest consumer of electricity; 37 out of 52 kWh/m^2 yr.
(b)   Day lighting created both glaring and gloomy areas.
(c)   Most lights were on regardless of occupation and automatic controls.

7.   Occupancy survey:

(a)   Did not include undergraduate students during the week of the questionnaire.
(b)   95% of staff questionnaires were completed.
(c)   Mostly a comfortable environment.
(d)   Third floor users experienced summer overheating.
(e)   Third floor users experienced unsatisfactory lack of ventilation in winter and summer.
(f)   Uneven air distribution and some downdraughts were experienced.
(g)   Noise from open plan areas caused great concern.
(h)   Office occupants rated the building in the bottom 10% of the Building Use Studies database.

8.   Helpful conclusions:

(a)   An innovative building aiming in the right direction for the *HM Government Carbon Plan 2011*.
(b)   Some summer overheating in a naturally ventilated building in the UK is bound to happen; most people treat it as normal and do not demand air conditioning.
(c)   The passive design maintained generally satisfactory conditions.
(d)   Combination of thermal mass, controllable natural ventilation, central heating, day lighting and solar shading were an endorsement of the passive design principle. Does this sound familiar? Of course it does, it is exactly the principal design for almost every home, commercial building, educational building, university, church, stately home and stone castle built prior to 1960 in the UK (CIBSE Heritage Group, *The Quest for Comfort*).

9.   Future possibilities:

(a)   Annual BEMS software updating with costly maintenance contract from supplier and staff training.

(b)   Continued investigations into why there remains a 40 kW, 100 kWh/day electrical demand when not occupied.

(c)   Summer overheating on 3rd floor will not be tolerated, leading to the installation of direct expansion refrigerated cooling units that will increase energy consumption and peak demand.

(d)   During the first 25 years of building use, staff turnover loses knowledge base; incoming facility managers and technicians bring different approaches to using a low energy building; complex ventilation control becomes inoperative due to wear, lack of understanding and maintenance; energy use increases.

(e)   By the *HM Government Carbon Plan 2011* target year 2020, the Queens Building design, BEMS, mechanical and electrical services require major refurbishment as the needs of the university change.

(f)   Fixed PC computer laboratories no longer needed by 2020; server and data cabling redundant; every user works from wireless network on portable computers emitting almost no heat; internal restructuring into cellular offices and small study rooms as lectures replaced by online learning software.

(g)   New cellular room design does not work with original ventilation strategy; VRV refrigeration direct expansion air conditioning units installed in every room, piped to rooftop condensing units; ventilation stacks demolished; electricity consumption increases.

Sustainability

In the perspective of global terms, *Homo sapiens'* occupation of planet Earth is unlikely to make living here unsustainable. We have lived here for 200,000 years or a lot longer, number around 7 billion and have a highly developed brain for problem-solving. Earth formed 4.5 billion years ago, its surface is 71% covered by water and weighs about 6×10^{24} kg. Anyone imagining that our puny scraping at its surface, burning of a few trees, drilling into superficial layers and emitting products of combustion will destroy Earth is forgetting history. Many of our modern buildings and means of transportation may not last 100 years and are soon to become recycled glass, concrete, steel, copper, aluminium and timber for new constructions at the whim of planners. Egypt's pyramids date back 5000 years; an advanced Maya civilization existed 4000 years ago while stone tools and evidence of the use of fire in China date back 1.3 million years, and their constructions remain standing. What happened on Earth during the 3 billion years before human occupation and anyone's ability to measure time is largely unknown. So, it is unlikely that the author and reader will cause the destruction of Earth by driving to work or play and in keeping warm. Animals search for food and eat until satisfied, but humans rationalize with their intelligence, gain knowledge and feel guilt for wrongdoing. We want to be good citizens and provide for future generations. 'Sustainable development meets the needs of the present without compromising the ability of future generations to meet their own needs' (Brundtland Commission of the United Nations, 1987). Pharaohs obviously intended their proud monuments to last forever, large-scale modern wars focussed on destroying what people had built, the current stock of buildings will endure while we keep the windows clean, supply the buildings with fuel and electricity, maintain their computers and software, while expecting to refurbish the structure, mechanical and electrical services every 25 years. Human expectations for the timescale of buildings have changed from hewn in stone to disposability. Today's principles of sustainability include reducing dependence upon fossil fuels, underground metals, minerals, synthetic chemicals, unnatural substances, encroaching less on the natural world and meeting human needs fairly, to each other, and efficiently; whether the current increasing use of reinforced concrete, carbon fibre polymer, silicon computer chips and photovoltaic solar cells, plastics, aluminium and glass complies with those ideals, remains to be seen.

Building shape

Shape has strong influence on energy demand and comfort conditions within a building. Energy demand is closely related to the exposed exterior surface of a building in relation to its floor area. Think about comparing the energy demand of underground buildings with their equivalent above ground. Ground temperatures surrounding caves remain static all year. Minimizing the surface area exposed to the weather may be helpful in lowering energy use. Shape and orientation depend upon location, climate and purpose of the building. A school used only in the daytime in northern Europe might have extensive tall glazing facing the sun to maximize natural warming, operational shading and controlled natural ventilation to control indoor air temperature. In the tropics where the sun is high in the sky all day, overhanging eaves shade the small glazed areas all day. Arid climates can have outdoor air temperatures of over 30°C every day with intense solar radiation making small shaded glazing preferable, and buildings are permanently cooled, 24/7. Ancient Mediterranean home designers used small shuttered windows and created internal shaded courtyard gardens having learnt to cope with their climate and allow an afternoon siesta time. Take note of those same practices today.

EXAMPLE 2.2

A major computer corporation designed a doughnut-shaped plan for their new headquarters in California recently. Test whether a square building plan shape could reduce its energy target. The outside diameter of the doughnut is 150 m, an inner diameter is 100 m and a single storey 4 m in height was designed.

For the doughnut shape:

$$Floor\ area = \frac{\pi}{4} \times \left(150^2 - 100^2\right) m^2$$

$$Ground\ Floor\ area = 9817\ m^2$$

$$Roof\ area = 9817\ m^2$$

$$Wall\ area = (\pi \times 150 \times 4 + \pi \times 100 \times 4) m^2$$

$$Wall\ area = 3142\ m^2$$

$$\frac{Exterior\ surface}{Floor\ area} = \frac{(3142 + 9817 + 9817)}{9817}$$

$$\frac{Exterior\ surface}{Ground\ floor\ area} = 2.32$$

For an alternative single-storey square plan shape:

$$Floor\ dimensions\ (99.08\ m \times 99.08\ m) = 9817\ m^2\ same\ as\ for\ the\ doughnut$$

$$Wall\ area = (4 \times 99.08 \times 4)\ m^2$$

$$Wall\ area = 1585\ m^2$$

$$\frac{Exterior\ surface}{Floor\ area} = \frac{(1585 + 9817 + 9817)}{9817}$$

$$\frac{Exterior\ surface}{Ground\ floor\ area} = 2.16\ a\ little\ less\ than\ the\ doughnut\ but\ not\ highly\ significant$$

For an alternative 10-storey square plan shape:

$$Floor\ dimensions\ 10 \times (31.33\ m \times 31.33\ m) = 9816\ m^2\ same\ as\ for\ the\ doughnut$$

$$Wall\ area = (10 \times 4 \times 31.33 \times 4)\ m^2$$

$$Wall\ area = 5013\ m^2$$

$$Ground\ floor\ area = (31.33 \times 31.33)\ m^2$$

$$Ground\ floor\ area = 982\ m^2$$

$$Roof\ area = 982\ m^2$$

$$\frac{Exterior\ surface}{Ground\ floor\ area} = \frac{(5013 + 982 + 982)}{982}$$

$$\frac{Exterior\ surface}{Ground\ floor\ area} = 7.1$$

$$\frac{Exterior\ surface}{Total\ floor\ area} = \frac{(5013 + 982 + 982)}{9816}$$

$$\frac{Exterior\ surface}{Total\ floor\ area} = 0.71$$

Conclusion: the multistorey alternative provides a very low exterior surface area per m² of total floor area but a much higher ratio per m² of ground floor area than either of the single-storey designs. The multistorey solution exposes a much larger external surface area to solar heat gains, wind, precipitation and external air temperature heat transfers than the more compact single-storey designs. It will use more energy even though the ratio of exterior per m² of total floor area is much lower. The low doughnut shape imitates the ancient principle of looking inward to a shaded atrium while minimizing the exterior exposure to the weather. This just proves that the latest computer brains still look to ancient principles. Experiment with other shapes to see if there is a better solution.

Small-scale generation

On-site generation of electricity may become financially worthwhile compared to purchases from the public supply or used as emergency stand-by power (*CIBSE Guide K Electricity in*

Buildings, 2005). The disadvantage is that the public supply has built-in plant redundancy, meaning that there is more generation capacity available than is being used at any one moment, so that maintenance can be conducted. Private generation must have 100% standby capacity if it is to guarantee never taking power from the public supply and avoid the utility company's fixed energy charges. That is the problem faced by private generators. When a private generator is running, it has to have a consistent demand for power. The public network is constantly monitored and controlled to match supply to demand but this is not the same situation in a private system. The principle of public generation was that power would be supplied at the lowest attainable cost from the lowest grades of fuel and with the largest-scale plant and distribution. Private generation has none of those advantages and pays a much higher price for plant and fuel as well as being completely responsible for capital cost, maintenance, repairs and reliability; a very demanding task that not even hospitals or public corporations really want to take on. On a household scale, paying for gas, petrol or diesel fuel plant is likely to be more costly than purchasing power. Solar panels provide free generation and can export surplus power to the public grid to create income for the homeowner but is the capital cost justifiable and is the supply of solar generation sufficiently reliable in the climate? That is the question and that depends on location. How long does a homeowner remain at one address? Rarely for long enough to justify over-capitalization of the property value and unlikely to recover the capital cost on resale of the home.

EXAMPLE 2.3

What does it cost to generate 3 kW for home use from a petrol generator that consumes 6 litres of unleaded petrol during a full load period of 4 hours? Petrol costs £1.4/litre. How does it compare to purchasing electricity?

$$Generation\ cost = 6\ l \times \frac{140\ p}{l} \times \frac{1}{3\ kW \times 4\ h}$$

$$Generation\ cost = 70 \frac{p}{kWh}$$

Purchased electricity costs 12–30 p/kWh peak rates making small-scale generation unattractive, and fuel has to be transported manually.

Larger power generators can be gas engine-driven to take advantage of piped natural gas and may be closer to the purchased cost for power. Engine heat output is used to produce hot water, making it into a combined heat and power system.

Questions

1. What are the differences between the designers and users of a building?
2. How many people use the computer building energy management system in a large office building, university campus and hospital every day and week?

    1. Everyone in the building.
    2. Specialist maintenance contractor.
    3. One person has the expertise and time to use it.
    4. Nobody.
    5. Everyone in the property and facilities management department.

3. Who is most interested in a macroscopic appreciation of a building?

    1. Owner.
    2. Facilities manager.
    3. Building services engineer.
    4. Employees.
    5. Architect.

4. Who is most concerned with the microscopic scale aspects of a building?

    1. Architect.
    2. Employees.
    3. Owner.
    4. Facilities manager.
    5. Building services engineer.

5. Rotherham Magistrates Court PROBE report (*BSJ*, March 1994) was constructed in 1994 as an air-conditioned, low energy public building. It has 10 courtrooms, gas-fired heating, refrigeration chillers and 7 air handling units; gross floor area, GFA, 5,450 m^2 and treated floor area, TFA, 4,350 m^2. An air leakage test revealed a flow through the building of 17 m^3/m^2 h, based on TFA m^2, at a static internal air pressure of 50 Pa. Metered electricity bills were 444 MWh/yr and gas 620 MWh/yr. Good practice data for an air-conditioned magistrates court was 31 kWh/m^2 yr for electricity and 125 kWh/m^2 yr for gas (*CIBSE Guide F*, 2004). Carbon dioxide atmospheric emission for electrical energy used was 0.52 kg CO_2/kWh and gas 0.2 kg CO_2/kWh (*A User Guide to iSBEM*). Download the report to understand more about the operation of the building, its services systems, lighting, ventilation and any user issues found. Identify three technical faults found by the PROBE team. Calculate the annual CO_2 emission from the building with either the manual method shown in Example 2.1, the workbook provided (asset rating), TM22 or commercial software. Find the EPC grade A–G this building would have by using only the information provided for this question. Estimate the energy usage to improve the EPC grade; recommend three energy-saving measures to improve the building's EPC grade with reasons, and the likelihood of adoption or rejection by the owners or tenants of the building.

6. Developers let the speculative office building 1 Aldermanbury Square, City of London, to the Standard Chartered Bank for 250 occupants in 1990 (Probe 2, *BSJ*, December 1995). It was a prestige air-conditioned 9-storey building; GFA 8,000 m^2; TFA, 7,000 m^2 with offpeak ice thermal storage to reduce peak electrical energy cost. Metered electricity bills were 2,597 MWh/yr and gas 224 MWh/yr. Good practice data for an air-conditioned prestige office building was 234 kWh/m^2 yr for electricity and 114 kWh/m^2 yr for gas (*CIBSE Guide F*). Use carbon dioxide atmospheric emission for electrical energy as 0.52 kg CO_2/kWh and gas 0.2 kg CO_2/kWh. Download the report to understand more about the operation of the building, its services systems, lighting, ventilation and any user issues found. Identify three technical faults found by the PROBE team. Calculate the annual CO_2 emission from the building with either the manual method shown in Example 2.1, the workbook provided (asset rating), TM22 or commercial software. Find the EPC grade A–G this building would have by using only the information provided for this question. Estimate the energy usage to improve the EPC grade; recommend three energy-saving measures to improve the building's EPC grade with reasons, and the likelihood of adoption or rejection by the owners of the building.

7. The Anglia Polytechnic University Learning Resource Centre, Chelmsford, had 750 work places and was built in 1994 (Probe 8 APU, *BSJ*, December 1996). GFA 6,018 m^2; TFA, 5,656 m^2; library, TV studio, conference room and café. A well-insulated building with triple glazing, brick/insulation/block walls, gas-fired central heating and mechanically controlled mixed mode natural ventilation. Metered electricity bills were 282.8 MWh/yr and gas 650.4 MWh/yr. Good practice data for a naturally ventilated educational library was 46 kWh/m^2 yr for electricity and 115 kWh/m^2 yr for gas (*CIBSE Guide F*). Use carbon dioxide atmospheric emission for electrical energy as 0.52 kg CO_2/kWh and gas 0.2 kg CO_2/kWh. Download the report to understand more about the operation of the building, its services systems, lighting, ventilation and any user issues found. Identify three technical faults found by the PROBE team. Calculate the annual CO_2 emission from the building with either the manual method shown in Example 2.1, the workbook provided (Asset rating), TM22 or commercial software. Find the Emission Performance Certificate, EPC grade A–G, this building would have by using only the information provided for this question. Estimate the energy usage to improve the EPC grade; recommend three energy-saving measures to improve the building's EPC grade with reasons, and the likelihood of adoption, or rejection by the owners of the building.

8. The Barclaycard HQ in Northampton (Probe, *BSJ*, March 2000) was known as the greenest corporate headquarters of the 1990s. It was a prestige mixed mode natural ventilation and air-conditioned 4-storey building for 2400 occupants, each had a PC; 2000 meals were served each day; estimated TFA 27,600 m^2. Perimeter heating was from two condensing gas-fired boilers serving low temperature hot water radiators fitted with thermostatic valves. An air leakage test revealed a flow through the building of 17.3 m^3/m^2 h at a static internal air pressure of 50 Pa. Estimated electricity use was 5,630 MWh/yr and gas 3,118 MWh/yr. Good practice data for an air-conditioned prestige office building was 234 kWh/m^2 yr for electricity and 114 kWh/m^2 yr for gas (*CIBSE Guide F*). Use carbon dioxide atmospheric emission for electrical energy as 0.52 kg CO_2/kWh and gas 0.2 kg CO_2/kWh. Download the report to understand more about the operation of the building, its services systems, lighting, ventilation and any user issues found. Identify three technical faults found by the PROBE team. Calculate the annual CO_2 emission from the building with either the manual method shown in Example 2.1, the workbook provided (asset rating), TM22 or commercial software. Find the EPC grade A–G this building would have by using only the information provided for this question. Estimate the energy usage to improve the EPC grade; recommend three energy-saving measures to improve the building's EPC grade with reasons, and the likelihood of adoption, or rejection by the owners of the building.

9. The Cheltenham & Gloucester Chief Office in Gloucester was built in 1989 for 930 occupants (Probe 3, C&G Chief Office, *BSJ*, February 1996). It was a prestige air-conditioned 4-storey building; GFA 19,900 m^2; TFA, 16,390 m^2. Gas-fired LPHW heating to perimeter convectors, air cooled chillers, air handling units and variable air volume distribution systems. An air leakage test was not conducted. Metered electricity bills were 6,048 MWh/yr and gas 1,655 MWh/yr. Good practice data for an air-conditioned prestige office building was 234 kWh/m^2 yr for electricity and 114 kWh/m^2 yr for gas (*CIBSE Guide F*). Use carbon dioxide atmospheric emission for electrical energy as 0.52 kg CO_2/kWh and gas 0.2 kg CO_2/kWh . Download the report to understand more about the operation of the building, its services systems, lighting, ventilation and any user issues found. Identify three technical faults found by the PROBE team. Calculate the annual CO_2 emission from the building with either the manual method shown in Example 2.1, the workbook provided (asset rating), TM22 or commercial software. Find the EPC grade A–G this building would have by using only the information provided

for this question. Estimate the energy usage to improve the EPC grade; recommend three energy-saving measures to improve the building's EPC grade with reasons, and the likelihood of adoption or rejection by the owners of the building.

10. A small health service building called the Woodhouse Medical Centre in Sheffield was constructed in 1989. It was intended as a low energy green building, single storey with brick/block walls and high thermal insulation and natural ventilation (Probe 6, Woodhouse Medical Centre, *BSJ*, August 1996). Single-storey; GFA 640 m^2; TFA, 640 m^2; LTHW gas-fired radiator heating system, each radiator had a thermostatic flow control valve. The occupants did not make use of the manually controlled ventilators, windows or openable roof lights; summer overheating led to the installation of room air conditioners. No air leakage test was conducted. Metered electricity bills were around 32,000 kWh/yr and gas 35,200 kWh/yr. Good practice data for a naturally ventilated health centre were taken as 55 kWh/m^2 yr for electricity and 174 kWh/m^2 yr for gas (*CIBSE Guide F*) due to incomplete data being available. Use carbon dioxide atmospheric emission for electrical energy as 0.52 kg CO_2/kWh and gas 0.2 kg CO_2/kWh. Download the report to understand more about the operation of the building, its services systems, lighting, ventilation and any user issues found. Identify three technical faults found by the PROBE team. Calculate the annual CO_2 emission from the building with either the manual method shown in Example 2.1, the workbook provided (Asset rating), TM22 or commercial software. Find the Emission Performance Certificate, EPC grade A–G, this building would have by using only the information provided for this question. Estimate the energy usage to improve the EPC grade; recommend three energy-saving measures to improve the building's EPC grade with reasons, and the likelihood of adoption, or rejection by the owners of the building.

11. The Cable & Wireless College (Probe 5, *BSJ*, June 1996) was constructed in 1993 as a teaching and residential educational building in Coventry. It had 2 lecture theatres, 20 classrooms, 22 technical training rooms, 168 study bedrooms, library, restaurant, swimming pool and sports buildings. It was single storey, had natural ventilation, GFA 12,019 m^2; TFA, 11,400 m^2; LTHW radiator heating from gas-fired boilers. No air leakage test conducted. Metered electricity bills were 2,132 MWh/yr and gas 4,560 MWh/yr. Good practice data for an education halls of residence is to be taken as 85 kWh/m^2 yr for electricity and 240 kWh/m^2 yr for gas as there is a wide range of building types on the campus (*CIBSE Guide F*). Use carbon dioxide atmospheric emission for electrical energy as 0.52 kg CO_2/kWh and gas 0.2 kg CO_2/kWh. Download the report to understand more about the operation of the building, its services systems, lighting, ventilation and any user issues found. Identify three technical faults found by the PROBE team. Calculate the annual CO_2 emission from the building with either the manual method shown in Example 2.1, the workbook provided (asset rating), TM22 or commercial software. Find the EPC grade A–G this building would have by using only the information provided for this question. Estimate the energy usage to improve the EPC grade; recommend three energy-saving measures to improve the building's EPC grade with reasons, and the likelihood of adoption or rejection by the owners of the building.

12. The Charities Aid Foundation, a U-shaped plan, three-storey brick building, was constructed in West Malling, Kent, in 1997 for 200 occupants (Probe 13, *BSJ*, February 1998). Designed as a mixed mode naturally ventilated building with openable windows and external solar shading, it was an example of solar architecture. Cooling was provided by water sprays into the exhaust air which then pre-cooled incoming warm outdoor air through heat exchangers; an adiabatic cooling system and low cost, with no mechanical refrigeration plant. Work areas were mainly open plan; GFA 3,900 m^2; TFA, 3,700 m^2; gas-fired heating with radiators

having thermostatic valves; two air handling units with cross-flow heat exchangers for heat recovery. A computer server room and the boardroom had direct expansion air conditioning units using variable refrigerant volume flow, VRV, for economy. No air leakage test was conducted but smoke pencils found leakages at windows. Metered electricity bills were 432,900 kWh/yr and gas 558,700 kWh/yr. Good practice data for an open plan naturally ventilated office building was 54 kWh/m^2 yr for electricity and 79 kWh/m^2 yr for gas (*CIBSE Guide F*). Use carbon dioxide atmospheric emission for electrical energy as 0.52 kg CO_2/kWh and gas 0.2 kg CO_2/kWh. Download the report to understand more about the operation of the building, its services systems, lighting, ventilation and any user issues found. Identify three technical faults found by the PROBE team. Calculate the annual CO_2 emission from the building with either the manual method shown in Example 2.1, the workbook provided (asset rating), TM22 or commercial software. Find the EPC grade A–G this building would have by using only the information provided for this question. Estimate the energy usage to improve the EPC grade; recommend three energy-saving measures to improve the building's EPC grade with reasons, and the likelihood of adoption or rejection by the owners of the building.

13. An air-conditioned green building for the Co-operative Retail Services HQ (Probe 17, *BSJ*, October 1998) had 4 storeys for 930 staff in Manchester. Built as crescent-shaped with an occupancy density of 5 m^2/person, it had a TFA of 17,300 m^2 and interior atria. It had chilled beams, displacement ventilation, gas-fired heating, chillers, air handling units and an ice thermal storage system to reduce peak electrical cost. An air leakage test revealed a flow through the building of 17.2 m^3/m^2 h at a static internal air pressure of 50 Pa. Metered electricity bills were 7,800 MWh/yr and gas 2,045 MWh/yr. Good practice data for an air-conditioned prestige office building was 234 kWh/m^2 yr for electricity and 114 kWh/m^2 yr for gas (*CIBSE Guide F*). Use carbon dioxide atmospheric emission for electrical energy as 0.52 kg CO_2/kWh and gas 0.2 kg CO_2/kWh. Download the report to understand more about the operation of the building, its services systems, lighting, ventilation and any user issues found. Identify three technical faults found by the PROBE team. Calculate the annual CO_2 emission from the building with either the manual method shown in Example 2.1, the workbook provided (asset rating), TM22 or commercial software. Find the EPC grade A–G this building would have by using only the information provided for this question. Estimate the energy usage to improve the EPC grade; recommend three energy-saving measures to improve the building's EPC grade with reasons, and the likelihood of adoption or rejection by the owners of the building.

14. The University of East Anglia built the Elizabeth Fry building, Norwich, in 1995 (Probe 14, *BSJ*, April 1998) for teaching. It was a low energy design, 4-storey building; GFA 3,250 m^2; TFA, 3,130 m^2; an air leakage test revealed a flow through the building of 6.53 m^3/m^2 h at a static internal air pressure of 50 Pa which corresponded to a natural infiltration rate of 0.97 air changes/h, which was a good standard. No refrigeration was installed; Termodeck air distribution through channels in the concrete floors distributed supply air to the rooms. There were cellular offices for academic staff, seminar rooms, a lecture theatre, dining room and kitchen. Walls were well-insulated concrete block/block; argon-filled triple-glazed windows with low emissivity glass and mid-pane venetian blinds. Heating came from gas-fired condensing boilers for the air handling unit heating coils. Metered electricity bills were 191MWh/yr and gas 96 MWh/yr. Good practice data for a naturally ventilated cellular office building, the nearest equivalent, was 33 kWh/m^2 yr for electricity and 79 kWh/m^2 yr for gas (*CIBSE Guide F*). Use carbon dioxide atmospheric emission for electrical energy as 0.6 kg CO_2/kWh and gas 0.2 kg CO_2/kWh. Download the report to understand more about the

operation of the building, its services systems, lighting, ventilation and any user issues found. Identify three technical faults found by the PROBE team. Calculate the annual CO_2 emission from the building with either the manual method shown in Example 2.1, the workbook provided (asset rating), TM22 or commercial software. Find the EPC grade A–G this building would have by using only the information provided for this question. Estimate the energy usage to improve the EPC grade; recommend three energy-saving measures to improve the building's EPC grade with reasons, and the likelihood of adoption or rejection by the owners of the building.

15. Gardener House, Homeowners Friendly Society, UK, was constructed in 1994 as an air conditioned, 2-storey low energy head office with air conditioning (Probe 7, Gardener House, *BSJ*, October 1996). Open plan office space for 120 occupants. Part of the ground floor was buried. Chilled beams and displacement ventilation provide the HFS with a degree of novelty to the air conditioning requested by client. GFA 4,300 m²; TFA, 3,800; stone clad exterior with double-glazed grey glass and internal venetian blinds. The building had open plan offices, boardroom, dining room and kitchen and gas-fired LTHW heating. An air leakage test revealed a flow through the building of 27 m³/m² hr at a static internal air pressure of 50 Pa with very leaky windows as found with smoke pencils. Metered electricity bills were 1,216 MWh/yr and gas 1,030 MWh/yr. Good practice data for an air-conditioned prestige office building was 234 kWh/m² yr for electricity and 114 kWh/m² yr for gas (*CIBSE Guide F*). Use carbon dioxide atmospheric emission for electrical energy as 0.52 kg CO_2/kWh and gas 0.2 kg CO_2/kWh. Download the report to understand more about the operation of the building, its services systems, lighting, ventilation and any user issues found. Identify three technical faults found by the PROBE team. Calculate the annual CO_2 emission from the building with either the manual method shown in Example 2.1, the workbook provided (asset rating), TM22 or commercial software. Find the EPC grade A-G this building would have by using only the information provided for this question. Estimate the energy usage to improve the EPC grade; recommend three energy-saving measures to improve the building's EPC grade with reasons, and the likelihood of adoption or rejection by the owners of the building.

16. Marston Book Services, Milton Park, Abingdon, had a 2-storey brick open office building constructed in 1996 as an office plus a large book warehouse. It was designed for 53 office staff and 46 for the warehouse (Probe 16, *BSJ*, August 1998). It was a low cost naturally ventilated office building plus a book warehouse; office TFA 962 m² having gas-fired LTHW radiators and openable windows. The warehouse was a single-storey, clear-span industrial building of GFA 5,028 m² on the ground plus 1,840 m² of mezzanine with gas-fired warm air heating. An air leakage test revealed a flow through the office building of 27.1 m³/m² h at a static internal air pressure of 50 Pa and 9.4 m³/m² h for the warehouse. Metered electricity bills for the office were 76,000 kWh/yr and gas 126,000 kWh/yr. Metered electricity bills for the warehouse were 332,000 kWh/yr and gas 240,000 kWh/yr. Good practice data for a naturally ventilated office building was 54 kWh/m² yr for electricity, 79 kWh/m² yr for gas and for a naturally ventilated warehouse was 34 kWh/m² yr for electricity and 187 kWh/m² yr for gas (*CIBSE Guide F*). Use carbon dioxide atmospheric emission for electrical energy as 0.52 kg CO_2/kWh and gas 0.2 kg CO_2/kWh. Download the report to understand more about the operation of the building, its services systems, lighting, ventilation and any user issues found. Identify three technical faults found by the PROBE team. Calculate the annual CO_2 emission from the building with either the manual method shown in Example 2.1, the workbook provided (asset rating), TM22 or commercial software. Find the EPC grade A–G this building would have by using only the information provided for this question. Estimate the energy

usage to improve the EPC grade; recommend three energy-saving measures to improve the building's EPC grade with reasons, and the likelihood of adoption or rejection by the owners of the building.

17. The University of Birmingham built the Orchard Learning Resource Centre in 1996 as a low energy, naturally ventilated library, academic offices and book archive (Probe 17,*BSJ*, July 2000); GFA and TFA 4,500 m^2; thermal insulation slightly better than the Building Regulations at the time; gas-fired LTHW radiators with TRVs. Ventilation was with both manually and automatically controlled ventilators. The building had 103 PCs. An air leakage test revealed a flow through the office building of 31.9 m^3/m^2 h at a static internal air pressure of 50 Pa. This was a very leaky building; smoke pencils identified leaks at structural junctions, where heating pipes passed through walls, doorways, beneath skirting boards, around windows and at automatically operated windows at high level that did not close securely. Metered electricity bills were 356 MWh/yr and gas 767 MWh/yr. Good practice data for an education library with natural ventilation was 46 kWh/m^2 yr for electricity, 115 kWh/m^2 yr for gas (*CIBSE Guide F*). Use carbon dioxide atmospheric emission for electrical energy as 0.46 kg CO_2/kWh and gas 0.2 kg CO_2/kWh. Download the report to understand more about the operation of the building, its services systems, lighting, ventilation and any user issues found. Identify three technical faults found by the PROBE team. Calculate the annual CO_2 emission from the building with either the manual method shown in Example 2.1, the workbook provided (asset rating), TM22 or commercial software. Find the EPC grade A–G this building would have by using only the information provided for this question. Estimate the energy usage to improve the EPC grade; recommend three energy-saving measures to improve the building's EPC grade with reasons, and the likelihood of adoption or rejection by the owners of the building.

18. Portsmouth University built the Portland Building, Portsmouth, for the School of Architecture in 1996 for 60 staff and 870–1300 students (Probe 18, *BSJ*, January 1999). It had a design suggesting a fortress but was E-shaped in plan, not as Henry VIII would have wished. There were lecture theatres, seminar rooms, staff offices and a library; GFA 6,230 m^2; TFA 6,000 m^2; four storeys. Ventilation was mixed mode natural and mechanical with gas-fired LTHW heating, some underfloor heating, and some direct expansion cooling units. Glare from windows was a problem for users. The lecture theatre had displacement ventilation that caused draughts and air flow whistling. Windows had both manual and automatic operation which provided poor control; motorized roof windows stuck partly open, closed slowly and let the rain in. There were automated external sun shades. Internal air temperature control had problems. It was a low cost, naturally ventilated office; office TFA 962 m^2 with gas-fired LTHW radiators and openable windows. An air leakage test revealed a flow through the building of 15.6 m^3/m^2 h at a static internal air pressure of 50 Pa. Electricity bills were estimated to be 300 MWh/yr and gas 600 MWh/yr due to meter errors. Good practice data for a naturally ventilated office building was used as being the nearest listed application, 54 kWh/m^2 yr for electricity, 79 kWh/m^2 yr for gas (*CIBSE Guide F*). Use carbon dioxide atmospheric emission for electrical energy as 0.52 kg CO_2/kWh and gas 0.2 kg CO_2/kWh. Download the report to understand more about the operation of the building, its services systems, lighting, ventilation and any user issues found. Identify three technical faults found by the PROBE team. Calculate the annual CO_2 emission from the building with either the manual method shown in Example 2.1, the workbook provided (asset rating), TM22 or commercial software. Find the EPC grade A–G this building would have by using only the information provided for this question. Estimate the energy usage to improve the EPC grade; recommend three energy-saving measures to improve the building's

EPC grade with reasons, and the likelihood of adoption or rejection by the owners of the building.

19. Tanfield House, Edinburgh, was an open plan, prestige 2-storey office building for the Standard Life administrative centre, built in 1995. It had internal atria, plant operated 24 hours/day, 7 days/week and the building was fully air conditioned. It was designed for 1300 office staff, had a roof garden, 450-seat restaurant and 306 basement car park spaces (Probe 1, *BSJ,* September 1995); GFA 20,000 m^2; TFA 19,780 m^2; 20 air handling units. There were 1–3 PCs for each workstation. The cool climate allowed a lot of free cooling through the air conditioning system. Openable windows had tinted glass and fixed external shading; they were rarely opened due to draughts. Internal solar shading had motorized sails beneath the dome roof glazing. Users had no local control over lighting or the HVAC. An air leakage test revealed a flow through the office building of 27.1 m^3/m^2 h at a static internal air pressure of 50 Pa. Metered electricity bills for the office were 6,329,600 kWh/yr and gas 6,547,180 kWh/yr. Good practice data for a prestige air-conditioned office building was 234 kWh/m^2 yr for electricity, 114 kWh/m^2 yr for gas (*CIBSE Guide F*). A separate building housed the IT servers and is not included in this data. Use carbon dioxide atmospheric emission for electrical energy as 0.52 kg CO_2/kWh and gas 0.2 kg CO_2/kWh. Download the report to understand more about the operation of the building, its services systems, lighting, ventilation and any user issues found. Identify three technical faults found by the PROBE team. Calculate the annual CO_2 emission from the building with either the manual method shown in Example 2.1, the workbook provided (asset rating), TM22 or commercial software. Find the EPC grade A–G this building would have by using only the information provided for this question. Estimate the energy usage to improve the EPC grade; recommend three energy-saving measures to improve the building's EPC grade with reasons, and the likelihood of adoption or rejection by the owners of the building.

3 Built environment

Learning objectives

Study of this chapter will enable the reader to:

1. relate human physiological needs to the internal and external environment;
2. calculate indoor thermal comfort equations;
3. know the instrumentation used for indoor environmental monitoring.

Key terms and concepts

air velocity 51; conduction, convection, radiation and evaporation 50; data logger 52; dry-bulb air temperature 52; dry resultant temperature 53; environmental temperature 54; globe temperature 52; humidity 55; mean radiant temperature 52; olf 50; operative temperature 54; pitot-static tube 54; sling psychrometer 51; thermistor anemometer 53; vane anemometer 53; vapour pressure 51; wet-bulb temperature 52; wet-bulb globe temperature 53; wind chill index 57.

Introduction

Identifies basic needs for comfort in different climates, means of heat transfer and how air quality is assessed. Explains how CO_2 level determines ventilation rate, comfort standards in rooms, measurements of atmospheric pressures and temperatures. Outlines external climate effects, environmental measurement instruments and calculation are shown.

Human comfort

One of people's basic needs is to maintain a constant body temperature while the metabolism regulates heat flows from the body to compensate for changes in the environment. We have become expert in fine-tuning the environmental conditions produced by the climate in relation to the properties of the building envelope to avoid discomfort. A simple tent or cave may be sufficient to filter out the worst of adverse weather conditions, but the ability of this type of shelter to respond to favourable heat gains or cooling breezes may be too fast or too slow to maintain comfort.

Outside of the tropics, latitudes beyond 23.5° from the equator, houses may be advantageously oriented towards the sun to take advantage of solar heat gains, which will be stored in the dense parts of the structure and later released into the rooms to help offset heat losses to the cool external air during winter. Buildings within the tropical zone require large overhanging roofs and shutters over the windows to exclude most summer solar radiation and shade the rooms. Thus, the building envelope acts to moderate extremes of climate, and by a suitable design of illumination and ventilation openings, together with heating, cooling and humidity controls, a stable internal environment can be matched to the use of the building.

The building services engineer is involved with every part of the interface between the building and its occupant. Visually, colours rendered by natural and artificial illumination are produced by combinations of decor and windows. The acoustic environment is largely attributed to the success achieved in producing the required temperatures with quiet services equipment, all of which is part of the thermal control and transportation arrangements. Energy consumption for thermal and lighting systems is the main concern, and close coordination between client, architect and engineer is vitally important.

Heat transfer between the human body and its surroundings takes place through conduction, convection and radiation. Points of conduction contact with the structure are made with furniture and the floor. Clothing normally has a substantial thermal insulation value and discomfort should be avoided. Heat removed from the body by natural convection currents in the room air, or fast-moving airstreams produced by ventilation fans or external wind pressure, is a major source of cooling. The body's response to a cool air environment is to restrict blood circulation to the skin to conserve deep tissue temperature, involuntary reflex action, shivering, if necessary, and in extreme cases inevitable lowering of body temperature. This last state of hypothermia can lead to loss of life and is a particular concern in relation to elderly people. Radiation heat transfer takes place between the body and its surroundings. The direction of heat transfer may be either way, but normally a minor part of the total body heat loss takes place by this method. Radiation between skin and clothing surfaces and the room depends on the fourth power of the absolute surface temperature, the emissivity, the surface area and the geometric configuration of the emitting and receiving areas. Thus, a moving person will experience changes in comfort level depending on the location of the hot and cold surfaces in the room, even though air temperature and speed may be constant. Some source of radiant heat is essential for comfort, particularly for sedentary occupations. Hot-water central heating radiators, direct fuel-fired appliances and most electrical heaters provide a combination of convection and radiated warmth. The elderly find particular difficulty in keeping warm when they are relatively immobile, and convective heating alone is unlikely to be satisfactory. A source of radiant heat provides rapid heat transfer and a focal point, easy manual control and quick heat-up periods. Overheating from sunshine causes discomfort and disability glare. Humid air is exhaled; further transfer of moisture from the body takes place by evaporation from the skin and through clothing. Maintenance of a steady rate of moisture removal from the body is essential; this is a mass transfer process depending on air humidity, temperature and speed as well as variables such as clothing and activity.

The quality of the air in a building depends upon the quantity, type and dispersal of atmospheric pollutants. Some of these, odorants, can be detected by the olfactory receptors in the nose. These are the odours, vapours and gases that ingress from the outdoor environment and are released from humans, animals, flora, furnishings and the structural components of the building. Solid particles of dust, pollen and other contaminants often have little or no smell. Cleaning fluids such as ammonia, cigarette smoke, hair spray, deodorants and perfumes can be most noticeable.

The inflow of diesel exhaust fumes, road tar, paint vapours and creosote creates unpleasantly noticeable pollution, even when of short duration. The presence of harmful pollutants such as carbon monoxide and radon gases is not detectable by the occupant. Indoor air quality may be said to be acceptable when not more than 50% of the occupants can detect any odour. Olfactory response adapts to pollutants over time, making those regularly exposed less able to identify odours that are immediately noticeable to a newcomer. The olf unit quantifies the biological effluent from one standard person or equivalent from other sources. 1 olf corresponds to a non-smoking sedentary person occupying $10\,m^2$ ventilated with an outside air supply of 10 l/s.

Outside air ventilation rates vary for different applications; 10 l/s per person is generally used, ranging within 5–20 l/s per person, or an air change rate for fume removal. Measurement of return air CO_2 content can be applied to ensure adequate air quality; outdoor air CO_2 is up to 400 ppm and up to 1200 ppm, 0.12%, while 0.5% is the maximum limit for humans. The designer needs to evaluate all aspects of the need for outdoor air ventilation. These include the heating and cooling plant loads that will be generated, the potential for energy recovery between the incoming fresh air and outgoing air at room temperature, avoidance of draughts in the occupied rooms, the variation of load with the occupancy level and the ability to utilize outdoor air to provide free cooling to the building when the outdoor air is between 10°C and 20°C.

Moving-air velocity in normally occupied rooms will be between 0 and 2 m/s where the upper figure relates to a significantly uncomfortable hot or cold draught. Still-air conditions are most unlikely to occur, all buildings leak air, convection currents from people, warmed surfaces and electrical equipment and computers promote air circulation. Room air movement patterns should be variable rather than monotonous and ventilation of every part of the space is most important.

Mean radiant temperature is a measure of radiation heat transfers taking place between various surfaces and has an important bearing on thermal comfort. Air temperature is that from a mercury-in-glass dry-bulb thermometer freely exposed to the air stream, usually in a sling psychrometer. Atmospheric vapour pressure is that part of the barometric pressure produced by the water vapour in humid air. Standard atmospheric pressure at sea level is 1013.25 mb, comprising about 993.0 mb from the weight of dry gases and 20.25 mb from the water vapour, depending on the values of barometric pressure, air temperature and humidity. Comfort assessment can be expressed as the percentage of people dissatisfied and their predicted mean vote. Dissatisfaction among less than one-third of occupants is achieved in indoor environments within the slightly cool to slightly warm categories. Personal variation in clothing, sedentary position and control over the microclimate are the final control mechanisms. People adapt to climates that are regularly encountered. To obtain values of temperature, velocity, carbon dioxide and humidity in the atmosphere, each is measured separately as 0–10 V or 0–20 mA analogue signals. The analogue-to-digital converter stores information in a data logger random access memory, and readings are accessed through a desktop or portable computer. Graphical output can show the history of the recorded temperature, or other variable, during a period of time. Such graphs are termed trends. The software can be programmed to take readings at, say, intervals of 2 s or longer, or only whenever a significant change takes place. Figure 3.1 shows the schematic arrangement of a data-logging system with a trend display of room air temperature.

Extremes of external climate are mainly of concern to construction workers and travellers in severe environments. There are two main indices: wind chill index and heat stress index. Wind chill index measures the cooling effect on the body of a moving air stream, leading to frostbite and hypothermia. Hot climates lead to an inevitable increase in body temperature, hyperthermia, with symptoms of fatigue, headache, dizziness, vomiting, irritability, fainting and

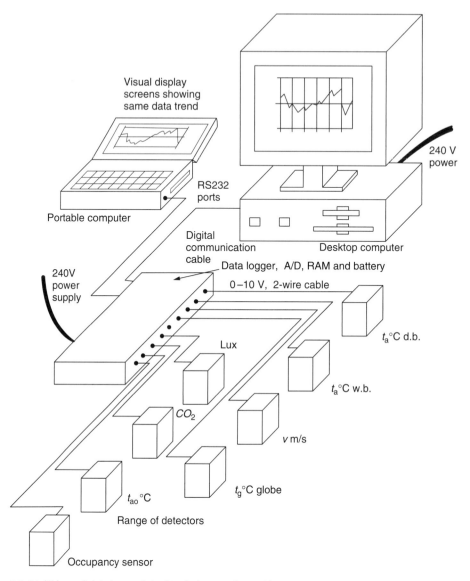

3.1 Multichannel data logger linked to desktop and portable computers.

failure of normal blood circulation to cope with the problem. Heat exhaustion of this sort can normally be counteracted by removal to a cool place.

Environmental measurements

Measurement of air temperature and humidity is accurately made by using both dry- and wet-bulb mercury-in-glass thermometers in a sling psychrometer, otherwise called a whirling hygrometer. The dry-bulb thermometer defines the air temperature t_a °C d.b. and evaporation of water from the cotton wick cools the wet-bulb thermometer; its reading is known as the wet-bulb temperature t_a °C w.b.

In order to find the mean radiant temperature t_r, the dry-bulb temperature t_a and the air velocity v m/s are measured:

$$t_r = t_g\left(1 + 2.35\sqrt{v}\right) - 2.35t_a\sqrt{v}$$

where t_g is the globe temperature measured at the centre of a blackened globe of diameter 150 mm suspended at the measurement location. The resultant temperature, t_{res} is found using a globe of diameter 100 mm and is calculated from:

$$t_{res} = t_g\left(1 + 3.17\sqrt{v}\right) - 3.17t_a\sqrt{v}$$

The wet-bulb globe temperature (WBGT) is used for the assessment of warm humid environments for health and safety at work conditions. It is found from the air wet bulb in weighted amounts to give greater emphasis to humidity. It is humidity that causes significant heat stress, while a hot dry air flow is efficient at removing bodily heat production. When the local air temperature approaches core body temperature, the sole remaining mechanism for bodily heat removal is evaporation from breath and skin; that is, when there is no handy swimming pool.

$$WBGT = 0.7t_aw.b. + 0.2t_g + 0.1t_ad.b.$$

Direct-reading air velocity measuring instruments are shown in Figures 3.2 and 3.3. Rotating vane anemometers are used to measure airstreams through ventilation grilles, where the rotational speed of the blades is magnetically counted. Thermistor and hot-wire anemometers use the air stream cooling effect on the probe. Duct air flow rates are found by inserting a pitot-static tube into the airway, taking up to 48 velocity readings and evaluating the air volume flow rate from the average air velocity found from all the readings. The term for air humidity is percentage saturation, and the most reliable method of measurement is to take dry- and wet-bulb air temperature readings using a sling psychrometer and refer to a psychrometric chart. Permanent monitoring and control require an electronic sensor utilizing a hygroscopic salt covering a coil

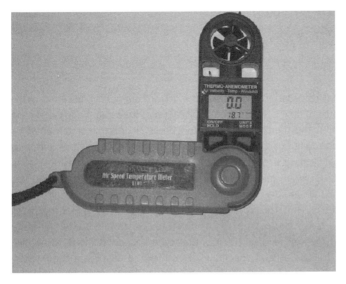

3.2 Thermistor anemometer.

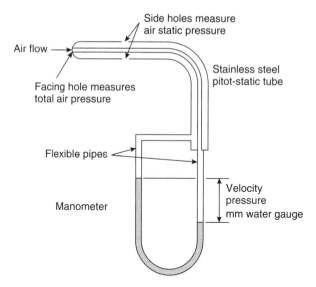

Side holes measure
air static pressure

Air flow

Facing hole measures
total air pressure

Stainless steel
pitot-static tube

Flexible pipes

Manometer

Velocity
pressure
mm water gauge

3.3 Pitot-static tube and U-tube manometer.

of wire. The output signal from this unit depends on the amount of moisture absorbed by the sensor. Non-touch temperature sensors receive infrared heat radiation emitted from all surfaces that are above absolute zero, and they are calibrated to give the surface temperature. Small areas can be surveyed with portable gun instruments and large scans can be performed by mobile television camera equipment, which produces temperature contour maps. Ground-level or aerial surveys are used to detect energy waste from non-existent, inadequate or damaged thermal insulation in homes, in factories or where there are buried pipes in district heating systems. Heat losses from buildings and the conditions required for thermal comfort both depend on the mean radiant temperature and the air temperature. The environmental temperature t_{ei} combines these two measures:

$$t_{ei} = 0.667t_r + 0.333t_{ai}$$

Operative temperature t_c is that recorded in the centre of a 40 mm diameter blackened globe freely exposed to room air. Sufficient time is needed for stabilization of the radiant and convective heat transfers to take place with the thermometer, thermocouple or thermistor temperature sensor in the globe. When taking readings, realize that room conditions may be variable due to movement of people, solar heat gains, air movement or the room temperature control system. During normal room conditions of low air velocity and where mean radiant and air temperatures are very close, operative, air dry-bulb and dry resultant temperatures, t_{res}, are substantially equal. Well-insulated modern buildings with small areas of glazing have the effect of raising room mean radiant temperature close to the air dry bulb. Dry resultant temperature is the temperature recorded by a thermometer at the centre of a 100 mm diameter blackened globe. Operative temperature is recommended for use as that specified for comfort conditions in the internal environment. Comfortable values range from 21°C for a residential living room through 18°C for a bedroom or lecture room to 16°C in passageways. It is related to the mean radiant, air and resultant temperatures and the air velocity v m/s by:

$$t_c = t_{res} = \frac{t_r + t_{ai}\sqrt{10v}}{1 + \sqrt{10v}}$$

In normally occupied rooms, the air temperature and the mean radiant temperature should be within a few degrees of each other. The air velocity within a habitable space should be barely discernible, in the region of 0.1 m/s. Under these conditions the dry resultant temperature at the centre of the room is:

$$t_c = t_{res} = 0.5t_{ai} + 0.5t_r$$

There should be no significant effect upon comfort conditions when t_c is within 1.5°C of its design value; this allows for some flexibility in the actual value of the mean radiant and air temperatures and the air velocity due to the occupants changing position within the room, or to weather variations.

EXAMPLE 3.1

An open plan office is designed for sedentary occupation and is to have general air movement not exceeding 0.25 m/s and an air temperature of 22°C d.b., in winter. It is expected that a globe temperature of 20°C would be found at the centre of the room volume. What would be the mean radiant, resultant, environmental and operative temperatures?

$$t_r = t_g \left(1 + 2.35\sqrt{v}\right) - 2.35t_a\sqrt{v}$$

$$t_r = 20 \left(1 + 2.35 \times \sqrt{0.25}\right) - 2.35 \times 22 \times \sqrt{0.25}$$

$$t_r = 17.6°C$$

$$t_{res} = t_g \left(1 + 3.17\sqrt{v}\right) - 3.17t_a\sqrt{v}$$

$$t_{res} = 20 \times (1 + 3.17 \times 0.25) - 3.17 \times 22 \times 0.25$$

$$t_{res} = 16.8°C$$

$$t_{ei} = 0.667t_r + 0.333t_{ai}$$

$$t_{ei} = 0.667 \times 17.6 + 0.333 \times 22$$

$$t_{ei} = 19.1°C$$

$$t_c = \frac{t_r + t_{ai}\sqrt{10v}}{1 + \sqrt{10v}}$$

$$t_c = \frac{17.6 + 22 \times \sqrt{10 \times 0.25}}{1 + \sqrt{10 \times 0.25}}$$

$$t_c = 20.3°C$$

The main comfort criteria for sedentary occupants in buildings in climates similar to that of the British Isles are: operative temperature 19–23°C depending on room use, a feeling of freshness, mean radiant temperature is slightly above air temperature. Air temperature and the mean radiant temperature should be approximately the same, as large differences cause either radiant overheating or excessive heat loss from the body to the environment, as would be experienced during occupation of a glasshouse through seasonal variations. Warmer temperatures become acceptable in summer where an upper limit of 26°C d.b. can be tolerated for short periods. Percentage saturation should be in the range of 40–70%. Maximum air velocity at the neck should be 0.1 m/s for a moving-air temperature of 20°C d.b. Hot and cold draughts to be avoided. Variable air velocity and direction are preferable to constant and this is achieved by changes in natural ventilation from prevailing wind, movement of people around the building, on–off or high–low thermostatic operation of fan-assisted heaters or variable-volume air conditioning systems. The minimum quantity of fresh air for room use that will remove probable contamination is 10 l/s per person. Mechanical ventilation systems should provide at least four air changes per hour to avoid stagnant pockets and ensure good air circulation. Incoming fresh air can be filtered to maintain a clean, dust-free, internal environment. The difference between room air temperatures at head and foot levels should be no more than 1°C. Ventilation air quantity can be determined by some other controlling parameter, for example, removal of smoke, fumes or dust, solar or other heat gains and dilution of noxious fumes. Living in tropical climates requires adaptation to constant high humidity and 30°C d.b. outdoors; indoor humidity of 40% and above in arid climates means that it is raining.

Questions

1. List and discuss the factors affecting thermal comfort.
2. State how extremes of heat and cold affect the workers on a site, what environmental measurements can be taken, and the corrective actions possible to ensure safe and healthy working conditions.
3. Describe with the aid of sketches how each of the following instruments functions: dry-bulb thermometer, wet-bulb thermometer, globe thermometer, vane anemometer, thermocouple, thermistor and infrared scanner.
4. An open-plan office is designed for sedentary occupation and is to have general air movement not exceeding 0.2 m/s and an air temperature of 22°C d.b. in winter. It is expected that a globe temperature of 20°C would be found at the centre of the room volume. What would be the mean radiant, resultant, environmental and operative temperatures?
5. A lecture theatre is designed for sedentary occupation and is to have general air movement not exceeding 0.5 m/s and an air temperature of 21°C d.b. in winter. It is expected that a globe temperature of 18°C would be found at the centre of the room volume. What would be the mean radiant, resultant, environmental and operative temperatures?
6. A conference room is designed for sedentary occupation and is to have general air movement not exceeding 0.35 m/s and an air temperature of 24°C d.b. in summer. It is expected that a globe temperature of 21°C would be found at the centre of the room volume. What would be the mean radiant, resultant, environmental and operative temperatures?
7. Survey the factors affecting thermal comfort and explain what they mean.
8. Where do indoor odours, vapours and gases come from?

    1. Radon gas emanating from the ground beneath the building.
    2. Carbon monoxide from traffic.

3. People, our clothes and what we put on our skin.
4. Passively acquired cigarette smoke prior to entry into the office building.
5. Last night's spicy meal.

9. Are any of these correct for biological effluent?

1. Is too complicated to be measured.
2. Comes from many sources within the working environment.
3. From one office worker in a $10.0\,m^2$ working space is standardized at 1.0 olf.
4. Is counteracted by plants within the occupied building, particularly with open atria.
5. We walk into the building with odours on our clothes.

10. Which of these are correct for excellent air quality in a building?

1. May need very high room air change rates.
2. May need outside air to be collected from the roof of a tall city centre building.
3. May be unachievable where the building is located in a polluted industrial area.
4. Can be improved with air filtering equipment.
5. Is mainly impractical due to its high cost.

11. How can heat leakages due to inadequate thermal insulation and damaged pipes or cables be detected?

12. State the factors that are taken into account when designing for the provision of ventilation with outdoor air.

13. List the atmospheric pollutants that are likely to be present within normally occupied buildings. Identify those pollutants that are used for the design of the ventilation system, the filtration equipment, acoustic insulation and general maintenance during occupation.

14. State why continuous logging is of value to the energy audit engineer, environmental system design engineer, building designer and building occupants, giving reasons for your statements.

15. Where does poor indoor air quality come from?

4 Energy economics

Learning objectives

Study of this chapter will enable the reader to:

1. understand the basis of energy auditing and design an energy audit;
2. understand degree days and their use;
3. calculate the economic thickness of thermal insulation for both hot surfaces and building structures;
4. analyse cash flow and financial return on investment in energy-saving measures.

Key terms and concepts

annual energy 64; carbon emission 71; capital repayment period 69; cash flow 70; cost per useful unit 63; degree day 64; economic thickness 68; energy audit 59; gross calorific value 61; load factor 65; overall efficiency 63; return on investment 69; net present value 69; internal rate of return 69.

Introduction

Buildings are such major consumers of primary energy and renewable sources such as hydroelectric, wind, solar and wave energy that accounting for energy use and calculating the consequent financial implications are of paramount importance to all who are involved in building design and operation. Logical methods of dealing with the calculations are introduced to enable the new user to cope with the complex conversion equations and calculation of energy costs per standard unit, the annual energy cost and the economic thickness of thermal insulation. Degree days are explained and their use as an accounting tool is explored. The financial implications of purchasing or leasing energy saving hardware equipment are investigated.

Energy audit

Management of the energy that is used for buildings has three major components:

1.   initial design;
2.   retrofitting energy-saving measures;
3.   maintenance practices.

Design engineers should provide heating, ventilating and air-conditioning systems that consume the minimum amount of fossil fuel energy in satisfying the needs of the site. The initial installation is designed in conjunction with the architecture and the client's requirements, in accordance with statutory legislation and in compliance with the standards of good engineering practice. The design includes the means of controlling the use of energy. Control can mean anything from switching building services plant on and off manually, up to a fully automated computer-based system that gives audible and visual alarms when something goes wrong. The best efforts of the design engineer are limited by the initial construction cost of the new installation, which is usually minimized. The installation of energy-saving systems often increases this cost. Some buildings are designed to be low-energy users. Most building services engineering systems are designed to provide thermal comfort for the conditions that are found in the building, for example, perimeter heating to overcome the heat loss through large areas of glazing and thermostatically controlled ventilation louvres in a naturally ventilated building in a cool climate.

   Energy-saving measures that are installed after the first few years of use of a new building, or during a major upgrade, can be justified for two primary reasons. Either the owner of the building has decided to refurbish it, and has found the capital funding that is needed for all the work, or the operating cost of the site is significantly greater than comparable facilities, and the owner or tenant is prepared to invest in measures that will reduce annual outgoings. The owner may be forced to provide leased office space that has competitive energy and maintenance costs. A building that has a labour-intensive maintenance and supervision workload, from steam boilers, manual switching of mechanical plant and lighting systems, unreliable water chillers, poorly maintained closed-circuit water conditions and highly stressed belt drives on fans, is not attractive to a new user.

   Many sites have maintenance practices that encourage the provision of breakdown repairs and replacements, rather than preventing breakdowns through good quality methods. The financial controller of the business may view the annual maintenance budget as expendable, through a lack of understanding about engineering equipment. This is understandable. The maintenance engineer has only to ask the finance director whether it is preferable for a company car to be taken for regular servicing or to wait for the car to break down on a motorway during inclement weather, because the engine has run out of oil, the engine cooling system has boiled dry and the brake pads have worn down to the metal! This is how the maintenance budget of some sites is managed, that is, breakdown maintenance only. Many building services are critical to the life safety of the users. These life safety systems are not just the emergency exit lighting, smoke spill ventilation fans, stairway air pressurization fans and electrical earth leakage circuit breakers, but also include the air conditioning to hospital operating theatres, lifts, outside air ventilation dampers, domestic hot water and cooling tower bacteria controls. Proficient maintenance practice helps to prevent breakdowns by:

1.   monitoring the condition of plant;
2.   optimizing the maintenance activity to replace items only when they are needed;

3. keeping the maintenance team well motivated;
4. planning expenditure;
5. comprehensive maintenance record keeping;
6. enabling a quick response to problems, such as the failure of a fan motor, before the tenants complain of experiencing poor quality air conditions. The building maintenance manager usually has about half an hour from when an air-conditioning fan ceases to function to when the tenants complain on the telephone. If the plant failure has been monitored through the building management system computer with audible and visual alarms, and an automatically sent message to the engineer's pager or mobile phone, the corrective response can be made within 5 minutes and the tenants provided with a briefing.

The energy audit engineer assesses the practical and financial viability of energy-saving measures for each site, as is appropriate. The purpose of the energy-saving analysis is to identify suitable investments in capital equipment that will reduce the use of energy and labour, so that the savings will provide a payback on the investment in a reasonable period. This period will vary from 1 year, for those only interested in this year's profits, to 3 years for those who rely upon their bank for capital funding, to 5 years for those who can source capital funds from an equity performance contracting partner, to the longer terms of 10 to 25 years when the user is a government department and is to retain ownership of the public buildings indefinitely. The retrofit energy-saving measures that are usually considered include the following:

1. thermal insulation of the building;
2. solar shading;
3. changing the fuel source for heating and cooling;
4. heat pumps;
5. heat reclaim;
6. co-generation of electricity with heating or cooling;
7. computer-based building management system;
8. digital control refrigerant circuit of the water chiller;
9. hot water, chilled water or ice thermal storage;
10. load shedding large electrical loads at critical times for short periods;
11. energy tariff change;
12. reducing the lighting system power usage;
13. variable speed drives of fan and pump motors;
14. reducing the usage of water by taps and in toilets;
15. economy air recycling ductwork and motorized damper controls;
16. air-to-air heat exchange between exhaust and incoming outside air ducts;
17. occupancy-sensing with infrared, acoustic or carbon dioxide sensing to control lighting and the supply of outside air;
18. air curtains at doorways;
19. oxygen sensing in the boiler flue gas to modulate the combustion air supplied to the burner;
20. replacement of old inefficient boilers and heating systems;
21. distribution of domestic hot water at 45°C with a mixing valve and temperature control;
22. replacement of steam-to-water heat exchangers and calorifiers with local gas-fired heating and domestic hot water systems;
23. thermal insulation of heating, cooling, steam pipes and heat exchangers;
24. recovery of the maximum quantity of condensate in a steam distribution system;
25. replacement and overhaul of steam traps and condensate pumping.

An energy audit of an existing building or a new development is carried out in the same way as a financial audit but it is not only money that is accounted. All energy use is monitored and regular statements are prepared showing final uses, costs and energy quantities consumed per unit of production or per square metre of floor area as appropriate. Weather data are used to assess the performance of heating systems. Monthly intervals between audits are most practical for building use, and in addition an annual one. Basic aims for an audit are to do the following:

1. establish total costs of energy purchased;
2. locate the principal energy-consuming areas;
3. notice any obvious losses or inefficient uses of heat, fuel and electricity;
4. take overall data to gain initial results quickly, which can be refined later and broken down into greater detail;
5. find where additional metering is needed;
6. take priority action to correct wastage;
7. survey buildings and plant use at night and weekends as well as during normal working hours;
8. initiate formal records monitored by the energy manager;
9. compare all energy used on a common basis (kilowatt-hours or megajoules);
10. list energy inputs and outputs to particular buildings or departments.

A vital part of auditing is enlisting the cooperation of all employee groups, and explaining the problem not just in financial terms but also in quantities of energy. A joint effort by all staff is needed. Posters, stickers and prizes for ideas can be used to stimulate interest. An overall energy audit will list each fuel, the annual quantity used and the cost for the year, including standing charges and maintenance; then a comparison is made with other fuels by converting to a common unit of measurement. Energy use performance factors enable comparisons to be made between similar buildings or items of equipment. These can be litres of heating oil per degree day, kilowatt-hours of electricity consumption per square metre of floor area, megajoules of energy per person per hour of building use, or other accounting ratios as appropriate. For example, car manufacturers may analyse energy used per car. As experience is gained in auditing a particular building, data can be refined to monthly energy use in conjunction with degree day figures for this period. This detailed analysis can be made for each building or department of a large site, each large room or factory area, each type of heating, air-conditioning or power-using system, each industrial process and each item of plant. The most serious deficiency in the acquisition of data is likely to be the lack of sufficient metering stations. Electricity, gas and other fuels are metered by the supply authority at the point of entry to the building or site; further metering is the responsibility of the site user. Frequently, no further meters are installed and capital expenditure is needed to obtain data.

The total heat energy content of a fuel is known as the gross calorific value (GCV) and is usually expressed in megajoules per kilogram (MJ/kg).

EXAMPLE 4.1

Heating oil has a specific gravity of 0.84 and a GCV of 44.8 MJ/kg. Find its heat content in kWh per litre.

$$GCV = 44.8 \frac{MJ}{kg} \times \frac{0.84\,kg}{1\,l} \times \frac{10^3\,kJ}{1\,MJ} \times \frac{1\,kWs}{1\,kJ} \times \frac{1\,h}{3600\,s}$$

$$GCV = \frac{44.8 \times 0.84 \times 10^3}{3600} \frac{kWh}{l}$$

$$GCV = 10.453 \frac{kWh}{l}$$

EXAMPLE 4.2

A site uses 48000 l of oil of GCV 44.0 MJ/kg and specific gravity 0.84 costing £50,000 and 2×10^6 kWh of electricity costing 10 p/kWh. The fixed charge for the electrical installation is £7000 and the servicing cost for the oil-fired heating system is £15,000. The period of use being considered is 1 year. Calculate the data for an energy audit.

$$Oil\ GCV = 44 \frac{MJ}{kg} \times \frac{0.84\,kg}{1\,l} \times \frac{10^3\,kJ}{1\,MJ} \times \frac{1\,kWs}{1\,kJ} \times \frac{1\,h}{3600\,s}$$

$$Oil\ GCV = 10.27\,kWh/l$$

$$Total\ cost\ of\ oil = £50000 + £15000$$

$$Total\ cost\ of\ oil = £65000$$

$$Oil\ energy = 48000\,l \times \frac{10.27\,kWh}{l}$$

$$Oil\ energy = 492960\,kWh$$

$$Cost\ per\ unit\ of\ oil\ energy = \frac{£65000}{492960\,kWh} \times \frac{100p}{£1}$$

$$Cost\ per\ unit\ of\ oil\ energy = 13.2\,p/kWh$$

$$Electricity\ cost = 2 \times 10^6 \times \frac{10p}{kWh} \times \frac{£1}{100p} + £7000$$

$$Electricity\ cost = £207000$$

$$Cost\ per\ unit\ of\ electricity = \frac{£207000}{2 \times 10^6\,kWh} \times \frac{100p}{£1}$$

$$Cost\ per\ unit\ of\ electricity = 10.35\,p/kWh$$

Energy cost may be presented per useful gigajoule as a means of comparison. Suppliers quote energy costs in the unitary system most convenient to their industry, and there is no obvious method of comparing the real cost of providing a specific amount of useful heat or power in the building. A decision can be made to reduce all costs to a common base unit and this may be the kilowatt hour or the gigajoule, where,

$$1\,GJ = 10^6\,kJ \times \frac{1\,kWs}{1\,kJ} \times \frac{1\,h}{3600\,s}$$

$$1\,GJ = \frac{10^6}{3600}\,kWh$$

If the overall efficiency of the energy conversion process is included in the unit cost, then the incurred cost of using that system of heat or power can be realistically assessed. The cost per useful kWh is the cost of providing 1 kWh of useful heat, or energy, at the place of use. The overall efficiency of a central heating system will include the following:

1. *Combustion efficiency of the fuel.* Regular maintenance is necessary with all fuel-burning appliances to ensure that the correct fuel-to-air ratio is maintained.
2. *Heat transfer efficiency of the appliance.* Both flue gas and water-side surfaces must be kept clean.
3. *Heat losses from distribution pipework.* All hot water pipes and surfaces must be adequately insulated unless they provide a useful heating surface in rooms. It is not good practice to allow heating system pipes to be bare metal as the uncontrolled heat transfer will lead to high fuel costs.
4. *Ability of the final heat emitter to transfer warmth to the occupants.* A hot water central heating radiator placed under a window sill should counteract down-draughts and provide a reasonably adequate air temperature at the window and at the inner surface of the outside wall; this has the effect of increasing heat flows through the window and wall. Placing the radiator on a warm internal wall will improve its useful heat output but at the loss of some warm usable space in the room as the window region will be colder.
5. *Thermal storage capacity of the heating system and building.* Large amounts of heat are stored in the water in heating systems and the dense fabric of buildings. An insensitive automatic control system, or the lack of such a system, will lead to wild swings above and below the desired resultant temperature and cause excessive fuel use.

The cost in use of a fuel or source of energy can be calculated from the basic price and the overall efficiency; for gas,

$$\text{Cost in use} = \text{unit}\frac{p}{kWh} \times \frac{100}{efficiency\%} + \frac{standing\ charge}{kWh}$$

When total kWhs used during the year are assessed, the annual standing charge can be apportioned to each kWh. This is not necessary if the standing charge has already been incurred by another use, for example, cooking and water heating and cost in use is being evaluated for an additional heating system.

EXAMPLE 4.3

A gas-fired central heating and hot water system is to be installed in a commercial property. The gas tariff is 5 p/kWh plus a standing charge of 50 p per day. The estimated annual heat energy that will be used by the occupants is 165 000 kWh. The annual maintenance works cost £250. Find the total energy bill and the average cost per kWh, for the year. Overall efficiency of the heating system is 70%.

$$\text{Metered input energy} = 165000 \times \frac{100}{70}\ \frac{kWh}{yr}$$

$$\text{Metered input energy} = 235714\frac{kWh}{yr}$$

The nearest whole number of kWh is the significant number. Do not waste time with too many decimal places. Fractions of a millimetre, watt, Pascal or kWh are of little or no significance to the reality of the calculation.

$$Cost = 5\frac{p}{kWh} \times 235714\frac{kWh}{yr} \times \frac{£1}{100p}$$

$$Cost = £11786$$

$$Standing\ charge = 365\frac{days}{yr} \times 50\frac{p}{day} \times \frac{£1}{100p}$$

$$Standing\ charge = £183$$

$$Annual\ bill = £11786 + £183 + £250$$

$$Annual\ bill = £12219$$

$$Cost\ per\ unit = \frac{£12219}{235714\,kWh} \times \frac{100p}{£1}$$

$$Cost\ per\ unit = 5.2\frac{p}{kWh}$$

The cost in use for other fuel or energy sources is calculated from the basic price. These are pence per litre for oils, £ per tonne for solid fuels and £ per kg refill for liquefied petroleum gas. The specific gravity of liquid fuel is used to convert volume to mass measurements. When the specific gravity of oil is 0.83, 1 litre of the oil weighs 0.83 kg. One litre of water at 4°C weighs 1 kg. Oil and paraffin are sold by the litre but its heat content, gross calorific value (GCV) is usually listed as around 45 MJ/kg. For oil costing 100 p/litre, GCV 45.8 MJ/kg used in a heating plant having an overall efficiency of 70% is:

$$Cost = 100\frac{p}{l} \times \frac{1l}{0.83\,kg} \times \frac{kg}{45.8\,MJ} \times \frac{100}{70} \times \frac{1\,MJ}{10^3\,kJ} \times \frac{1\,kJ}{1\,kWs} \times \frac{3600\,s}{1\,h}$$

$$Cost = 13.5\frac{p}{kWh}$$

Annual energy costs

Annual fuel costs can be estimated from the energy cost per useful unit, length of heating season, operational hours of the system, mean internal building temperature, design external temperature and degree days for the locality. Dynamic simulation software utilizes detailed hourly weather data, building usage, installed systems, thermal storage from admittances and building design. The design steady-state building heat loss is known as the design external air temperature, ranging from −1°C to −5°C. Throughout the heating season, the heat loss will fluctuate with the cyclic variations in ambient temperature. Fortuitous heat gains will reduce fuel consumption provided that the automatic controls can reduce heating system performance sufficiently and avoid overshooting the desired room temperatures.

   Degree days are recorded temperature data that facilitate the production of a climatic correction or load factor for the calculation of heating system operational costs and efficiency over months or yearly time intervals. They are applied to normally occupied buildings where the heat loss from the warm interior is balanced by the gains of heat from the sun, the occupants, lighting, cooking and hot water usage at an external air temperature of 15.5°C; this is known as the base temperature. The value taken for the base temperature is an estimate of the conditions under which there will be no net heat loss from a traditionally constructed residence; thus no fuel will be consumed at this and higher outside temperatures. Calculation of the actual base temperature for a particular building may reveal another value; consequently, care is needed in the application of degree day data, and correction factors may be included for other than traditionally constructed dwellings: for example, highly insulated or commercial structures and where internal heat gains from electrical equipment are high. The standard method of use is to assess the daily difference between the base temperature and the mean value of the external air temperature during each 24-hour period. A modified calculation is made when the base temperature is below the external mean temperature, as this would indicate a net heat gain to the building. Degree day data are not used for air-conditioning cooling-load calculations as they are not appropriate. Typical fluctuations in external air temperature are relative to base temperature, such as maximum and minimum air temperatures of 10°C and 6°C, the 24-hour mean is 8°C. There is a difference of 7.5°C per day, 7.5 degree days are added to the accumulated total for that month. The maximum possible number of degree days for a particular location and period of heating system operation can be found as shown in Example 4.4.

EXAMPLE 4.4

A house is continuously occupied during a 30-week heating season. The design external air temperature is −1.0°C. Find the maximum possible number of degree days.

$$Number\ of\ days = 30\,weeks \times \frac{7\,days}{week}$$

$$Number\ of\ days = 210$$

$$Maximum\ temperature\ difference = (15.5 - 1)°\,C$$

$$Maximum\ temperature\ difference = 16.5°C$$

$$Maximum\ degree\ days = 210\,days \times 16.5°C$$

$$Maximum\ degree\ days = 3465\,degree\ days$$

The load factor is the ratio of actual to maximum degree days and is used to find the average rate of heat loss from a building over the heating season:

$$Load\ factor = \frac{degree\ days\ for\ locality}{maximum\ posible\ degree\ days}$$

EXAMPLE 4.5

Assess the rate of heating plant power used during a heating season when there were 2460 degree days, the steady state design heat loss was 24.5 kW at an outdoor design air temperature of −1°C.

$$Load\ factor = \frac{2460}{3465}$$

$$Load\ factor = 0.71$$

Seasonal average heater output power $= 0.71 \times 24.5\,kW$

Seasonal average heater output power $= 17.4\,kW$

The heating system plant has an average output of 17.4 kW, plus any domestic hot water demand and wasted heat emission from pipework. This gives an approximation to what can be expected for energy use. Degree days can be used to monitor fuel consumption and check that it is not being used wastefully. Incorrectly serviced fuel-burning appliances would show an increasing use of energy per degree day rather than a constant rate. Deterioration of the performance of an automatic control system or lack of proper manual regulation of ventilation openings would also result in a departure from expected ratios. A graph of energy consumption against degree days should be linear for a building, and any major divergence will show that corrective action is needed. Estimation of annual fuel costs is now a matter of finding the number of kWh or GJ consumed in the building and then the cost of providing this useful amount of heat.

EXAMPLE 4.6

A hospital is heated for 24 hours per day, 7 days per week for 30 weeks a year. Its steady-state heat loss at −1°C outside is 2850 kW; a gas-fired boiler with a hot water radiator central heating system has an overall efficiency of 78%. Estimate the annual fuel cost for heating the building if there are likely to be 2240 degree days in that locality. Maximum degree days are 3465. Gas costs 5 p/kWh.

$$Load\ factor = \frac{2240}{3465}$$

$$Load\ factor = 0.65$$

$$Annual\ energy\ used = heat\ loss \times load\ factor \times running\ hours \times \frac{100}{efficiency}$$

$$Annual\ energy = 2850\,kW \times 0.65 \times \frac{24\,h}{day} \times \frac{210\,days}{year} \times \frac{100}{78} \times \frac{1\,GWh}{10^6\,kWh}$$

$$Annual\ energy\ input = 11.97\frac{GWh}{yr}$$

$$Gas\ bill = 11.97\frac{GWh}{yr} \times \frac{5\,p}{kWh} \times \frac{£1}{100\,p} \times \frac{10^6\,kWh}{1\,GWh}$$

$$Gas\ bill = £598500$$

EXAMPLE 4.7

An initial energy audit of a hospital revealed the data shown in Table 4.1. The data were for a month that had 260 degree days, and the energy manager required energy use performance factors of total cost per square metre of floor area, heating system energy used per degree day and electrical energy used per person per hour. Gas consumed was for the heating system and cost 5 p/kWh plus £750 per month standing charge and £550 per month for maintenance. Electricity cost was 10 p/kWh and maintenance costs amounted to £550 per month.

Table 4.1 Energy audit data for Example 4.7.

Location	Electricity (kWh)	Gas (kWh)	Floor (m²)	Usage (h)	Occupants
Medical	12000000	13000000	45000	670	2300
Administration	1500000	1000000	3500	350	220
Engineering	1500000	250000	1000	400	23
Totals	13650000	14250000	49500	1420	2543

Energy use performance factor of total cost per square metre of floor area for the month:

$$Electricity\ bill = 13650000\ kWh \times \frac{10\,p}{kWh} \times \frac{£1}{100\,p} + £550$$

$$Electricity\ bill = £1365550$$

$$Gas\ bill = 14250000\ kWh \times \frac{5\,p}{kWh} \times \frac{£1}{100\,p} + £750 + £550$$

$$Gas\ bill = £713800$$

$$Energy\ cost\ per\ m^2 = \frac{£1365550 + £713800}{49500\ m^2}$$

$$Total\ energy\ cost = £42\ per\ m^2\ floor\ area$$

$$Gas\ energy = \frac{14250000\ kWh}{260\ degree\ days}$$

$$Gas\ energy\ used\ in\ the\ month = 54808\frac{kWh}{degree\ day}$$

Electrical energy used per person per hour:

$$Total\ occupation = \sum(occupants \times usage\ hours)$$

$$Total\ occupation = (2300 \times 670) + (220 \times 350) + (23 \times 400)$$

$$Total\ occupation = 1627200\ person\ hours$$

$$Electrical\ energy\ used = \frac{13650000\ kWh}{1627200\ person\ hours}$$

$$Electrical\ energy\ used = 8.4\,kWh\ per\ person\ per\ hour$$

You may wish to evaluate the energy use performance factors for various locations for comparison. We used energy use standards in the evaluation of the PROBE reports with *CIBSE Guide F* in Chapter 2.

Economic thickness of thermal insulation

A balance needs to be made between the capital cost of thermal insulation of buildings or hot surfaces and the potential reduction in fuel costs in order to obtain the lowest total cost combination of these two cash flows. Capital cost is normally expected to be recovered from fuel cost savings during the first two or three years of use; however, longer periods than this are needed for major structural items, such as cavity fill and double glazing, and there will be additional benefits, such as improved thermal storage capacity, reduced external noise transmission, fewer draughts and added value to the property, that do not fit easily into a financial treatment of their worth. For flat surfaces, the cost of heat loss per square metre through the structure is:

$$Energy\ cost = U \frac{W}{m^2 K} \times (t_1 - t_2)K \times \frac{1\,kW}{10^3\,W} \times load\ factor \times \frac{hours}{year} \times \frac{energy\,p}{kWh} \times \frac{£1}{100p}$$

$$Energy\ cost = U \times (t_1 - t_2) \times load\ factor \times annual\ hours \times \frac{p}{kWh} \times \frac{1}{100} \times \frac{1}{10^3} \frac{£}{m^2 yr}$$

The cost of fuel usage for a range of thermal transmittances U can be calculated for a particular structure. This is usually a decreasing curve for increasing insulation thickness as each additional layer reduces the thermal transmittance by progressively smaller amounts. If the cost $£/m^3$ of the thermal insulation as installed is known, then the cost for each thickness per square metre of surface area can be found from:

$$Insulation\ installed\ cost = \frac{£}{m^3} \times \frac{thickness\ m}{repayment\ time\ years}$$

Data from these equations can be drawn on a graph. Addition of the two curves produces a total cost curve. The lowest point on this curve gives the optimum insulation thickness; if its lower part is fairly flat, then any one of a number of commercially available thicknesses will be economic.

EXAMPLE 4.8

Expanded polystyrene board is to be added to the internal face of a wall having an initial thermal transmittance of 3.30 W/m²K in thicknesses of 25, 50, 75, 100, 125 and 150 mm. Insulated wall thermal transmittances will be 0.96, 0.56, 0.40, 0.31, 0.25 and 0.21 W/m²K. The insulation costs £30 per cubic metre fitted and the capital recovery period is to be 3 years. Fuel costs 5 p/kWh and is used at an overall system efficiency of 70%. Internal and external design temperatures are 21°C and −1°C respectively, the load factor is 0.608 and the building is to be heated for 3000 h per year. Use the information provided to find the economic thickness of insulation.

$$Energy\ cost = U \times (21 + 1) \times 0.608 \times 3000 \times \frac{5p}{kWh} \times \frac{1}{100} \times \frac{1}{10^3} \frac{£}{m^2 yr}$$

$$Energy\ cost = 2 \times U \frac{£}{m^2 yr}$$

$$Insulation\ cost = \frac{£30}{m^3} \times \frac{thickness\ m}{3\ years}$$

Table 4.2 Cost data for Example 4.8.

Thickness I (m)	0.025	0.050	0.075	0.100	0.125	0.150
U (W/m²K)	0.96	0.56	0.40	0.31	0.25	0.21
Insulation cost (£)	0.25	0.5	0.75	1	1.25	1.5
Energy cost (£)	1.92	1.12	0.8	0.62	0.5	0.42
Total cost (£)	2.17	1.62	1.55	1.62	1.75	1.92

$$Insulation\ cost = 10 \times thickness \frac{£}{m^2 yr}$$

Results in Table 4.2 show that the total cost curve can be drawn by adding the fuel and insulation cost curves for each insulation thickness. The economic thickness is 75 mm.

Accounting for energy-economizing systems

Once the capital cost and fuel cost savings have been assessed for thermal insulation, improved glazing, on-site generation, ventilation control, energy-saving systems or improved controls, the capital repayment period or return on capital investment can be calculated in simple terms:

$$Capital\ repayment\ period = \frac{capital\ cost}{annual\ cost\ saving}$$

$$Investment\ return = \frac{annual\ cost\ saving}{capital\ cost} \times 100\%$$

Further refinements such as discounted cash flow, loan interest charges, tax allowances and grants can be included to improve accuracy. Cash flow statements for limited companies are handled differently from government departments, non-profit organizations and home-owners. Allowances for capital expenditure and taxation on increased company profitability due to energy economies can markedly improve estimates of payback times. A company making a purchase costing £500,000 which would save £200,000 in the first year's energy bill, would appear to take 2.5 years for capital recovery, but the cash flow projection may be as shown in Table 4.3 when a discount rate of 8%, inflation is 3% and interest of 10% are included as shown. A government capital grant allowance of 25% of the purchase cost is made a year in arrears. Company tax is 40% paid a year in arrears. Figures in parentheses are outward cash flows from the business. Energy costs are indicated to increase by 3% per year, so savings increase by the same amount. This investment commences cash generation during the second year of equipment use and would start to provide funds after year 3 for further investment. Certain items of equipment can be leased rather than purchased and this releases cash earlier but calls for continuous payments to the leasing company. Cash flow is always positive to the company, but leasing payments are made for 10 years and then at a reduced rate after that period. Self-contained items of plant such as heat pumps or electricity generators may be leased. Use the workbook (*DCF Analysis*) provided for capital and leasing cash flow projections. This workbook shows 5 years of cash flow and demonstrates the cash investment has a Net Present Value of £550,189, meaning the investment is worth that much money as cash in hand today. The Internal Rate of Return of the investment is 39%. The company has to decide where else an equal or higher rate of return could be achieved. NPV and IRR are standard workbook functions.

Table 4.3 Cash flow forecast for the purchase of an energy-economizing system.

		Year 1	Year 2	Year 3
A	Cash balance brought forward	0	(315000)	(53000)
B	Capital purchase	(500000)	0	0
C	Energy saving	200000	210000	220500
D	Capital allowance 25% × B	0	125000	0
E	Cash balance (A − B + C + D)	(300000)	20000	167500
F	Interest, say, 10% × E × 0.5	(15000)	1000	8375
G	Tax, 1 year in arrears, 40% × (C + F)	0	(74000)	(84000)
H	Net cash flow (C + D + F − B − C)	(315000)	262000	144475
I	Cash balance (A + H)	(315000)	(53000)	91475

The effect on gas consumption of thermal insulation in houses

Dwelling energy use is reduced by the addition of thermal insulation, double glazing, use of passive solar radiation, ventilation control and draught-proofing providing increased standards of thermal comfort. However, the full potential saving due to the extra insulation may not be reflected in the fuel bills as expected. Field measurements (British Gas, 1980) have shown a correlation between domestic gas consumption, design heat loss, occupancy and degree days for the locality, where Y are annual degree days, Q the design heat loss in kW and N the occupancy:

$$Annual\ gas\ use = 29.3056 \times \left(61 + \frac{70 \times Y \times Q}{2222} + 59 \times N\right) kWh$$

EXAMPLE 4.9

A large house in the Thames region has a design heat loss of 28 kW and up to six occupants. Added thermal insulation reduces the design heat loss to 20 kW. Estimate the probable energy, greenhouse gas and cost savings for the gas-fired central heating and hot water system. Gas costs 5 p/kWh, total degree days for the year are 2230.

$$Existing\ gas\ use = 29.3056 \times \left(61 + \frac{70 \times 2230 \times 28}{2222} + 59 \times 6\right) kWh$$

$$Existing\ gas\ use = 69808\ kWh$$

$$Insulated\ gas\ use = 29.3056 \times \left(61 + \frac{70 \times 2230 \times 20}{2222} + 59 \times 6\right) kWh$$

$$Insulated\ gas\ use = 53337\ kWh$$

$$Gas\ saving = (69808 - 53337)\ kWh$$

$$Gas\ saving = 16471\ kWh$$

$$CO_2\ emission\ saving = 16471\ kWh \times \frac{0.194\ kg\ CO_2}{kWh}$$

$$CO_2\ emission\ saving = 3195\ kg\ CO_2$$

$$Gas\ cost\ saving = 16471\ kWh \times \frac{5p}{kWh} \times \frac{£1}{100p}$$

$$Gas\ cost\ saving = £823$$

Questions

1. State the function of an energy audit. What data are collected? How are the data presented? What is likely to be the most serious barrier to data collection?
2. Explain the term 'degree day' and state its use.
3. How is the load factor calculated and how is it used?
4. A factory uses 20000 l of oil for its heating and hot water systems, 160000 kWh of electrical power and 300000 kWh of gas for furnaces in a year. Fixed charges are £800 for the oil, £700 for the electrical equipment and £1200 for gas equipment. Use the data provided in this chapter and current energy prices to produce an overall energy audit based on the gigajoule unit and find the average cost of all the energy used.
5. Calculate the annual cost of a gas-fired heating system in a house with a design heat loss of 30 kW at $-2°C$ for 16 h per day, 7 days per week for 30 weeks in the year. Use the data provided in this chapter and the current fuel price.
6. Find the total annual cost of running a gas-fired heating and hot water system in a house with four occupants if its design heat loss is 32 kW. Maintenance charges amount to £160 per year.
7. Discuss the statement, 'Economic thickness of thermal insulation of houses is no longer the relevant argument.'
8. A city centre building in Leeds has a predicted energy consumption of 1500000 kWh per year for only the space heating system. The design engineer is to recommend the energy source and system type to be used on the basis of minimizing the greenhouse gas emissions. The average seasonal usage efficiency of the alternative systems are 95% for electrical heating systems of various types, 75% for gas-fired radiator heating system, 65% for coal-fired radiator heating system and 75% for an oil-fired ducted warm air heating system. How might renewable energy sources be used? Calculate the carbon emission in tonnes per year and make a suitable recommendation to the client.
9. Which of these adequately describe an energy audit of a building?

    1. Points out what building operators could do to save energy.
    2. Are only for cosmetic appearance of doing something to reduce greenhouse emissions.
    3. Identify and quantify viable energy-saving investments.
    4. Concentrate on finding almost zero cost short-term payback energy-saving opportunities.
    5. Only analyses technical projects and not financial investment criteria.

10. How can discounted cash flow calculation aid analysis of an energy-saving project?
11. State what Net Present Value of an investment decisions means.
12. State how Internal Rate of Return of a cash flow is an aid to making the decision to invest.
13. Which of these is not a correct multiple?

    1. $kJ = 10^3\ J$.
    2. $MWh = 1000\ W \times 1\ h$.
    3. $1\ GJ = 10^6\ kJ$.
    4. $1\ mm = 10^{-3}\ m$.
    5. $1\ GW = 1000\ MW$.

14. Which is degree day load factor not relevant to?

    1. Calculation of heating system kW load for design.
    2. Ratio of degree days from meteorological data.
    3. Minimum outside air temperature for design.
    4. Seasonal weather variability.
    5. Maximum possible Degree Days for the locality.

5 Ventilation and air conditioning

Learning objectives

Study of this chapter will enable the reader to:

1. recognize the physiological reasons for fresh air ventilation of buildings;
2. calculate fresh air requirement;
3. understand the basic design criteria for air movement control;
4. describe the combinations of natural and mechanical ventilation;
5. understand mixed mode ventilation;
6. know the leakage testing of buildings;
7. discuss why air conditioning is used in various climates;
8. describe the working principles of air conditioning systems;
9. calculate ventilation air quantities;
10. understand psychrometric cycles for humid air;
11. calculate air conditioning heating and cooling plant loads;
12. describe the various forms of air conditioning systems;
13. state where reciprocating piston, screw and centrifugal compressors are suits;
14. understand the coefficient of performance;
15. explain the states of refrigerant occurring within a vapour compression refrigeration cycle;
16. explain the operation of refrigeration equipment serving an air conditioning system;
17. comprehend the absorption refrigeration cycle;
18. explain how ventilation rates are measured;
19. choose suitable materials for air conditioning ductwork;
20. understand the relationship between CFCs and the environment;
21. know the uses of CFCs, and good practice and handling procedures;
22. be able to discuss the problem of sick building syndrome (SBS);
23. know the symptoms, causes and possible cures for SBS;
24. relate the daily cyclic variation of air temperatures to the need for air conditioning;

25. recognize how solar-powered cooling can be applied;
26. relate ventilation and air conditioning to CO_2 emission reduction;
27. understand offpeak thermal storage cooling application.

Key terms and concepts

absorption 102; air change 77; air conditioning systems 83; air flow rate 84; air velocity 75; anemometer 105; biocide treatment 107; building air tightness 77; carbon dioxide 74; centrifugal compressor 100; chlorofluorocarbon 74; Coanda effect 75; coefficient of performance 101; dew-point temperature 91; dual duct 96; duct size 84; ductwork materials 106; dynamic thermal analysis 81; energy recovery 80; energy-saving systems 83; evaporative cooling 81; fan coil 98; humidity control 89; ice thermal storage 105; induction 97; latent heat gains 89; low-cost cooling 81; mechanical ventilation 76; natural ventilation 75; packaged units 98; pollutants 75; pressure–enthalpy diagram 101; psychrometric chart and cycles 91; recirculated air 76; refrigeration 100; screw compressor 100; sensible heat gains 86; shading 78; sick building syndrome 106; single duct 83; smoke 82; solar cooling 105; stack effect 75; supply air moisture content 87; supply air temperature 87; total environmental loading 107; vapour compression 100; variable volume 96; ventilation rate measurement 77.

Introduction

The reasons for ventilation lead into an understanding of the necessary combinations of natural, mechanical and mixed mode systems. Air conditioning means full mechanical control of air movement, temperature and comfort conditions through the building. Calculation of air changes and air flow rates from basic human requirements, and then for the removal of heat gains, is fundamental. Sizing of air ducts and the calculation of heater and cooler loads utilizing the psychrometric chart of humid air properties is shown. Systems of air conditioning ranging from small self-contained units to large commercial applications are described. Appropriate refrigeration elements are developed, from thermodynamic principles, into complete installations in a form that is easily understood. Both vapour compression and vapour absorption cycles are explained. Sick building syndrome has been attributed to air conditioning, but has been found to be due to various possible causes, none of which is singly identifiable as a cause. Chlorofluorocarbons were widely used in thermal insulation and refrigeration until their potential for environmental damage resulted in the Montréal Protocol agreement. The effects of this agreement on the building services industry are discussed. How solar-powered cooling might be applied is outlined. Building air tightness is explained and related to PROBE.

Ventilation requirements

Ventilation of occupied spaces is determined from:

1. allowable CO_2 level;
2. bodily heat production, about 100 W per person during sedentary occupation;
3. moisture exhaled and evaporated from the skin, about 40 W per person for sedentary occupation;
4. body odour;

5.   removal of pollutants;
6.   control of air temperature and humidity;
7.   occupant density and activity.

Recommended fresh air supply per person varies from 5–20 l/s depending upon which of the four indoor air quality standard is adopted from low/moderate/medium/high standards (*CIBSE Guide A4*). A minimum of 10 l/s per person is required for most non-domestic applications. A single person workstation occupies around 5–10 m^2 floor area when half of the walkway is included, so an outdoor air supply of 10 l/s per person equates to 1–2 l/s m^2 floor area. CO_2 concentration may be allowable in the range 350–1600 ppm and 900 ppm meets the medium standard. Outdoor air CO_2 concentration can be taken as 400 ppm so indoor sources can raise the room total to 900 ppm. A sedentary occupant needs around 1 l/s of outdoor air to remove their CO_2 emission so that is not a major issue for ventilation design.
    The ventilation system should not produce monotonous draughts but preferably variable air speed and direction. Facilities for manual control of ventilation terminals allow some freedom of choice over their environment. Careful location of ventilation grilles and control of both the temperature and velocity of moving air in mechanical systems can ensure that neither cool nor hot draughts are caused. The maximum air velocity that can be perceived at neck level is related to its temperature; air movement of 0.32 m/s at 24°C should avoid annoyance. Grille manufacturers' data reveal the length of the jet of air entering the room for particular air flow rates. Additionally, the air jet may rise or fall depending on whether it is warmer or cooler than the room air. Moving air currents tend to attach themselves to a stationary surface and follow its contours, this is known as the Coanda effect. When this boundary layer flow either comes to the end of, say, a wall or hits a bluff body, such as a beam or luminaire, it may suddenly be detached and cause turbulent flow; this appears as a draught, an uncomfortable air movement. An air jet entering a room should be allowed to mix with the room air before entering the occupied space. This can be done either above head height with cooling systems, or in circulation spaces with heating systems and displacement ventilation inlets at low level. The design criteria for ventilation systems are:

1.   correct fresh air quantity, minimum normally 10 l/s per person;
2.   avoidance of hot or cold draughts by design of the air inlet system;
3.   some manual control over air movement;
4.   mechanical ventilation to provide a minimum of four air changes per hour to ensure adequate flushing of all parts of rooms;
5.   air change rates that can be increased to remove solar and other heat gains;
6.   air cleanliness achieved by filtration of fresh air intake and recirculated room air.

Ventilation systems

The five possible combinations of natural and mechanical ventilation are:

1.   *Natural inlet and outlet*: utilizing openable windows, air bricks, louvers, doorways and chimneys. Up to about three air changes per hour may be provided, depending upon prevailing wind direction and strength, the stack effect of rising warm air currents, and adventitious openings around doors and windows.
2.   *Natural inlet, mechanical outlet*: mechanical extract fans in windows or roofs and ducted systems where the air is to be discharged away from the occupied space owing to its contamination with heat, fumes, smoke, water vapour or odour. This system can be used

in dwellings, offices, factories or public buildings. A slight reduction in air static pressure is caused within the building, and external air flows inwards. This inflow is facilitated by air inlet grilles, sometimes situated behind radiators or convector heaters. There is no filtration of incoming impurities. This system is used particularly for toilet or kitchen extraction, smoke removal from public rooms and heat or fume removal from industrial premises. Heating of the incoming air is essential for winter use.

3.   *Mechanical inlet, natural outlet*: air is blown into the building through a fan convector or ducted system to pressurize the internal atmosphere slightly with a heated air supply. The air leaks out of the building through adventitious openings and permanent air bricks or louvers. This system can be used for offices, factories, large public halls or underground boiler plant rooms.

4.   *Mechanical inlet and outlet*: where natural ventilation openings would be unable to cope with large air flow rates without disturbing the architecture or causing uncontrollable draughts, full mechanical control of air movement is assumed. This may augment natural ventilation at times of peak occupancy or solar heat gain. When a building is to be sealed from the external environment, then a full air-conditioning system is used. Figure 5.1 shows the basic arrangement of a single-duct system.

5.   *Mixed mode: any combination of the previous four systems.* A low energy fan may be used to assist natural movement of air through a building when needed. Natural and mechanical ventilation may coexist or these may be used at alternate times depending upon need. Examples are: an academic building having few occupants out of teaching hours might have mechanical systems switched off and allow natural ventilation, passive heating, cooling and illumination to provide satisfactory conditions; a below ground car park or road tunnel is naturally ventilated but has speed-controlled exhaust fans switched from CO and smoke sensors that only run when needed.

The designers of new buildings bear a heavy responsibility for the future environmental condition of the planet. A building that is a large user of fossil fuel primary energy to power its mechanical systems, usually air conditioning, is a charge on society for 50 or more years. Such buildings

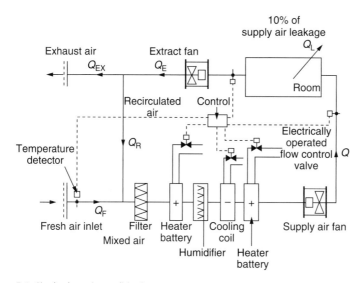

5.1 Single-duct air conditioning system.

are unlikely to have their mechanical heating, cooling, lifts, lighting and computer power supply networks removed for reasons of economy by future tenants. Once the pattern of energy use is set for a building by the architect and original client, it remains that way until it is demolished.

Building air tightness is measured by replacing an external door with a fixed panel, air duct, flow meter and a variable speed supply fan to pressurize the building by up to 50 Pa. This is enough not to cause structural damage or blow out sanitary appliance water seals, but does overcome prevailing wind pressure. Interior doors are wedged open, windows, doors and ventilation openings sealed with polythene sheet and tape. Air flow into the building at a stable interior pressure is the rate of leakage and expressed as $m^3/h\,m^2$ floor area. Some mixed mode ventilation buildings cannot maintain 50 Pa, so a lower value is used and then the flow for 50 Pa is calculated. Sources of air leaks are sought out with smoke pencils while running the fans in reverse to create a suction pressure indoors as incoming smoke is easy to identify. An airtight standard is around $9\,m^3/h\,m^2$ floor area, the average is $18\,m^3/h\,m^2$ and a leaky building has $36\,m^3/h\,m^2$. Uncontrolled air leakage occurs between window and door frames and the structure, at service penetrations, windows and doors not closing tightly, construction and maintenance faults. Such leakages were found to be highly significant and greater than the designers intended (CIBSE PROBE reports).

EXAMPLE 5.1

What is meant by tight, average and leaky building air flow rates for rooms 3 m high? Refer to the PROBE reports for information and descriptions. Relate the matter to the notion of zero carbon buildings.

$$Floor\ area = 1\,m^2$$

$$Volume\ of\ room = 1\,m^2 \times 3m$$

$$Volume\ of\ room = 3\,m^3$$

$$Tight\ building\ air\ leakage = 9\,\frac{m^3}{h\,m^2}$$

$$Tight\ building\ air\ leakage = 9\,\frac{m^3}{h} \times \frac{1\ air\ change}{3m^3}$$

$$Tight\ building\ air\ leakage = 3\,\frac{air\ change}{h}$$

$$Average\ building\ air\ leakage = 18\,\frac{m^3}{h} \times \frac{1\ air\ change}{3m^3}$$

$$Average\ building\ air\ leakage = 6\,\frac{air\ change}{h}$$

$$Leaky\ building\ air\ leakage = 36\,\frac{m^3}{h} \times \frac{1\ air\ change}{3m^3}$$

$$Leaky\ building\ air\ leakage = 12\,\frac{air\ change}{h}$$

Answer is 3, 6 and 12 air changes/h for tight, average and leaky building ventilation rates from uncontrolled sources. These are very high air change rates that cannot be controlled and represent significant heating and cooling demands in addition to the controlled ventilation, solar and internal heat gains. There will not always be such a high static air pressure difference as that tested. Prevailing wind is highly variable. Uncontrolled ventilation varies throughout the year and on a minute-by-minute basis due to wind, occupant behaviour, window, door and ventilation system manual and automatic control. PROBE reports were all on modern buildings constructed since the late 1980s. Describing a building as zero carbon, as is the current trend, seems generous when uncontrolled ventilation may cause 3–12 air changes/h with unheated or uncooled outdoor air.

The increasing need for comfortably habitable buildings in the temperate climate of the British Isles since the 1960s led to demands for air conditioning. Warm weather data for the UK shows that an external air temperature of 25°C d.b. is normally only exceeded during 1% of the summer period of June to September. Compare this to those areas of the world within 45° of latitude from the equator, where the outside air temperature exceeds 30°C d.b.; people living and working in such areas would probably consider the use of air conditioning in the UK a waste of energy resources. These four UK summer months total 120 days and include weekends, a public holiday and many people's annual holidays from their places of work. One day per summer is likely to have an outdoor air temperature that exceeds 25°C d.b. When a long, hot summer is experienced in southern England, there can be several days which exceed 25°C d.b., but this may not be repeated for a few years. In the rest of the UK, lower outdoor air temperatures are experienced most of the time. The UK has a basically maritime climate, with the Atlantic Ocean, the English Channel and the North Sea providing humid air flows at air temperatures that are modified by the evaporative cooling effect of the seas. Long, hot summers are produced by strong winds from Eastern Europe. These winds have travelled across the hot and dry land mass of northern Europe and then been reduced in temperature due to the evaporation of water from the North Sea. Such evaporation of sea water is aided by high wind speed. Sea water is evaporated into the wind and a change of phase occurs in the water. In order to evaporate water into moisture that is carried by the air stream, the water is boiled into steam, a change of phase. This phase change can only take place with the removal of sensible heat from the air and the water, as the latent heat of evaporation has to be provided, just as in a steam boiler. Removal of sensible heat from the air lowers its dry-bulb temperature. This is the principle of evaporative cooling that is employed in cooling towers and in the evaporative cooling heat exchangers that are used for indoor comfort cooling in hot, dry climates. The occasional long, hot summers in the UK coincide with the increase of outside air wet-bulb temperature that is associated with the evaporative cooling effect from the sea. The result is warm and humid weather in south-east England. It is the combined effect of the higher dry- and wet-bulb outdoor air temperatures which produces the humid air that many people find uncomfortable. The building designer has to decide how to address periods of discomfort due to warm outdoor weather conditions and explain the likely outcome to the client.

The principal source of indoor thermal discomfort in the UK is the manner in which modern buildings are designed and then filled with people and heat-producing equipment. Buildings that were constructed of masonry and brick, with small, openable windows, overhanging eaves from pitched tiled roofing, hinged wooden external or internal slatted shutters; permanent ventilation openings from air bricks, chimneys and roof vents, and north light windows for industrial use, have a large thermal storage mass to smooth out the fluctuations of the outdoor air temperature. Such buildings, designed in the late nineteenth and early twentieth centuries, are better able to provide shaded and cool habitable zones than the exposed glazing high-rise buildings that have been popular since the 1960s. Deep core floor plan buildings that are unreachable by outdoor air to provide the required ventilation, and exterior façades which are deliberately exposed to solar heat gain with no provision for external shading screens, trees or shadows cast from surrounding buildings, are designed to create interior discomfort. Current buildings are also packed to capacity with productive workers who each need a personal computer and permanent artificial illumination in order to carry out their tasks. The other mistake is to employ people to work in these glass boxes only during daylight hours that coincide with the peak cooling load times. The Mediterranean countries close many of their businesses during the afternoons, close the wooden window shutters and sleep during the hottest part of the day, coming out in the late afternoon to start work again.

Energy conservation engineers can assist the building maintenance manager to make the best use of what has been built. There will be many technical methods that are available to reduce the use of energy by the mechanical and electrical systems within the building, but at best, these are likely to reduce energy use by a maximum of 25%, often much less can be achieved. Retrofitting a building with energy-saving systems may accompany a change of use or an overall refurbishment. There has been a strong move in the direction of low-energy-use buildings in the UK in recent years. Greenfield sites in the mild climate of the British Isles have provided building designers with more options than are available in more extreme climates. The outdoor air temperatures that are used for the design of the heating and cooling systems in the UK are within the range of −3°C d.b. to 30°C d.b. The minimum overnight outdoor air temperature may drop to −17°C d.b. once per year during a severe winter. The extreme low and high outdoor air temperatures that can be experienced are of little importance to the designers of commercial and most industrial buildings, unless indoor air temperatures are of critical importance for the condition of products, human or animal safety, or industrial processes. The users of most buildings are expected to withstand colder or warmer indoor temperatures for a few days per year and not complain too much. In extreme cases, workplace agreements are made to allow work to be stopped for short periods so that the occupants of the building can move away from the workstation for recuperation. An example is a production factory where 10-minute breaks in an 8-hour shift are agreed for production-line workers to go outside the building when the indoor air temperature exceeds 30°C d.b., which it does frequently during long, hot summers, due to the minimalist thermal insulation value of the building's walls and roofing.

The design team of a new building need to develop a method of solving the complex issues surrounding the task of providing a building which satisfies:

1. the basic needs of the owner;
2. architectural and local planning design philosophy;
3. the requirements of legislation;
4. access, spatial, visual, aural and thermal comfort needs of the occupants;
5. use of energy during the 50-year service period of the building;
6. sources of energy that are available;
7. maintainability of the whole complex;
8. the *HM Government Carbon Plan, 2011* (DECC, 2011a).

Those matters which relate to the need, or otherwise, for air conditioning, can be summarized as:

1. local design weather conditions;
2. indoor design set points for zone air temperature and relative humidity;
3. allowable variation in the indoor design air conditions;

4. number of occasions during each year when divergence from the specified indoor design conditions is allowable;
5. time periods when the building will be occupied by the main users and the service personnel;
6. sources of primary energy available for the building, their long-term reliability, storage requirements and safety considerations;
7. renewable energy available for the site;
8. usage of the building;
9. means by which the building can be heated;
10. outdoor air ventilation quantity;
11. peak energy demand for heating and cooling;
12. location and sizes of plant and service shaft spaces available;
13. whether natural ventilation or assisted natural ventilation is appropriate;
14. whether mechanical cooling systems are needed to maintain the specified peak design conditions;
15. need to provide accurate indoor environmental conditions for equipment, material storage and handling, or an industrial process;
16. whether low-cost cooling systems can be used;
17. whether there is a real need to provide a mechanical means of air conditioning;
18. process requirement for closely conditioned air within the building;
19. energy recovery strategies available;
20. maintainability of the mechanical services and how replacement plant can be provided without major structural works becoming necessary.

These considerations all have an impact on the building design team's decision-making process. The local weather conditions that create the maximum heating and cooling loads during the occupied part of the day, or night, will determine the size of the heating and cooling plant that are required. Occupancy times can sometimes be varied in order to minimize the plant capacity that is needed, for example, by using the Mediterranean region off-peak working principle, however unpalatable this may seem to northern European practices. The architectural and engineering designers have the option of experimenting with the thermal insulation value of the exterior envelope of the building in varying the area of glazing, walling and roofing and the thermal transmittance of each component. Life-cycle costing of each alternative thermal design will reveal the total cost of constructing and using the building for 50 years, with reasonable assumptions on price changes each year and the cost of refitting the building every 20 years because of improvements in technology.

Selection of the indoor design air set points for temperature and relative humidity will be determined by legislative and comfort standards. Short-term variances within an allowable range can minimize the use of energy at peak times. It may be possible to avoid the installation of mechanical cooling systems where the users of the building agree to accept regular, short-term divergence from the standard design conditions. The provision of ceiling-mounted or portable fans with comfort breaks and refrigerated drinks dispensers might allow a building to avoid the installation of mechanical cooling, depending upon the number and frequency of divergences from the normal standards. Allowing the indoor air temperature to rise to 25°C d.b. in summer before switching on the air-conditioning chiller during the afternoon, and whether the building is occupied, the time is before 3.30 p.m. and the outdoor air temperature is above 25°C d.b; all these conditions can be accessed through the computer-based building management system, if one is provided. Automatic control programming can be set to minimize the use of electrical energy by the mechanical cooling system as a deliberate strategy by the designers and operators of the building. Dynamic thermal analysis software is used to model buildings and their services systems to assess the indoor air conditions that are expected to occur.

The provision of outdoor air ventilation is a legislative requirement based on the activities of and the numbers of people within the enclosure. The minimum quantity of outdoor air must be maintained for each person to comply with the standards. Outdoor air ventilation does more than provide air for breathing, it also flushes the building with outdoor air to remove heat and to control the concentrations of odours and atmospheric pollutants that are produced within the building. Toilet exhaust air flows often partly match the inflow rate of outdoor air in commercial buildings. Any balance of flow between the exhaust quantity and inflow of outdoor air may be allowed to leak though doors, permanent ventilators and other openings in the structure, such as gaps around window frames. The inclusion of CO_2 sensors in the return or exhaust air ducts allows the building management system to minimize the opening of the outside air intake motorized dampers and consequently reduce the heating or cooling load of the air-conditioning plant and save energy. CO_2 in air that is removed from the occupied space is a direct assessment of the number of occupants and their activity level, allowing energy use to match the instantaneous load on the building. Other means of assessing room occupancy are available from infrared or ultrasonic sensors. If the designers of the building are able to know the divergence in the patterns of occupancy of rooms or spaces, then decisions can sometimes be taken to diversify the provision of lighting, heating and cooling zones to minimize the use of mechanical and electrical plant and systems, with benefits in the initial plant capacity and in the use of energy in the long term. Where incoming outdoor unconditioned air can be pre-conditioned, an air-to-air flat plate heat exchanger may be used. This is a compact box from the size of a suitcase up as part of the air handling plant, having counter-flow passages for incoming and outgoing air separated by aluminium foil plates. Incoming winter air is warmed by the room air being exhausted; in summer, room air at the correct temperature cools the outdoor ventilation air stream.

Natural and mixed mode ventilation can be used in commercial atria and industrial buildings where the stack effect of height within the building is used to create air movement. Low-level outside air intakes and roof-level air extractors or openable ventilation units can be mechanically controlled to match the heating and cooling load on the building to the flow of air through the spaces. The avoidance of draughts around sedentary occupants will always be the main challenge. Low-level radiant heating from warmed floors or overhead radiant panels can offset high rates of air movement. Anyone who has sat within cathedral ceiling spaces will recognize the potential discomfort problems in cold weather within intermittently used high thermal-mass buildings, for example, stone churches and sports halls. The use of natural ventilation in the UK is accompanied by the problem of allowing the uncontrolled ingress of the moisture that is present in the outside air. Any surface within the building that is below the dew-point temperature of the space will accumulate damp and create long-term mould growth. This becomes a comfort, health and maintenance cost if damage to the building is to be avoided. In tropical climates where the outdoor air temperature rises above 20°C d.b. on most days of the year, frequently in the 30s with high humidity, the use of outdoor air natural ventilation is often out of the question. This is due to the combination of intense solar radiation heat gains in these regions and the lack of a natural cooling function by the outdoor air.

Hot dry climates in the Middle East and southern continents make use of evaporative low cost cooling. Figure 5.2 shows an Australian type where outdoor air of above 25°Cd.b. and around 10%–30% saturation is drawn inwards through wetted vertical panels and blown directly into occupied rooms as quickly as possible. Operational costs are for a 400 Watt axial fan, low water consumption, water bacteria control and maintenance cleaning. Larger units serve commercial premises and may have an indirect gas-fired heat exchanger to provide winter heating. The supply

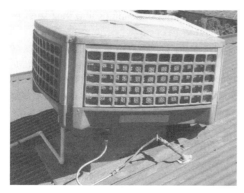

5.2 Ducted evaporative cooler used in hot climates.

air has a high humidity but is at reduced dry-bulb air temperature. All cooling is created from the evaporation of water, the latent heat transfer that provides sensible cooling of the conditioned space through high air flow. Air is released from the building through screened window and door openings or exhaust fans in large applications. Air handling units for commercial applications may use an evaporative cooling section to cool and humidify outgoing exhaust air. The cooled exhaust air then passes through a flat plate heat exchanger or thermal wheel that reduces the temperature of the incoming outdoor air; this is indirect evaporative cooling. This avoids the problems of increasing indoor air humidity which would create discomfort in mild European and tropical climates and bacterial contamination of the water.

In the tropics, the constant high moisture content of the outdoor air makes it unsuitable for natural ventilation practice in commercial buildings. Buildings in the UK that have an internal atrium have restored a historical precedent in turning the building inside-out. The exterior surfaces need less glazing as the occupants' view is directed inwards towards a planted open space that has natural daylight. The atrium can be used to return conditioned air back to the air-handling plant room without the need for a return air fan, collect exhaust air and expel it through roof openable vents and can facilitate the removal of smoke during an emergency. Heat produced by the occupants, fluorescent lighting, computers and electrical equipment assists the upward flow of air away from the occupied zones. Atria are used in all climates from sub-zero through 40°C d.b. environments and fully provide indoor office, retail shopping malls, casino, hotel and entertainment spaces throughout the world.

The options available to the designers in the selection of the method of controlling the environmental conditions within the building include the following:

1. *Natural ventilation.* Applicable when the external climate, the use and design of the building permit it. Mild climate localities usually close to the coast where the sea is warm; internal air conditions are allowed to vary widely and are directly related to the external weather conditions; central heating system; cooling may be provided from packaged direct expansion refrigeration units within each zone; manually operated internal or external shading blinds; passive solar architecture that may include thermal storage walls; chilled-water beams or flat panels may be installed at high level within offices to provide limited cooling; chilled water in the beams and panels is maintained at a temperature that is above the room air dew-point so that there is no condensation on the exposed surfaces or within the ceilings, avoiding dehumidification and control of the zone relative humidity; chilled beams and panels provide no ventilation air.

2. *Mixed mode ventilation.* This is a development of natural ventilation. Mechanically operated ventilation louvers and exhaust air fans improve the control of air flow through the occupied spaces; incoming outside air may be cooled with a water spray evaporative cooler in hot, dry climates; evaporative cooling is a low-cost means of cooling the incoming outside air, which is exhausted by either natural openings of doors and windows, or by exhaust air fans in confined zones; incoming outdoor air may be cooled through a specific temperature range, say, 10 K, to provide limited cooling by means of direct expansion or chilled water refrigeration plant.

3. *Mechanical ventilation which only passes outside air through the zone.* This usually applies to moderate climates such as the UK where minimal cooling is required; in climates where the outside air temperature exceeds 30°C d.b. the flow rate of outside air is likely to be insufficient to provide enough cooling for zone temperature control in an air-conditioning system; where it is possible to locate the exhaust air duct alongside the incoming outside air duct, within the ceiling space or a plant room, an air-to-air flat-plate heat exchanger is used to transfer heat between the incoming and outgoing air streams; the outgoing exhaust air is already at the correct zone temperature, of around 22°C d.b. (it cannot be recycled as it is vitiated with carbon dioxide, odours and atmospheric pollutants); heat transfer works throughout the year (in winter, the incoming outdoor air is preheated by up to 10 K from the outgoing exhaust air; in summer, the incoming outdoor air is precooled by up to 10 K from the outgoing exhaust air); heat transfer efficiency is around 55%; a similar heat transfer can be obtained from a recuperative heat wheel that transfers heat from the outgoing exhaust air to the incoming outside air, generating up to a 10 K temperature change in the outside air stream.

4. *Mechanical ventilation with recirculated room air.* The maximum quantity of conditioned room air is recirculated to save energy use at the heating and refrigeration plant; the outside air motorized dampers are modulated from closed to fully open to control the zone air temperature without the use of the mechanical refrigeration plant for as long a time as possible, providing low-cost cooling to the building; when the refrigeration plant has to be used, the outside and exhaust air motorized dampers are moved to their minimum outside air positions, often around 10% open, allowing the maximum use of recirculated room air; a range of ducted air-conditioning systems are in use, including single duct, dual duct, induction units, fan coil units and variable air volume systems.

A single-duct system extracts from the room, exhausts some to the atmosphere and as much as possible is recirculated to reduce running costs of heating and cooling plants. Incoming fresh air is filtered and mixed with that recirculated; it is then heated by a low-, medium- or high-pressure hot-water or steam finned pipe heat exchanger or an electric resistance element. The heated air is supplied through ducts to the room. The hot water flow rate is controlled by a duct-mounted temperature detector in the extract air, which samples room conditions. The electrical signal from the temperature detector is received by the automatic control box and corrective action is taken to increase or reduce water flow rate at the electrically driven motorized valve at the heating coil. During summer operation, chilled water from the refrigeration plant is circulated through the cooling coil and room temperature is controlled. A temperature detector in the fresh air duct varies the set value of the extract duct air temperature; higher in summer, lower in winter; minimizing energy costs. A low-limit temperature detector will override the other controls to avoid injection of cold air to the room. The building is slightly pressurized by extracting only about 95% of the supply air volume, allowing some conditioned air to leak outwards. Energy savings are maximized by recirculating as much of the conditioned room air as possible. Room air recirculation with economy-cycle motorized dampers can, sometimes, be retrofitted to existing systems as an energy conservation measure. In mild climates, such as in the UK, full outside air systems are also used. These have no recirculation air ducts; either a flat-plate heat exchanger or run-around pipe coils can be installed to preheat and precool the incoming outside air to save energy. Such heat exchangers are around 55% efficient, which is not as good as recirculation.

EXAMPLE 5.2

A room 15 m × 7 m × 2.8 m high is to have a ventilation rate of 11 air changes per hour. Air enters from a duct at a velocity of 8.5 m/s. Find the air volume flow rate to the room and the dimensions of the square duct.

$$Air\ flow\ rate\ Q = N\frac{air\ changes}{h} \times V\frac{m^3}{air\ change} \times \frac{1h}{3600s}$$

$$V = room\ volume\ for\ one\ air\ change\ m^3$$

$$Q = \frac{NV}{3600}\ \frac{m^3}{s}$$

$$Q = \frac{11 \times 15 \times 7 \times 2.8}{3600}\ \frac{m^3}{s}$$

$$Q = 0.9\frac{m^3}{s}$$

$$Also,\ Q\frac{m^3}{s} = duct\ cross-sectional\ area\ A\ m^2 \times air\ velocity\ v\frac{m}{s}$$

$$A = \frac{Q}{v}\ m^2$$

$$A = \frac{0.9}{8.5}\ m^2$$

$$A = 0.106\ m^2$$

Length of side of square duct = l m

$$l = \sqrt{A\,m^2}$$

$$l = \sqrt{0.106\,m^2}$$

$$l = 0.325\ m$$

EXAMPLE 5.3

A lecture theatre has dimensions 25 m × 22 m × 6 m high and has 100 occupants; 8 l/s of fresh air and 25 l/s of recirculated air are supplied to the theatre for each person. A single-duct ventilation system is used. If 10% of the supply volume leaks out of the theatre, calculate the room air change rate and the air volume flow rate in each duct.

$$\text{Supply air flow rate } Q = (8 + 25)\frac{l}{s \ person} \times \frac{1\,m^3}{10^3 l} \times 100 \ people$$

$$\text{Supply air flow rate } Q = 3.3\frac{m^3}{s}$$

$$Q = \frac{NV}{3600}\frac{m^3}{s}$$

$$N = \frac{Q \times 3600}{V}\frac{air \ changes}{h}$$

$$N = \frac{Q \times 3600}{V}\frac{air \ changes}{h}$$

$$N = \frac{3.3 \times 3600}{25 \times 22 \times 6}\frac{air \ changes}{h}$$

$$N = 3.6\frac{air \ changes}{h}$$

$$\text{Leakage from the theatre} = 10\% \times Q\frac{m^3}{s}$$

$$\text{Leakage from the theatre} = 10\% \times 3.3\frac{m^3}{s}$$

$$\text{Leakage from the theatre} = 0.33\frac{m^3}{s}$$

$$\text{Air extracted from the theatre} = (3.3 - 0.33)\frac{m^3}{s}$$

$$\text{Air extracted from the theatre} = 2.97\frac{m^3}{s}$$

$$\text{Outside air supply to the theatre} = \frac{8 \times 100}{10^3}\frac{m^3}{s}$$

$$\text{Outside air supply to the theatre} = 0.8\frac{m^3}{s}$$

$$\text{Recirculated air flow} = supply \ air - outside \ air$$

$$\text{Recirculated air flow} = (3.3 - 0.8)\frac{m^3}{s}$$

$$\text{Recirculated air flow} = 2.5\frac{m^3}{s}$$

$$\text{Exhaust air flow} = return \ air \ from \ theatre - recirculated \ air \ at \ AHU$$

$$\text{Exhaust air flow} = (2.97 - 2.5)\frac{m^3}{s}$$

$$\text{Exhaust air flow} = 0.47\frac{m^3}{s}$$

EXAMPLE 5.4

There are 35 people in a gymnasium, each producing CO_2 at a rate of 10×10^{-6} m^3/s. If the CO_2 level is not to exceed 0.15%, 1500 ppm the recommended maximum, find the air supply rate necessary. The outdoor air CO_2 concentration is 0.04%, 400 ppm a likely global average. Explain what this means to the ventilation designer.

$$Outdoor\ air\ flow\ needed = \frac{rate\ of\ CO_2\ production}{(room\ level - outdoor\ air\ level)}$$

$$Outdoor\ air\ flow\ needed = \frac{35 \times 10 \times 10^{-6}}{(0.15 - 0.04) \times 10^{-2}}\ \frac{m^3}{s}$$

$$Outdoor\ air\ flow\ needed = 0.32\ \frac{m^3}{s}$$

This is a very small air flow of outdoor air through a gymnasium where strenuous activity by 35 athletic people is to take place. The outdoor air of 320 litre/s may be able to be provided by natural ventilation through controlled low and high level vents to outside. Additional recirculated air flow to control the air temperature with heating or cooling systems is provided if the climate required mechanical air treatment.

Removal of heat gains

Ventilation air flow and in-room cooling units such as chilled-water beams, floor and ceiling panels and packaged air conditioners are used to remove excess heat gains from buildings. Sensible heat gains are those that affect room air temperature while latent heat gains add moisture into the conditioned room. Sensible heat gains result from solar radiation, conduction from outside to inside during hot weather, warm ventilation air, lighting, electrical machinery and equipment, people and industrial processes. Such heat gains affect the temperature of the air and the building construction. Latent heat gains result from exhaled and evaporated moisture from people, moisture given out from industrial processes and humidifiers. Latent heat gains do not directly affect the temperature of the surroundings but take the form of transfers of moisture and are measured in weight of water vapour transferred or its latent heat equivalent in watts. The latent heat of evaporation of water into air at a temperature of 20°C and a barometric pressure of 1013.25 mb is 2453.61 kJ/kg. The latent heat (LH) evaporating 60 g of water in this air is:

$$LH = 60g \times \frac{1\,kg}{10^3 g} \times 2453.61\ \frac{kJ}{kg}$$

$$LH = 147.22\,kJ$$

If this evaporation takes place over, say, 1 h, the rate of latent heat transfer will be:

$$LH = 147.22\frac{kJ}{h} \times \frac{1\,h}{3600\,s} \times \frac{10^3 J}{kJ} \times \frac{Ws}{J}$$

$$LH = 40.9\,W$$

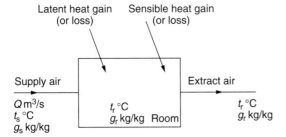

5.3 Schematic representation of heating, cooling and humidity control of an air-conditioned room.

This is the moisture output from a sedentary adult. Removal of sensible heat gains to control room air temperature is carried out by cooling the ventilation supply air and increasing the air change rate to perhaps 20 changes per hour. Figure 5.3 shows this scheme. The temperature and moisture content of the supply air increase as it absorbs the sensible and latent heat gains until it reaches the desired room condition. The net sensible heat flow will be into the room in summer and in the outward direction in winter.

Rooms that are isolated from exterior building surfaces have internal heat gains from people and electrical equipment, producing a net heat gain throughout the year. The heat balance is as follows:

net sensible heat flow into room = sensible heat absorbed by ventilation air

SH = air mass flow rate $\times$ specific heat capacity $\times$ temperature rise

SH is the sensible heat. Specific heat capacity of humid air is 1.012 kJ/kg K. The volume flow rate of air is normally used for duct design. Density of air at 20°C d.b. and 1013.25 mb is $\rho = 1.205\,kg/m^3$.

$$Q\frac{m^3}{s} = \text{air mass flow rate}\frac{kg}{s} \times \frac{m^3}{kg}$$

The supply air temperature $t_s°C$ can have any value and, as density is inversely proportional to the absolute temperature, from the general gas laws. Room design air temperature is $t_r°C$.

$$SH\ kW = Q\frac{m^3}{s} \times 1.205\frac{kg}{m^3} \times \left(\frac{273+20}{273+t_s}\right) \times (t_r - t_s)K \times 1.012\frac{kJ}{kg\,K} \times \frac{kWs}{kJ}$$

For summer cooling, t_r is greater than t_s during a net heat gain. For winter heating, t_r is less than t_s as there is a net heat loss, and so the temperature difference $(t_s - t_r)$ must be used in the equation. It is more convenient to rewrite the equation in the form,

$$Q = \frac{SH\ kW}{(t_r - t_s)} \times \frac{(273+t_s)}{357}\frac{m^3}{s}$$

EXAMPLE 5.5

An office 20 m × 15 m × 3.2 m high has 30 occupants, 35 m² of windows, 25 (2 x 30) W fluorescent tube light fittings, a photocopier with a power consumption of 1500 W and conduction heat gains during summer amounting to 2 kW. Solar heat gains are 600 W/m² of window area. The sensible heat output from each person is 110 W. The room air temperature is not to exceed 24°C when the supply air is at 13°C. Calculate the supply air flow rate required and the room air change rate. State whether the answers are likely to be acceptable.

Table 5.1 shows the results.

$$Q = \frac{29.3 \, kW}{(24-13)} \times \frac{(273+13)}{357} \frac{m^3}{s}$$

$$Q = 2.13 \frac{m^3}{s}$$

Some engineers prefer to calculate the mass flow rate of air flow through the air-conditioned space. This is easily found by multiplying the volume flow rate by the density of air at that location, thus,

$$Q = 2.13 \frac{m^3}{s} \times 1.205 \frac{kg}{m^3}$$

$$Q = 2.57 \frac{kg}{s}$$

Air mass flow rate does not change with temperature, volume flow does.

$$Q = \frac{NV}{3600} \frac{m^3}{s}$$

$$N = \left(\frac{3600 \times Q}{V} \right) \frac{air \; changes}{h}$$

$$N = \left(\frac{3600 \times 2.13}{20 \times 15 \times 3.2} \right) \frac{air \; changes}{h}$$

$$N = 8 \frac{air \; changes}{h}$$

Table 5.1 Summary of sensible heat gains in Example 5.5

Source	Quantity (W/m² × m²)	SH (W)
Windows	600 × 35	21000
Occupants	110 × 30	3300
Lights	25 × 2 × 30	1500
Photocopier		1500
Conduction		2000
	SH gain	29300
		29.3 kW

Between 4 and 20 air changes per hour are likely to create fresh air circulation through an office without causing draughts and should be suitable for the application.

EXAMPLE 5.6

A room has a heat demand in winter of 32 kW and a supply air flow rate of 3.5 m³/s. The room air temperature is to be maintained at 22°C. Calculate the supply air temperature to be used.

For winter, the equation is:

$$Q = \frac{SH\,kW}{(t_s - t_r)} \times \frac{(273 + t_s)}{357} \frac{m^3}{s}$$

This can be rearranged to find the supply air temperature t_s required:

$$Q \times 357 \times (t_s - t_r) = SH \times (273 + t_s)$$

$$Q \times 357 \times t_s - Q \times 357 \times t_r = 273 \times SH + SH \times t_s$$

$$Q \times 357 \times t_s - SH \times t_s = 273 \times SH + Q \times 357 \times t_r$$

$$t_s \times (Q \times 357 - SH) = 273 \times SH + Q \times 357 \times t_r$$

$$t_s = \frac{273 \times SH + Q \times 357 \times t_r}{Q \times 357 - SH}$$

In this example:

$$t_s = \frac{273 \times 32 + 3.5 \times 357 \times 22}{3.5 \times 357 - 32}$$

$$t_s = 29.75°C$$

The heated or cooled air supply may also have its humidity modified by an air conditioning plant so that the percentage saturation of the room air is controlled. A water spray or steam injector can be used for humidification. The refrigeration cooling coil lowers the air temperature below its dew-point to condense the moisture out of the air. Figure 5.4 shows how the properties of humid air are determined from a psychrometric chart. If g_r and g_s are the moisture contents of the room and the supply air respectively, then a heat balance can be written for the latent heat gain absorbed by the previously calculated supply air flow rate Q:

Latent heat gain to room = latent heat equivalent of moisture removed

by the conditioned air

$$LH\,kW = moisture\ mass\ flow\ rate \frac{kg}{s} \times latent\ heat\ of\ evaporation \frac{kJ}{kg}$$

$$LH = air\ mass\ flow \frac{kg}{s} \times moisture\ absorbed \frac{kg\ H_2O}{kg\ dry\ air} \times 2453.61 \frac{kJ}{kg\ H_2O}$$

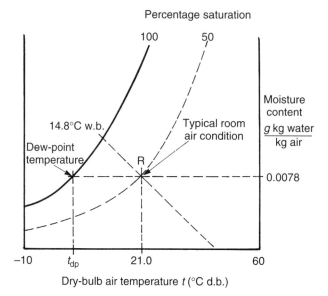

Percentage saturation

5.4 The psychrometric chart.

$$LH = Q\frac{m^3}{s} \times 1.205\frac{kg}{m^3} \times \left(\frac{273+20}{273+t_s}\right) \times (g_r - g_s)\frac{kg\ H_2O}{kg\ dry\ air} \times 2453.61\frac{kJ}{kg\ H_2O}$$

$$Q = \frac{LH}{(g_r - g_s)} \times \frac{(273+t_s)}{866284}\frac{m^3}{s}$$

The denominator can be rounded to 860 000 with an error of less than 1.0%.

EXAMPLE 5.7

The people in the office in Example 5.5 each produce 40 W of latent heat. Find the supply air moisture content to be maintained, given that the room air is to be at 50% saturation and the corresponding moisture content g_r is 0.008 905 kg H_2O/kg air.

$$LH = 30\ people \times 40\frac{W}{person} \times \frac{1\ kW}{10^3\ W}$$

$$LH = 1.2\ kW$$

$$t_s = 13°C$$

$$Q = 2.13\frac{m^3}{s}$$

$$Air\ mass\ flow\ rate = 2.13\frac{m^3}{s} \times 1.205\frac{kg}{m^3}$$

$$Air\ mass\ flow\ rate = 2.57\frac{kg}{s}$$

$$Q = \frac{LH}{(g_r - g_s)} \times \frac{(273 + t_s)}{860000} \frac{m^3}{s}$$

$$(g_r - g_s) = \frac{1.2}{2.13} \times \frac{(273 + 13)}{860000} \frac{kg\, H_2O}{kg\, dry\, air}$$

$$g_s = (g_r - 0.000187) \frac{kg\, H_2O}{kg\, dry\, air}$$

$$g_s = (0.008905 - 0.000187) \frac{kg\, H_2O}{kg\, dry\, air}$$

$$g_s = 0.008718 \frac{kg\, H_2O}{kg\, dry\, air}$$

Psychrometric cycles

Heating and cooling processes in air-conditioning equipment can be represented on the psychrometric chart in the following manner.

Heating is performed with a low or high pressure hot water, possibly steam, finned pipe heating coil, electric resistance heater or fuel-fired heat exchanger as shown in Figure 5.5. SE is the specific enthalpy, total heat content, of the air, as read from the chart.

Cooling is performed by passing chilled water, brine solution or refrigerant through a finned pipe coil. When the coolant temperature is below the air dew-point, condensation occurs and the air will be dehumidified. Figure 5.6 shows the two possible cycles.

Mixing of two airstreams occurs when the fresh air intake joins the recirculated room air. The quantity of each air stream is regulated by multi-leaf dampers operated by electric or pneumatic motors under the direction of an automatic control system. Varying the intake of fresh air between the minimum amount during peak summer and winter conditions and 100% when free atmospheric cooling, can be achieved during mild weather and summer evenings and can result in minimizing the energy costs of the heating and refrigeration plants. Figure 5.7 shows the operational process.

The mass flow balance for the junction is:

Mass flow of stream 1 + mass flow of stream 2 = mixed mass flow

$$Q_1 \times \rho_1 + Q_2 \times \rho_2 = (Q \times \rho) \frac{kg}{s}$$

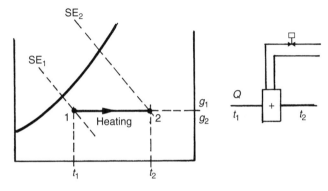

5.5 Air heating depicted on a psychrometric chart.

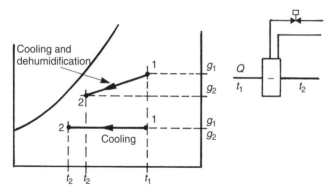

5.6 Cooling and dehumidification.

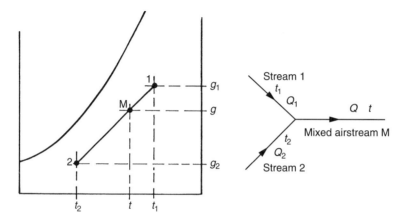

5.7 Psychrometric cycle for the mixing of two airstreams.

The enthalpy balance, taking the specific heat capacity and density as constants, can be represented:

$$Q_1 \times t_1 + Q_2 \times t_2 = (Q \times t)kW$$

Dividing through by Q gives:

$$\frac{Q_1}{Q} \times t_1 + \frac{Q_2}{Q} \times t_2 = t$$

The mixed air temperature and moisture content lie on the straight line connecting the two entry conditions and can be found by the volume flow rate proportions as indicated by this equation.

In winter, incoming fresh air with low moisture content can be humidified by steam injection, water sprays, evaporation from a heated water tank or a spinning disc atomizer. A pre-heater low-pressure hot water coil usually precedes the humidifier to increase the water-holding capacity of the air. This also offsets the reduction in temperature of the air owing to transference of some of its sensible heat into latent energy which is needed for the evaporation process. Figure 5.8 shows such an arrangement.

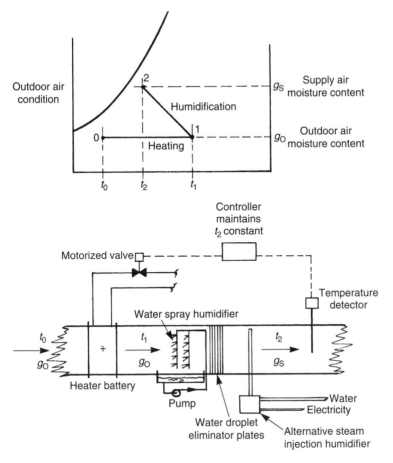

5.8 Preheating and humidification.

A temperature sensor in the humidified air is used as a dew-point control by modulating the pre-heater power to produce air at consistent moisture content throughout the winter. For comfort air conditioning, the room percentage saturation will be 50% ± 10%. This permits a wide range of humidifier performance characteristics. The humidification process often follows a line of constant wet-bulb temperature. The water spray temperature is varied to alter the slope of the line on the psychrometric chart. A complete psychrometric cycle for a single-duct system during winter operation is shown in Figure 5.9. The preheating and humidification stages have been omitted, as close humidity control is deemed not to be needed in this case. A typical summer cycle is shown in Figure 5.10.

Some reheating of the cooled and dehumidified air will be necessary because of the practical limitations of cooling coil design. Part of the boiler plant remains operational during the summer. Reheating can be avoided by using a cooling coil bypass which mixes air M and air C to produce the correct supply condition. Heating and refrigeration plant capacities are found from the enthalpy changes and specific volume, read from the chart, and the air volume flow rate and the specific volume of the supply air $v_s \frac{m^3}{kg}$ from a psychrometric chart or data:

$$\text{Heat transfer rate} = Q \frac{m^3}{s} \times \frac{kg}{v_s m^3} \times (SE_1 - SE_2)\frac{kJ}{kg} \times \frac{kWs}{kJ}$$

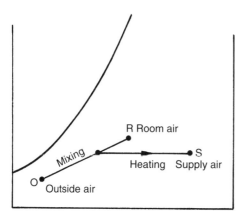

5.9 Winter psychrometric cycle for a single-duct system.

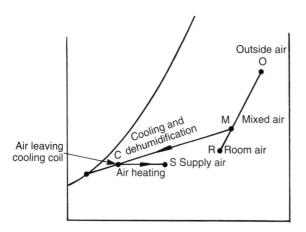

5.10 Summer psychrometric cycle for a single-duct system.

EXAMPLE 5.8

Outside air at $-5°C$ d.b., 80% saturation enters a pre-heater coil and leaves at 24°C d.b. The air volume flow rate is 6.5 m^3/s. Find: (a) the outdoor air wet-bulb temperature and specific volume; (b) the heated air moisture content and percentage saturation; and (c) the heating coil power.

From the CIBSE psychrometric chart:

(a) $-5.9°C$ and 0.7615 m^3/kg;

(b) 0.00198 kg H_2O/kg air and 10%;

(c) SE_2 at 24°C d.b. and 0.00198 kg H_2O/kg is 29.0 kJ/kg.

   SE_1 at $-5°C$ d.b. and 0.00198 kg H_2O/kg is -0.073 kJ/kg

$$Heater\ duty = 6.5\frac{m^3}{s} \times \frac{kg}{0.7615\ m^3} \times (29 + 0.073)\frac{kJ}{kg} \times \frac{kWs}{kJ}$$

$$Heater\ duty = 248.2\ kW$$

EXAMPLE 5.9

A cooling coil has water passing through it at a mean temperature of 10°C, an air flow of 10.25 m³/s air enters the coil at 28°C d.b., 23°C w.b. and leaves at 15°C d.b. Find the leaving air wet-bulb temperature and percentage saturation. Calculate the refrigeration capacity of the coil in equivalent tonnes refrigeration capacity, ton(r), when 1 ton(r) is equal to 3.527 kW(r).

Plot the cooling and dehumidification line on the psychrometric chart in the manner shown in Figure 5.5 with a target point of 10°C on the saturation curve. The air leaves the cooling coil at 15°C d.b., 14.2°C w.b. and 91% saturation.

Specific enthalpy of the air entering the coil is 68 kJ/kg and is 40 kJ/kg on leaving. Specific volume of the air entering the coil is 0.874 m³/kg.

$$Refrigeration\ capacity = 10.25\frac{m^3}{s} \times \frac{kg}{0.874\ m^3} \times (68 - 40)\frac{kJ}{kg} \times \frac{kWs}{kJ}$$

$$Refrigeration\ capacity = 328.4\ kW\,(r)$$

$$Refrigeration\ capacity = 328.4\ kW\,(r) \times \frac{1\ ton\,(r)}{3.517\ kW\,(r)}$$

$$Refrigeration\ capacity = 93.4\ ton\,(r)$$

Refrigeration capacity is commonly rated in ton(r) as this is the energy needed to freeze a US tonne of ice over a period of 24 hours,

$$1\ TR = \frac{2000\ lb}{24\ h} \times \frac{144\ Btu}{lb} \times \frac{1\ kWh}{3412\ Btu}$$

$$1\ TR = 3.517kW\,(r)$$

EXAMPLE 5.10

6 m³/s of recirculated room air at 22°C d.b., 50% saturation is mixed with 1.5 m³/s of incoming fresh air at 10°C d.b., 6°C w.b. Calculate the mixed air dry-bulb temperature. Plot the process on a psychrometric chart and find the mixed air moisture content.

$$Q_1 = 6\,m^3/s,\, Q_2 = 1.5\,m^3/s,\, Q = supply\ air\ flow\ 7.5\,m^3/s,$$

$$t = \frac{Q_1}{Q} \times t_1 + \frac{Q_2}{Q} \times t_2$$

$$t = \frac{6}{7.5} \times 22 + \frac{1.5}{7.5} \times 10$$

$$t = 19.6°C$$

From the chart,

$$g = 0.0076 \frac{kg\ H_2O}{kg\ dry\ air}$$

Air-conditioning systems

Single-duct system

The single-duct system (Figure 5.11) is used for a large room such as an atrium, a banking hall, a swimming pool, or a lecture, entertainment or operating theatre. It can be applied to groups of rooms with a similar demand for air conditioning, such as offices facing the same side of the building. A terminal heater coil under the control of a temperature sensor within the room can be employed to provide individual room conditions.

A variable air volume (*VAV*) system has either an air volume control damper or a centrifugal fan in the terminal unit to control the quantity of air flowing into the room in response to signals from a room air temperature sensor. Air is sent to the terminal units at a constant temperature by the single-duct central plant, according to external weather conditions. A reduction in the demand for heating or cooling detected by the room sensor causes the damper to throttle the air supply or the fan to reduce speed until either the room temperature stabilizes or the minimum air flow setting is reached. Air flow from the diffuser is often blown across the ceiling to avoid directing jets at the occupants. As a result of the Coanda effect, the air stream forms a boundary layer along the ceiling and entrains room air to produce thorough mixing and temperature stabilization before it reaches the occupied part of the room. When the *VAV* unit reduces air flow, there may be insufficient velocity to maintain the boundary layer, and in summer cool air can dump or drop from the ceiling onto the occupants, resulting in complaints of cool draughts.

Dual-duct system

In order to provide for wide-ranging demands for heating and cooling in multi-room buildings, the dual-duct system, as shown in Figure 5.12, is used. Air flow in the two supply ducts may, of necessity, be at a high velocity (10–20 m/s) to fit into service ducts of limited size. Air turbulence

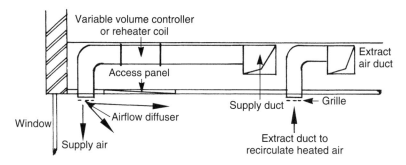

5.11 Single-duct all-air installation in a false ceiling.

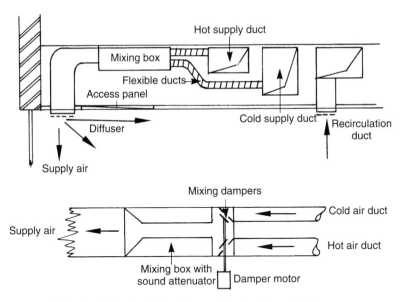

5.12 Dual-duct installation in a false ceiling and detail of the mixing box.

and fan noise are prevented from entering the conditioned room by an acoustic silencer. In summer, the hot duct will be for mixed fresh and recirculated air, while the cold duct is for cooled and dehumidified air. The two streams are mixed in variable proportions by dampers controlled from a room air temperature detector. In winter, the cold duct will contain the untreated mixed air and the air in the hot duct will be raised in temperature in the plant room. The system is used for comfort air conditioning as it does not provide close humidity control. It reacts quickly to changes in demand for heating or cooling when, for example, there is a large influx of people or a rapid increase in solar gain.

Induction

Induction is a less costly alternative to the all-air single- and dual-duct systems for multi-room applications. The central air-conditioning plant handles only fresh air, perhaps only 25% of the supply air quantity for an equivalent single-duct system. All the humidity control, and also some of the heating and cooling for the building, are achieved by conditioning the fresh air intake in the plant room. Primary fresh air is injected through nozzles into the induction unit in each room. These units may be in the floor, in the ceiling void or under the window sill. Because of the high-velocity jets, the local atmospheric pressure within the unit is lowered and air is induced into it from the room. The induced air may enter at three or four times the volume flow rate of primary air and it flows through a finned pipe bank and dust filter before mixing with primary air and being supplied to the room. The secondary air flow rate can be manually adjusted using a damper. Either hot or chilled water is passed through the room coil, depending upon demand. A two-, three- or four-pipe distribution system will be used. The two-pipe system requires a change-over date from heating to cooling plant operation, but a three-way valve can blend hot and chilled water from the three-pipe arrangement. The third alternative has separate hot- and chilled-water pipe coils and pipes. Extract ductwork and a fan remove 90% of the primary air supply and exhaust it to the atmosphere. All recirculation is kept within the room and this greatly

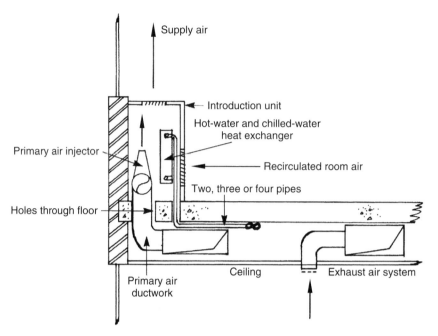

5.13 Induction unit installation in a multi-storey building.

reduces duct costs and service duct space requirements. Figure 5.13 shows a typical installation in an office.

Fan coil units

Heating and cooling loads that prove to be too great for induction units can be dealt with by separate fan and coil units fitted into the false ceiling of each room or building module. Better air filtration can be achieved than with the induction unit. A removable access hatch below the unit is required to facilitate motor and filter maintenance. Care is taken to match the fan-generated noise to the required acoustic environment. As with the other systems, the extracted air can be taken through ventilated luminaires to remove the lighting heat output at source and avoid overheating the room. The supply and extract ducts only carry the fresh air. All recirculation is confined to the room. A typical layout is shown in Figure 5.14.

Packaged unit

A packaged unit is a self-contained air-conditioning unit comprising a hermetically sealed refrigeration compressor, a refrigerant evaporator coil to cool the room air, a hot-water or electric resistance heater battery, filter, a water- or air-cooled refrigerant condenser and automatic controls. Packaged units can either be completely self-contained, needing only a supply of electricity, or piped to central heating and condenser cooling-water plant. Small units are fitted into an external wall and have a change-over valve to reverse the refrigerant flow direction to enable the unit to cool the internal air in summer and the external air in winter. Heat rejected from the condenser is used to heat the internal environment in winter. In this mode of operation it is called a heat pump. A separate ventilation system may be needed. Compressor and fan

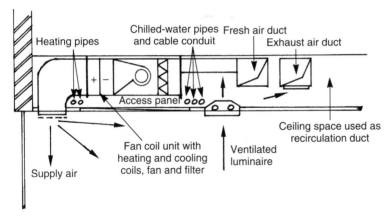

5.14 Fan coil unit installation in a false ceiling.

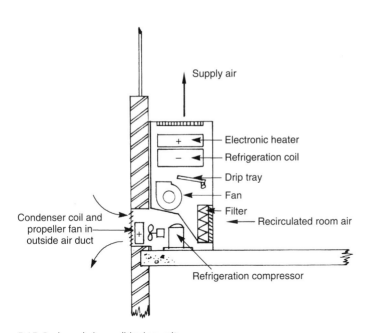

5.15 Packaged air-conditioning unit.

noise levels are compared with the acceptable background acoustic environment. Maintenance requirements are filter cleaning, bearing lubrication and replacement of the compressor when it becomes too noisy or breaks down.

Split system units have a separate condenser installed outside the building. Two refrigerant pipes of small diameter connect the internal and external equipment boxes. This allows greater flexibility in siting the noise-producing compressor. Ducted models provide conditioning and ventilation and are often on flat roofs. Figure 5.15 shows a typical through-the-wall installation.

Vapour compression refrigeration

The electrically driven vapour compression refrigeration system is the principal type used. Its rival, the absorption cycle, burns gas to produce cooling but has a coefficient of performance of around 1, whereas vapour compression has a coefficient of performance in the range 2–3, and so it is cheaper to operate. Compressor types are as follows.

1. *Single- or multi-cylinder reciprocating piston compressor with spring-loaded valves:* domestic refrigerators and small air conditioners have hermetically sealed motor-compressor units which are sealed for their service period, i.e. about 10 years. Condensing units comprise a sealed compressor, refrigerant condenser, liquid receiver, pipes and controls. Refrigerant pipework is installed on site from this unit to a finned-pipe forced-draught air-cooling coil in the air-conditioning system. Large air-cooling plant comprises a multi-cylinder in-line or V-formation compressor, a shell and tube refrigerant to a water evaporator producing chilled water, and a shell and tube refrigerant to a water condenser where the refrigerant vapour is condensed into liquid and the heat given out is carried away by a water circuit to a cooling tower on the roof.
2. *Centrifugal compressors* are used in large chilled-water plants where the noise and vibration produced by the reciprocating type would be unacceptable. A centrifugal impeller of small diameter is driven through a step-up gearbox from a three-phase electric motor. The lack of vibration and compactness of the very high-speed compressor make siting the plant easier.
3. *A screw compressor* has two meshed gears, which compress the refrigerant in the spaces between the helical screws. One gear is driven by an electric motor through a step-up gearbox. The compressor operates at high speed and has very low noise and vibration levels.

A simplified diagram of a vapour compression refrigeration plant is shown in Figure 5.16.

Refrigerants commonly used are non-toxic fluids with high latent heat. Small to medium-sized refrigeration plant with a capacity of up to about 175 kW and motor cars use refrigerant R134a which boils at $-29.8°C$ in the atmosphere. In a typical system it will be evaporated at $5°C$ under a pressure of 3.6 bars and condensed at $40°C$ at 9.6 bars. Larger plant uses fluorinated hydrocarbon R22 ($CHClF_2$), which has a greater refrigerating effect per kilogram but

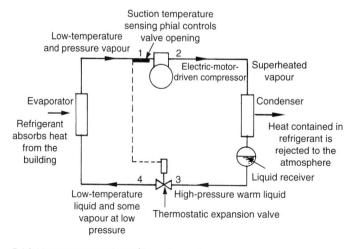

5.16 Vapour compression refrigeration system.

is more expensive and has atmospheric ozone depletion potential (ODP). Other refrigerants in use are ammonia, NH_3, for industrial systems such as in food production and cold stores, CO_2, hydrocarbons such as propane and halocarbon HFCs. The use of CO_2 as a refrigerant appears to be contrary to the Kyoto Protocol reducing global warming emissions but refrigerants are not allowed to be discharged into the atmosphere; putting CO_2 to good use may be a benefit to the world. CO_2 is said to have a low global warming potential, in this context, zero ODP, safe, non-toxic and inflammable, as previously discussed; often used for supermarket refrigeration. All refrigerants have safety risks, require trained installers and service technicians, display warning signs, sensors for leakage, alarms, plant ventilation, first aid availability, safety and breathing apparatus. Large-scale ammonia leakage requires area evacuation. Coefficient of performance (COP) is an expression of cycle efficiency and is found from:

$$COP = \frac{heat\ absorbed\ by\ refrigerant\ in\ the\ evaporator\ W}{power\ consumption\ by\ the\ compressor\ W}$$

COP can be expressed as an instantaneous theoretical ratio, an ideal performance from manufacturers' test data, a measured performance on the installed plant, or as a seasonal average from kWh consumed by the compressor and produced as refrigeration energy, including real weather and load variations for that site. There can be large differences between all these COPs. The vapour compression cycle can be represented on a pressure-enthalpy diagram for the refrigerant as shown in Figure 5.17. Referring to Figures 5.16 and 5.17, compression 1–2 raises the temperature of the refrigerant dry superheated vapour from about 20°C to 60°C, where it can then be cooled and condensed at a sufficiently high temperature to reject the excess heat from the building to the hot external environment.

Vapour condenses at 40°C and collects in the liquid receiver. This warm high-pressure liquid passes through an uninsulated pipe so that it is sub-cooled to below its saturation temperature, about 20°C, at the expansion valve located alongside the evaporator. The pressure rise produced through the compressor is dissipated in friction through the fine orifice in the valve. Such a sudden pressure drop is almost an adiabatic thermodynamic process. This is represented by the vertical line 3–4 on the pressure–enthalpy diagram. Some heat loss from the valve body takes place, so that 3–4 will be slightly curved. Condition 4 is at the lower pressure of the evaporation process, where the refrigerant temperature has dropped to 5°C. Some of the refrigerant liquid has flashed into vapour. Liquid and flash vapour mixture flows through the evaporator, where it is completely boiled into vapour and is then given a small degree of superheat (path 4–1). It then enters the compressor as dry low-pressure superheated vapour at 20°C. A suction

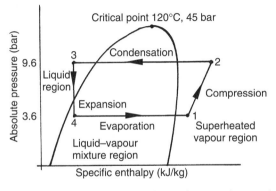

5.17 Pressure enthalpy diagram for a refrigerant showing the vapour compression cycle.

temperature-sensing phial controls the refrigerant flow rate by means of a liquid-filled bellows on the thermostatic expansion valve. This matches refrigerant flow to the refrigerating effect required by the air-conditioning system and ensures that liquid droplets are not carried into the compressor, where they could cause damage. Large plant has refrigerant pressure controllers, which reduce compressor performance by unloading some cylinders. Lubricating oil contaminates the refrigerant leaving the compressor and is separated gravitationally and returned to the crankcase. The pressure–enthalpy diagram depicts the reverse Carnot cycle. In the other direction, the cycle is for an internal combustion engine. The theoretical ideal coefficient of performance is given by:

$$Ideal\ COP = \frac{T_1}{T_2 - T_1}$$

T_1 is the evaporator absolute temperature, K, and T_2 is the condenser absolute temperature. With the temperatures previously used, a summer ideal COP may be:

$$Ideal\ cooling\ COP = \frac{(273 + 5)}{(273 + 40) - (273 + 5)}$$

$$Ideal\ cooling\ COP = 7.94$$

In winter, the same system may be operated as a reverse cycle heat pump to warm indoor air at 30°C and refrigerate outdoor air at 0°C. Some heat pumps are used to heat domestic hot water at 65°C, in winter its ideal COP is:

$$Ideal\ water\ heating\ COP = \frac{(273 + 0)}{(273 + 65) - (273 + 0)}$$

$$Ideal\ water\ heating\ COP = 4.2$$

Friction losses from fluid turbulence, heat transfers to the surroundings, mechanical and electrical losses in the compressor all reduce COP to a range of 2–3 in commercial equipment when electricity consumption of fans and pumps adds to the running energy.

Figure 5.18 shows the installation of a chilled-water refrigeration plant serving one of the cooling coils in an air-conditioning system in a large building. Each air-handling system has filters, heaters, humidifiers and recirculation ducts appropriate to the design. There are several separate air-handling plants, each serving its own zone. Zones are decided by the similarity of demand for conditioning. The south-facing orientation has a cyclic requirement for cooling that is distinct from that of the other sides of a building. Internal areas require cooling throughout the year. These differing needs are often met by having separate zones for each area.

Absorption refrigeration cycle

An example of a two-drum absorption refrigeration cycle is shown in Figure 5.19. The input heat source may be gas, steam or hot water from a district heating scheme, or rejected heat from a gas turbine electricity-generating set. The generator (1) contains a concentrated solution of lithium bromide salt in water. Pure water is boiled off this solution and condenses on the cooling-water pipes, which are connected to an external cooling tower. The generator drum pressure is sub-atmospheric at 0.07 bar, with boiling and condensation taking place at 38°C. Water leaves

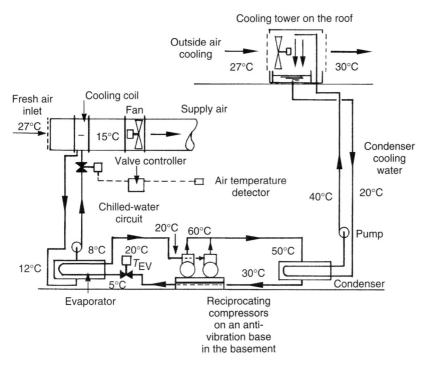

5.18 Refrigeration plant serving an air-conditioning system, showing typical fluid temperatures.

the condenser (2), and then passes through an expansion valve (3), where its pressure and temperature are lowered to 0.01 bar and 7°C. It then completely evaporates while being sprayed over water pipes in the evaporator. These pipes are the chilled-water circuit at 6–10°C, which supplies the refrigeration for the air-conditioning cooling coils. Water vapour in the evaporator drum (4) is sucked into a weak lithium bromide solution by the salt's affinity for water. Latent heat given up by the water vapour as it condenses into the solution is removed from the cooling tower by cooling-water pipes. The weak solution (5) is pumped back into the higher-pressure generator drum to complete the cycle. This pump is the only moving part of some systems. Concentrated salt solution (6) is passed down to the absorber via a heat exchanger and pressure-reducing valve to replace the salt removed. The production of chilled water is equivalent to about half the heat input to the generator.

The gas domestic refrigerator works on an absorption system using liquid evaporation to provide the cooling effect. Figure 5.20 shows the features of modern equipment. A solution of ammonia in water is heated in the boiler (A) by a small gas flame. This is the only energy input. Ammonia gas is driven off the solution and then condensed to a liquid in the air-cooled condenser (B) outside the refrigerator cabinet. The liquid ammonia then passes, with some hydrogen, into the evaporator (C) inside the refrigerator cabinet. This is the ice box. The ammonia completely evaporates while absorbing the heat from the cabinet. The two gases are then led into the absorber (D), where the ammonia is absorbed by a weak solution trickling down the absorber. The strong ammonia solution produced is then driven back into the boiler, while the hydrogen gas, which is not absorbed, passes to the evaporator. The weak solution trickling down the absorber is provided from the boiler. Both types of absorption refrigerator provide cooling from a source of heat, they have few or no moving parts and they only require low-power pumps.

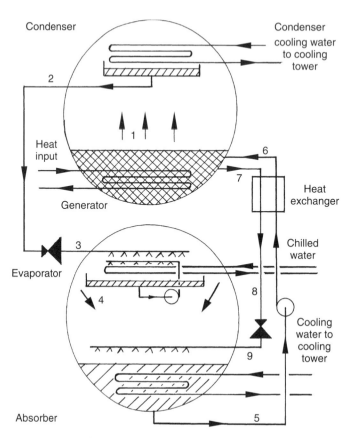

5.19 Two-drum absorption refrigeration cycle.

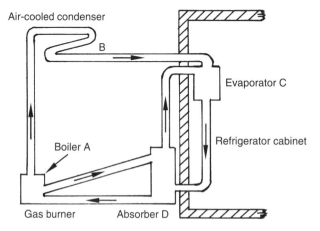

5.20 Gas-fired domestic absorption refrigeration system.

This makes the cycle suitable for solar power input. The equipment has no vibration and very quiet in operation.

Solar-powered cooling

Cooling for air conditioning usually comes from vapour compression refrigeration systems that consume peak-hours electricity from the public supply. Peak demand kW and peak period energy kWh are always the most expensive and put pressure on generators to keep up with the demand. Daytime electrical energy cost can be reduced by making ice during the night in large thermal storage tanks by using the off-peak electricity rate of charge. Daytime chilled water to the air conditioning cooling coils comes from heat exchangers in the ice store system rather than refrigeration compressor water chilling machines. There may be times when all the ice has melted, cooling is still required and the refrigeration chillers have to be restarted. This should be avoided as it incurs high energy cost and a step up in the peak demand kW taken from the public supply that attracts a monthly charge lasting all year. One way to minimize electrical demand kW is to use an absorption refrigeration machine to chill water. An absorption cycle only needs two water pumps and a source of heat to operate it. Gas-fired absorption water chillers are quiet, free of vibration, slightly larger in dimensions than an equivalent vapour compression reciprocating chiller but use very little electrical energy. Its heat source can be gas-firing, steam or hot water. Flat plate, concentrating parabolic tracking or evacuated tube solar collectors generate hot water up to 100°C and can be the heat source for an absorption water chiller. When there is insufficient solar energy available, such as during rain, storms, cool seasons and night time, ice thermal storage can be designed to maintain required cooling.

Herein lies the problem. Air conditioning cooling coils take chilled water at 6°C from a refrigeration plant where the refrigerant evaporates at 5°C. The 5°C value is as high a temperature as possible to maximize efficiency and minimize electrical energy use and running costs. Making ice overnight or with lower-cost off-peak energy requires the refrigerant to be evaporated at a much lower temperature of −5°C. Salt water brine is used as the heat transfer medium from the evaporator to the ice thermal storage tanks. The ice-making brine chiller is a separate machine from the daytime water chiller. This adds significantly to plant cost and complexity.

In conclusion, solar-powered absorption cooling can be possible in an appropriate climate and where the demand for cooling is mainly daytime and relates closely to the weather. Peak sunshine and peak cooling load should coincide. Arrays of solar collectors are costly and demand a large area of ground or roof. A solar-powered cooling system is more expensive to install than vapour compression refrigeration. It requires back-up systems when solar energy is insufficient. It does have the potential to reduce electrical energy costs and CO_2 emissions. Solar energy is free but the systems have maintenance needs. Whether such investment is justified depends on the reasons for doing it and the medium-term financial payback requirement.

Ventilation rate measurement

Measurement of room and building ventilation rate may be required for research into the energy consumption of heated buildings or to carry out commissioning. A pressurization test on the complete building finds its natural ventilation leakage as previously described and demonstrated in the PROBE reports. Smoke is used to visualize air movement. Air velocity through each ventilation grille and any obvious gaps around doors and windows is measured using a suitable anemometer: rotating vane for large grilles: thermistor, mini-vane or pitot-static tube for small airways. The air flow rates into and out of the room are calculated from the airway areas and the average air velocities through them. Non-toxic tracer gas such as nitrous oxide or helium can be released into the room and thoroughly mixed with portable fans to fill the complete volume. Samples of room air are taken at intervals and passed through an analyser, which measures the concentration of tracer gas. The room air change rate is calculated from two known concentrations and the time interval between them. This technique can be used for naturally ventilated buildings and produces accurate results. The katharometer measures air electrical conductivity and gives an output of percentage concentration of tracer in air. An infrared analyser uses a source of infrared radiation and passes it down two tubes to receiving photocells. One tube contains a reference gas and the other the sample of room air. The different gases absorb different amounts of radiation, and the variation in the signals from the photocells is calibrated as the percentage of tracer gas in the air.

Materials for ventilation ductwork

Materials used for ventilation ductwork are mostly thin-gauge galvanized mild steel sheet ducts which are the most popular because of their low cost and ease of manufacture off-site. Prefabricated ducts and fittings allow rapid site erection. Circular, rectangular, flat or spirally wound circular ducts are generally used. Joints are pop-riveted and sealed with waterproof adhesive tape, hard-setting butyl bandage or heat-shrunk plastic sleeves. Large ducts have bolted angle-iron flanges, which also act as support brackets. Stiffening steel strips or tented sheets are used to reduce the drumming effect on flat duct sides caused by air turbulence. Bare metal is painted with metal oxide or zinc chromate paint. Ducts are thermally insulated with resin-bonded glassfibre boards or expanded polystyrene. Kitchen exhaust hoods and ducts are stainless steel unless other metals are preferred for ornamental purpose. Builders' works ducts of brick, concrete, suspended ceiling spaces and plasterboard may be used but keeping these clean and free of debris is a problem as can be witnessed in older buildings. Below ground labyrinths for outdoor air intakes are used to cool incoming air naturally: these bare concrete caves represent a dust, biological material, insect and rodent trap that needs maintenance cleaning. Ducts are tested with air pressure consisting of the design operational pressure plus 250 Pascal (Pa) (1 Pa = N/m^2) applied after installation. The maximum allowed leakage rate is 1% of the system design air flow rate and leaks must not be audible.

Sick building syndrome (SBS)

Indoor environments may be made artificially close to the warm Spring day that most people would like to inhabit. What we breathe indoors is not a genuine atmosphere and is polluted with synthetic particles and vapours plus other contributory factors such as tobacco smoke, body odour, deodorants, vapours from cleaning fluids, photocopiers, paints and furnishings, dust, airborne bacteria, noise, flickering lamps, glare from artificial illumination and the sun, carpet dust and odour, furniture, polyvinyl chloride, food and drink smells, wet clothing, animal odours, waste basket materials, scents, skin preparations, hair treatments, clothing odours, paper smell and dust, formaldehyde, traffic fumes, heated dust from computer systems, volatile chemicals, bacteria grown in stagnant water in humidifiers, treated water aerosols distributed from showers, washing facilities or fountains, dog and cat odours, fish tanks, insect and rodent droppings, decaying insects and rodents. There may be more sources. Are you feeling sick yet? It is enough to make the stoutest lungs and olfactory systems complain. And there is another thing. Do not go to hospital; that is where all the sick people go! Seriously though, we make each other sick.

Air temperature, humidity and air movement, which will seem either stagnant or too draughty, can rarely please more than 95% of the occupants and frequently please a lot fewer. The total environmental loading upon the occupants may rise to an unacceptable level, which can be low for those who are hypersensitive, i.e. physically and psychologically unable to fight off such a bombardment of additional foreign agents to the body. Sick building syndrome (SBS) is epitomized by the occupants exhibiting a pattern of lethargy, headaches, dry eyes, eye strain, aching muscles, upper respiratory infections, catarrh and aggravated breathing problems such as asthma, upon returning to their workplace after the weekend. Apparent causes are sealed windows, air conditioning, recirculated air, recirculating water humidifiers, high-density occupation, low negative ion content, smoking, air ductwork corrosion, airborne micro-organisms, dust, and excrement from dust mites in carpets. SBS can be defined as a combination of health malfunctions that noticeably affect more than 5% of the building's population. This means that there should be sick house syndrome as well. Perhaps there is, or perhaps we are more tolerant at home. Cases of formaldehyde vapour irritation after cavity insulation have been noted. The pattern of house occupation is different, and variation of climatic controls is easily achieved. When we enter somewhere new to us, we notice the unique odours and characteristics but soon become accustomed. Relief from discomfort can be gained by operating windows, temperature control, sun blinds, air grilles, by a brisk walk or by going outdoors to stimulate the body to sweat toxins out.

It has been easy to blame air conditioning for SBS but the cause is more complex and has much to do with the standards we demand of our buildings, the psychological influence of having to go into the workplace at all, and the total internal environment created. Naturally ventilated buildings often have a higher bacteria and dust count than air-conditioned buildings which use filters and have sealed windows. Outbreaks of Legionella diseases have been attributed to the growth of bacteria in stagnant water in wet cooling towers. These bacteria are distributed on air currents and breathed in by those susceptible to infection, sometimes with fatal results. Dry heat exchangers are preferred for discharging surplus building heat gains back into the external atmosphere but they are rather large. Adequate cleanliness and biocide dosing of recirculated cooling-tower water is mandatory. The cure for SBS requires the following actions:

1. Measure pollutants to identify causes.
2. Remove recirculating water humidifiers from air-conditioning plant and replace only with direct steam or water injection.
3. Allow individuals to have control over local air movement, direction and temperature.
4. Clean recirculating water systems such as wet cooling towers and remaining humidifiers and treat them with biocides.
5. Ensure that fresh-air ventilation ductwork, filters, heating and cooling coils and grilles are internally and externally clean and fully functional.
6. Inspect and clean air-conditioning systems and potential dust-traps regularly.
7. Appraise the lighting system to maximize natural illumination and reduce glare.

Air temperature profile

The recommended upper limit for the room temperature for normally occupied buildings is 27°C. External design air temperature for comfort in offices in London may be chosen as 29°C d.b., 20°C w.b.; higher outdoor air temperatures occur. The indoor limit of 27°C will be exceeded in naturally ventilated buildings in the UK and in warmer locations. The elevation of indoor temperature above that of the outdoor air is caused by a combination of the infiltration of external air, solar radiation and indoor heat gains. In parts of the world where high solar radiation intensity and continuously higher external temperatures are common, for example, Sydney with 35°C d.b. and 24°C w.b., the necessity for controlled air circulation and refrigeration can be recognized. Internal comfort temperature is a combination of mean radiant and air temperatures. Intense solar radiation through glazing during the summer can lead to the mean radiant temperature being higher than the air temperature. The temperature of the air in an office, factory or residence may need to be kept to an upper limit of, say, 26°C d.b. in order to counteract radiation heating. Such conditions are tolerable, but not comfortable, for sedentary work. Considerable discomfort is experienced when strenuous physical activity is conducted.

Indoor air temperature fluctuates through each 24-hour period owing to the position of the sun relative to the building. South-facing rooms that have a large area of glazing are likely to be exposed to the greatest indoor air temperatures. Figure 5.21 shows the variation of outdoor air, indoor air, window glass and predicted outdoor air temperatures for a south-facing academic office in Southampton on a Sunday. Windows and doors remained closed throughout the weekend. The office had only natural ventilation, no mechanical cooling, open light grey slatted venetian blinds and had been used normally for the preceding week. The exterior wall had a 70% glazed area.

The general profile of the outdoor air temperature t_{ao} that is expected can be calculated from a sine wave (Jones, 1985, p. 113):

$$t_{ao} = t_{max} - \left(\frac{t_{max} - t_{min}}{2} \right) \times \left(1 - \sin \left(\frac{\theta \pi - 9\pi}{12} \right) \right)$$

The 24-hour clock time is θ hours: that is, a time between 0 and 24 h. The predicted outdoor air temperature curve has been calculated for maximum and minimum values, t_{max} and t_{min}, of 28°C d.b. and 13°C d.b. on an hourly basis for 24 hours. This corresponds to the conditions after a week of warm sunny weather in June in Southampton. This has been a common occurrence.

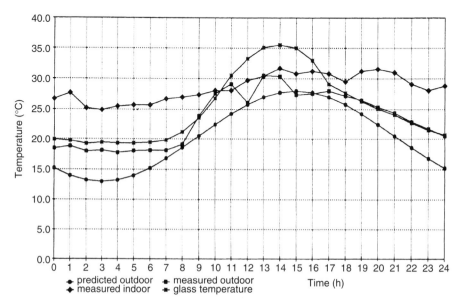

5.21 Temperature profiles of a south-facing office.

A thermocouple temperature logger measured the outdoor air, indoor air and the internal surface temperature of the outer pane of the double-glazed window, hourly. Figure 5.21 shows that the measured outdoor air temperature follows the general shape of the predicted sine wave. The thermocouple that was adhering to the glass showed a combination of two factors: first, that the air temperature in the cavity between the panes of double glazing rose to 35°C; second, that the glass absorbed some of the incident solar radiation and was raised in temperature. The internal room air temperature was measured in a shaded location at just above desk height. The room air temperature remained between 25°C d.b. and 31°C d.b. During normal use, the same office produced an internal air temperature of 25°C d.b. when the outdoor air temperature peaked at 27°C d.b.

While such an example does not prove conclusively that all south-facing rooms in the UK need to be air conditioned for human thermal comfort, it does give some evidence to strengthen the argument in favour of mechanical cooling. Working in air temperatures that move above 24°C d.b. in naturally ventilated spaces that have significant solar radiation can be noticeably uncomfortable. Whether the performance of human productivity or effectiveness becomes impaired is arguable. Low-cost cooling systems can be designed that make use of cool parts of a building to lower the temperature of the areas that are exposed to solar radiation. Heat pump systems, mechanical ventilation and evaporative water-cooling towers can be used to limit room air temperatures, without the need to involve high-cost refrigeration equipment.

EXAMPLE 5.11

A south-facing office in Basingstoke has natural ventilation and a large area of glazing. The maximum and minimum outdoor air temperatures are expected to be 26°C d.b. and 12°C d.b. on a summer day. Calculate the outdoor air temperature that is expected at 1600 h. Find the indoor air temperature that will be generated if solar radiation heat gains raise the indoor air by 1.5°C from that of the outdoor air. Comment upon the thermal comfort conditions that are provided for the office and make recommendations.

$\theta = 16\ h$ for the time of $1600\,h\,GMT$

$$t_{ao} = t_{max} - \left(\frac{t_{max} - t_{min}}{2}\right) \times \left(1 - \sin\left(\frac{\theta\pi - 9\pi}{12}\right)\right)$$

$$t_{ao} = 26 - \left(\frac{26 - 12}{2}\right) \times \left(1 - \sin\left(\frac{16\pi - 9\pi}{12}\right)\right)$$

$$t_{ao} = 26 - 7 \times (1 - \sin 1.833)$$

The 1.833 is in radians and not degrees. Switch on the radians mode of the calculator then press the SIN X key to find the answer of sin 1.833 = 0.966. Alternatively, in degree mode, multiply 1.833 by 360 and then divide the answer by 2π. This produces an angle of 105°. Press the SIN key and find:

$\sin 105 = 0.966$

There are 2π radians in $360°$.

$$t_{ao} = 26 - 7 \times (1 - 0.966)$$

$$t_{ao} = 25.8°C \ d.b.$$

$$Indoor \ air \ temperature = (25.8 + 1.5)° \ C \ d.b.$$

$$Indoor \ air \ temperature = 27.3°C \ d.b.$$

The calculated indoor air temperature exceeds that recommended. While this may prove acceptable for reasonable thermal comfort, there is solar radiation and possible glare during the afternoon in summer. External solar shading or interior blinds are recommended if seating cannot be relocated away from direct solar glare. Sedentary office workers would benefit from measures to increase the throughput of cooler air from the north side of the building or by means of a mechanical cooling system.

Questions

1. A banking hall is cooled in summer by an air-conditioning system that provides an air flow rate of $5 \, m^3/s$ to remove sensible heat gains of $50 \, kW$. Room air temperature is maintained at $23°C$. Derive the formula for calculating the supply air temperature and find its value.
2. A room has a sensible heat gain of $10 \, kW$ and a supply air temperature of $10°C$ d.b. Find the supply air rate required to keep the room air down to $20°C$ d.b.
3. Figure 5.22 shows a west-facing window in a warm climate at a latitudes of $35°$ south around midday. Explain how solar control is being achieved here and how and if it may be applied to commercial buildings in other latitudes. Computer simulation may be popular for such analysis but how could you physically model shading for design assessment at almost no cost?
4. Ten people occupy an office and each produces $50 \, W$ of latent heat. The supply air flow rate is $0.5 \, m^3/s$ and its temperature is $12°C$ d.b. If the room is to be maintained at $21°C$ d.b. and 50% saturation, calculate the supply air moisture content.
5. The cooling coil of a packaged air conditioner in a hotel bedroom has refrigerant in it at a temperature of $16°C$. Room air enters the coil at $31°C$ d.b. and 40% saturation and leaves at $20°C$ d.b. at a rate of $0.5 \, m^3/s$.

    1. Is the room air dehumidified by the conditioner?
    2. Find the room air wet-bulb temperature and specific volume.
    3. Calculate the total cooling load in the room.

6. A department store has 340 people in an area of $35 \, m \times 25 \, m$ that is $4 \, m$ high. Smoking is permitted. Calculate the fresh air quantity required to provide $12.5 \, l/s$ per person. If the air change rate is not to be less than 5 changes per hour, find the following:

    1. supply air quantity;
    2. percentage of fresh air in the supply duct;
    3. extract air quantity if 85% of the supply air is to be mechanically withdrawn;
    4. recirculated air quantity;
    5. ducted exhaust air quantity.

7. Air enters an office through a $250 \, mm \times 200 \, mm$ duct at a velocity of $5 \, m/s$. The room dimensions are $5 \, m \times 3 \, m \times 3 \, m$. Calculate the room air change rate.

5.22 Shaded window.

8. Show two methods of allowing fresh air to enter a room where extract ventilation is by mechanical means and the incoming air is not to cause any draughts.
9. Discuss the relative merits of centrifugal and axial flow fans used in ventilation systems for occupied buildings.
10. Sketch and describe the arrangements for natural and mechanical ventilation of buildings. State two applications for each system.
11. Describe the operating principles of four different systems of air conditioning. State a suitable application for each.
12. State, with reasons, the appropriate combinations of natural and mechanical ventilation for the following: residence, city office block, basement boiler room, industrial kitchen, internal toilet accommodation, hospital operating theatre, entertainment theatre.
13. Explain, with the aid of sketches, how the external wind environment affects the internal thermal environment of a building.
14. A 4-storey commercial building is to be mechanically ventilated. Air-handling plant is to be sited on the roof. Each floor has dimensions 20 m × 10 m × 3 m and is to have 6 air changes per hour. Of the air supplied, 10% is allowed to exfiltrate naturally and the remainder is extracted to roof level. The supply and extract air ducts run vertically within a concrete service shaft and the limiting air velocity is 10 m/s. Estimate the dimensions required for the service shaft. Square ducts are to be used and there is to be at least 150 mm between the duct and any other surface.
15. List the procedure for the design of an air-conditioning system for an office block.
16. A lecture theatre has dimensions 15 m × 15 m × 4 m and at peak occupancy in summer has sensible heat gains of 30 kW and latent heat gains of 3 kW. Room and supply air temperatures are to be 23°C d.b. and 14°C d.b. respectively. Room air moisture content is to be maintained

at 0.008 kg H_2O/kg air. Calculate the supply air volume flow rate, the room air change rate and the supply air moisture content.

17. To avoid draughts, a minimum supply air temperature of 30°C d.b. is needed for the heating and ventilation system serving a public room. The room has an air temperature of 21°C d.b. and a sensible heat loss of 18 kW. It is proposed to supply 2 m^3/s of air to the room. Calculate the supply air temperature that is required. If it is not suitable, recommend an alteration to meet the requirements.

18. Describe the operation of the vapour compression refrigeration cycle and sketch a complete system employing chilled-water distribution to cooling coils in an air-conditioning system.

19. Discuss the uses of the absorption refrigeration cycle for refrigerators and air-conditioning systems.

20. Show how refrigeration systems can be used to pump heat from low-temperature sources, such as waste water, outdoor air arid solar collectors, to produce a usable heat transfer medium for heating or air-conditioning systems.

21. A gymnasium of dimensions 20 m × 12 m × 4 m is to be mechanically ventilated. The maximum occupancy will be 100 people. The supply air for each person is to comprise 20 l/s of fresh air and 20 l/s of recirculated air. Allowing 10% natural exfiltration, calculate the room air change rate, the air flow rate in each duct and the dimensions of the square supply duct if the limiting air velocity is 8 m/s.

22. Study the commercial or academic building that you are familiar with. Sketch and describe how it is ventilated, heated and cooled. Do the systems perform satisfactorily? Are you actively involved with controlling the systems? If you were to redesign the HVAC systems for low energy use in compliance with the *HM Government Carbon Plan, 2011*, what would you recommend? Write an illustrated report of your recommendations with economic justification, CO_2 and other greenhouse gas emission reductions and whether such a project may ever be implemented and, if so, when.

6 Heat demand

Learning objectives

Study of this chapter will enable the reader to:

1. identify and use the thermal conductivity and resistivity of building materials;
2. calculate the thermal resistance of a composite structure;
3. use surface and cavity thermal resistance values;
4. calculate the thermal transmittance of walls, roofs, floors and windows;
5. use the proportional area method to calculate the average U value of a thermally bridged structure;
6. calculate the fabric and ventilation building heat loss components for a building;
7. define hot water storage power requirements;
8. calculate total heating plant power.

Key terms and concepts

boiler power 119; hot water heat load 119; intermittent heat load 118; proportion area method 115; steady-state heat loss 118; surface resistance 114; thermal conductivity 114; thermal resistance 114; thermal storage of structure 119; thermal transmittance 114; U value 114; Y value 119.

Introduction

The terms and techniques for handling thermal properties of building materials are introduced to enable calculation of the thermal resistance, thermal transmittance and use of admittance of composite building elements. The *Chartered Institution of Building Services Engineers (CIBSE Guide A, Environmental Design)* data are used throughout, sample values are used, readers need to refer to *Guide A* for further detailed information and methods. Calculation of building heat loss allows load estimation for heating equipment.

Thermal resistance of materials

Building materials resist the passage of heat to outdoors by trapping still air pockets and with low conductivity. The thermal resistance of a slab of homogeneous material is calculated by dividing its thickness by its thermal conductivity:

$$Thermal\ resistance\ R = \frac{material\ thickness\ d\ m}{thermal\ conductivity\ \lambda W/mK}$$

R is the thermal resistance (m² K/W), d is the thickness of the slab (m); λ is the thermal conductivity (W/mK). Resistance to heat flow by a material depends on its thickness, density, water content and temperature. Insulating materials are usually protected from moisture and the possibility of physical damage as they are of low density and strength. The thermal conductivity of masonry can be found from the bulk dry density and the moisture content, which depends on whether it is exposed to the climate or is in a protected position. Some thermal insulation products combine radiation, convection and conduction resistances, such as foil-faced foam or hollow blocks; the manufacturer's stated R value is used as they cannot be calculated from thickness.

EXAMPLE 6.1

Find the thermal resistance of a 110 mm thickness of brickwork inner leaf of thermal conductivity 0.62 W/m K.

$$R = 0.11m \times \frac{mK}{0.62\,W}$$

$$R = 0.18\frac{m^2K}{W}$$

Thermal transmittance (U value)

Thermal transmittance is found by adding the thermal resistances of adjacent material layers, boundary layers of air, cavities, and then taking the reciprocal. Boundary layer or surface film thermal resistances, R_{si} for interior surfaces, result from the near-stationary air layer surrounding each part of a building, with an allowance for the radiant heat transfer at the surface. Heat transmission across cavities depends upon their width, ventilation and surface emissivity's, R_a. The external surface resistance R_{se} depends upon the building's exposure to weather; sheltered, up to the third floor of buildings in city centres; normal, most suburban and rural buildings, fourth to eighth floors of buildings in city centres; exposed for buildings on coastal or hill sights, floors above the fifth in suburban or rural districts, and floors above the ninth in city centres.

EXAMPLE 6.2

Old stock houses often have external walls consisting of 105 mm brick, 50 mm cavity, 105 mm brick and 13 mm dense plaster with a severe exposure. Find the U value, given R_{si} 0.12, R_a 0.18, R_{se} 0.03 m² K/W, brick λ 0.84, plaster λ 0.5 W/m K.

$$R = \left(0.12 + \frac{0.013}{0.5} + \frac{0.105}{0.84} + 0.18 + \frac{0.105}{0.84} + 0.03 \right)$$

$$R = 0.606 \frac{m^2 K}{W}$$

$$U = \frac{1}{0.606}$$

$$U = 1.65 \frac{W}{m^2 K}$$

EXAMPLE 6.3

Calculate the thermal transmittance of the wall in Example 6.2 if the cavity is filled with urea formaldehyde foam λ 0.04 W/m K.

$$R = \left(0.12 + \frac{0.013}{0.5} + \frac{0.105}{0.84} + \frac{0.05}{0.04} + \frac{0.105}{0.84} + 0.03 \right)$$

$$R = 01.676 \frac{m^2 K}{W}$$

$$U = \frac{1}{1.676}$$

$$U = 0.6 \frac{W}{m^2 K}$$

Elements of buildings that are bridged by a material of noticeably different thermal conductivity, such as timber, steel and concrete frames, dense concrete and stone columns and walls, concrete and steel lintels, or such as window and door frames, all create a lower resistance path through an insulated structure. For example, a dense wall connecting a first floor to the ground is a thermal bridge having a unique longitudinal thermal transmittance in W/m K, which is Watts per metre length per degree Kelvin. The real thermal transmittance of a structure is not only a combination of all its elements but depends upon whether thermal insulation covers adequately, has gaps in it due to inaccurate fitting, the effect of penetrations for services and how much damage was done during the installation of electrical cables, switch boxes and services pipes. Calculated theoretical U values may not always be realized in practice. *CIBSE Guide A3* shows how detailed calculations are made for composite structures. Bridged structures can be calculated by combining the R or U values of the elements using the proportional area method. If U_1 and P_1 are the thermal transmittance and the insulated wall and proportion of the gross wall area, and U_2 and P_2 are the same parameters for the bridging material, the overall U value is given by:

$$U = P_1 \times U_1 + P_2 \times U_2$$

Where,

$$P_1 = \frac{area\ of\ material\ 1}{combined\ areas\ of\ materials\ 1\ and\ 2}$$

Alternatively, the overall R value for the structure combines those heat transfers which happen in series with each other, and those which happen as parallel pathways. Heat flow takes the path of least resistance, meaning that three-dimensional heat flow paths take place. Where major components of a building are used repeatedly, complete façade panels may be laboratory tested rather than relying on theoretical calculated R values. For series resistances;

$$R = R_{si} + P_1 \times R_1 + R_a + P_2 \times R_2 + R_{se}$$

EXAMPLE 6.4

In a concrete-framed commercial building the external walling of brick has a U value of $1.8\,W/m^2K$. The building has a gross perimeter of 180 m and is 3.6 m high. Sixty dense concrete pillars 180 mm wide penetrate the walling from inside to outside. The exposure is normal and the wall thickness is 300 mm. Find the overall U value from R_{si} 0.12, R_{se} 0.06 m^2 K/W, concrete pillar λ 1.40 W/m K.

For the brick wall panel, including surface resistances,

$$U_1 = 1.8 \frac{W}{m^2 K}$$

$$Total\ R_1 = \frac{1}{1.8} \frac{m^2 K}{W}$$

$$Total\ R_1 = 0.56 \frac{m^2 K}{W}$$

Subtract surface resistances to find brick wall R_1 only;

$$R_1 = (0.56 - 0.12 - 0.06) \frac{m^2 K}{W}$$

$$R_1 = 0.38 \frac{m^2 K}{W}$$

For the concrete pillar:

$$R_2 = \left(0.12 + \frac{0.3}{1.4} + 0.06 \right)$$

$$R_2 = (0.12 + 0.22 + 0.06)$$

The resistance of the concrete pillar only is 0.22.

$$R_2 = 0.4 \frac{m^2 K}{W}$$

$$U_2 = \frac{1}{0.4}$$

$$U_2 = 2.5 \frac{W}{m^2 K}$$

The brick and concrete heat flow paths are significantly different, so an overall R value is needed.

$$\text{Pillar surface area } A_2 = 60 \times 0.18 \times 3.6\,m^2$$

$$A_2 = 38.8\,m^2$$

$$\text{Gross wall area } A = 180 \times 3.6\,m^2$$

$$A = 648\,m^2$$

$$\text{Brick proportion } P_1 = \frac{648 - 38.8}{648}$$

$$P_1 = 0.94$$

$$\text{Column proportion } P_2 = \frac{38.8}{648}$$

$$P_2 = 0.06, \text{ or } 6\%$$

$$\text{Overall } R = R_{si} + P_1 \times R_1 + P_2 \times R_2 + R_{se}$$

$$R = 0.12 + 0.94 \times 0.38 + 0.06 \times 0.22 + 0.06\frac{m^2 K}{W}$$

$$R = 0.55\frac{m^2 K}{W}$$

Overall value of U is:

$$U = \frac{1}{R}$$

$$U = \frac{1}{0.55}\frac{W}{m^2 K}$$

$$U = 1.8\frac{W}{m^2 K}$$

Using the proportional areas with U values gives:

$$U = P_1 \times U_1 + P_2 \times U_2$$

$$U = 0.94 \times 1.8 + 0.06 \times 2.5$$

$$U = 1.8\frac{W}{m^2 K}$$

In this case, the concrete columns make no significant difference.

Where only one leaf of a structure containing a cavity is bridged, the resistance of each leaf is calculated separately using the proportional areas as appropriate and then the resistances can be added. The centre line of the cavity can be chosen as the dividing line between the two leaves and half the air space resistance added into each side of the structure. Heat bridges are thermal routes having a lower resistance than the surrounding material that cause distortions to the otherwise uniform temperature gradients. Precise calculation of overall thermal transmittance may require the use of a finite-element computer program that investigates the two- or three-dimensional heat conduction process taking place. The thermal transmittance of a flat roof is calculated in the manner outlined for walls. Heat flow through solid ground floors in contact with the earth depends on the thermal resistance of the floor slab and the ground which, in turn, is largely determined by its moisture content. The thermal conductivity of the earth can vary from 0.70 to 2.10 W/m K which is about the same as for a concrete slab, and so the U values given can be used for floors of any thickness. Dense floor-finishing materials will not influence thermal resistance but floor covering does and also reduces the floors' ability to absorb and store solar radiation from windows. The thermal resistance of insulation placed under the screed of a solid floor, or on netting between the joists of a suspended timber floor, can be added to the reciprocal of the U value of the uninsulated floor and the new thermal transmittance can be calculated. An insulation material placed vertically around the edge of a concrete floor slab which has a thermal resistance of at least 0.25 m² K/W and a depth of 1 m, for example, a 10 mm thickness of expanded polystyrene, will reduce the U value of the floor.

The thermal transmittance of windows depends on glazing, frame types and exposure. If a low-emissivity reflective metallic film is applied to the inside surface of the glass, then the internal surface resistance value can be significantly increased, resulting in a lower U value and reduced heat and light transmission from outside. Glass and metal window frames, in themselves, offer negligible resistance to heat flow, but when resistive materials are used, the overall U value can be found using the proportional area method.

Heating plant load

Heat loss occurs by convection and radiation from the outside of the building and by infiltration of outdoor air. Heating equipment is sized on the basis of steady-state heat flows through the building fabric, with an estimation of the effect of non-steady influences relating to the thermal storage capacity of the structure, adventitious heat gains from people, lighting and machines, and the intermittency of heating system operation. The steady-state heat loss through the building fabric can be assessed by:

$$\text{Heat loss} = \left(\sum (A \times U) + 0.33 \times N \times V\right) \times (t_c - t_{ao})$$

$\sum (A \times U)$ is the sum of the products of the area and thermal transmittance of each room surface; heat flows to adjacent rooms that are warmer than the outdoor air are found by using the appropriate temperature difference between them. 0.33 is the ventilation coefficient, N is the air changes per hour and V is the volume of the heated space, t_c is the operative and t_{ao} the outside air temperature. This is the simplified CIBSE equation and does not take into account the type of radiative or convective heating system.

EXAMPLE 6.5

An office building 20 m long by 10 m wide and 3 m high is to have a hot water panel radiator heating system that will maintain an operative temperature of 21°C at the centre of the room at an external air temperature of −4°C. There are 10 windows each of area 2 m² and two doors each of area 4 m². The roof can be taken as being flat. Infiltration of outside air amounts to 1.0 air changes per hour. Thermal transmittances are as follows: windows, 5.7 W/m² K; walls, 0.5 W/m² K; doors, 5.7 W/m² K; roof, 0.3 W/m² K; floor, 0.6 W/m² K. Find the total rate of heat loss from the building under steady-state conditions.

$$\sum (A \times U) = 20 \times 5.7 + 8 \times 5.7 + 152 \times 0.5 + 200 \times 0.3 + 200 \times 0.6$$

$$\sum (A \times U) = 415.6 \, W/K$$

$Heat\ loss = (415.6 + 0.33 \times N \times V) \times (t_c - t_{ao})$

$Heat\ loss = (415.6 + 0.33 \times 1 \times 20 \times 10 \times 3) \times (21 - 4)\,W$

$Heat\ loss = 15.34\,kW$

Where a building is occupied only occasionally, for example, a traditional heavyweight stone church or a brick-built assembly hall, or allowed to go cold during weekends, the heating system is used intermittently and steady-state heat loss calculations are inappropriate. Admittance factors, $Y\,W/m^2\,K$, are used to evaluate the heat flow into the thermal storage of the structure, rather than through it and are two to three times the value of thermal transmittance. Thermal storage capacities of buildings are classified as ranging from light, as in a caravan, to heavy, as for massive concrete structures. Thermal storage is very useful at moderating indoor air temperature fluctuations as weather changes, but takes a lot of energy to achieve. Solar radiation through clear glazing warming bare concrete surfaces is a very useful way of reducing fuel consumption but has a downside during cold cloudy weather from low temperature radiation to outdoors. Heater power required for a building is found from the sum of the design steady state demand for space and hot water heating, pipe heat emission that is not useful to the occupied spaces and an allowance for preheating from a cold start. A cold building absorbs warmth into its structure, using admittance values, until stable temperatures exist across the structure, then steady state heat transfer is said to exist, but does it? Weather rarely stabilizes; the interior of buildings are infrequently occupied for 24/7, unless they are hospitals or call centres, so steady state heat transfer calculation may not represent typical conditions. Dynamic thermal simulation software using hourly weather data is used to more accurately represent reality.

EXAMPLE 6.6

Calculate the heating power required for a commercial building having a peak heat loss of 900 kW, a low-pressure hot-water radiator heating system, pipe heat losses of 10%; preheat capacity addition 25%, hot water services, HWS, for 450 occupants and a 3-hour heating period for the hot-water storage cylinder. Water is to be stored at 65°C, daily consumption is expected to be 65 litres per person; 1 litre of water weighs 1 kg; specific heat capacity of water is 4.186 kJ/kg K and the temperature of mains cold water is 10°C.

$Heating\ system\ capacity = 900 \times 1.1 \times 1.25\,kW$

$Heating\ system\ capacity = 1237.5\,kW$

$$HWS\ capacity = \frac{450\ people}{3\,h} \times \frac{65\,l}{person} \times \frac{1\,kg}{1\,l} \times \frac{4.186\,kJ}{kg\,K} \times (65 - 10)\,K \times \frac{1\,h}{3600\,s} \times \frac{1\,kWs}{1\,kJ}$$

$HWS\ capacity = 624\,kW$

$Plant\ heating\ capacity = (1237.5 + 624)\,kW$

$Plant\ heating\ capacity = 1861.5\,kW$

Thermal transmittance measurement

The current stock of buildings creates the need for the energy engineer to be able to discover the value of the thermal transmittance of existing structures. The building has design *U*values that were calculated in accordance with standard practice and regulations at the time, but what is the reality of the designer's intentions? Does the design *U* value exist in the components that have been constructed? When the components of the building, such as walls, windows, corners of walls, and interfaces between walls and floors, are taken together as an integrated package, were the design *U* values achieved? Has the process of construction destroyed the designer's work? Such possibilities have a lasting influence upon the energy consumption of the building. A large proportion of the energy consumed in the UK is used to keep the inside of buildings warm. Monitoring the quality of in-situ thermal insulation and for the retrofitting of additional insulation to existing structures is an important part of energy management. The built thermal transmittance of a structure can be calculated from measurements of air and surface temperatures from a thermocouple temperature instrument or multichannel data logger. It is not necessary to know the constructional details of the wall, roof, floor, door or window in order to discover its *U* value. The visiting surveyor will not wish to drill holes through brick, concrete and timber to measure the thickness of each material. Even if this is done, the quality of the materials remains largely unknown and assumptions about the water content and the integrity of each layer would have to be made. The constructional detail is unknown. There may be air spaces, vapour barriers and layers of thermal insulation in place, but these are hidden from view.

Figure 6.1 represents a cross-section through the unknown structure. It could be an external wall, internal wall, roof, floor, glazing or door. All that can be realistically assessed are the temperatures on either side, at nodes 1 and 4, and on the surfaces at nodes 2 and 3. A shielded surface-contact thermocouple probe can be used to measure each surface temperature. An exposed thermocouple junction or a sling psychrometer can be used to find the air temperatures. The values for the inside and outside surface film resistances, R_{si} and R_{se} m^2 K/W, are assumed to be their normal, tabulated values for the appropriate applications. The heat transfer equations that describe the heat flow through the structure, Q W, are as follows.

For the whole structure:

$$Q = U \frac{W}{m^2 K} \times A m^2 \times (t_1 - t_4) K$$

For the interior film:

$$Q = \frac{1}{R_{si}} \frac{W}{m^2 K} \times A m^2 \times (t_1 - t_2) K$$

For the unknown structure:

$$Q = \frac{1}{R} \frac{W}{m^2 K} \times A m^2 \times (t_2 - t_3) K$$

R m^2 K/W is the resistance of the unknown parts of the construction. For the exterior film:

$$Q = \frac{1}{R_{se}} \frac{W}{m^2 K} \times A m^2 \times (t_3 - t_4) K$$

Heat flow is considered to be under steady-state conditions, that is, it remains at a stable rate over several hours and certainly while the measurements are taken. This is the same assumption

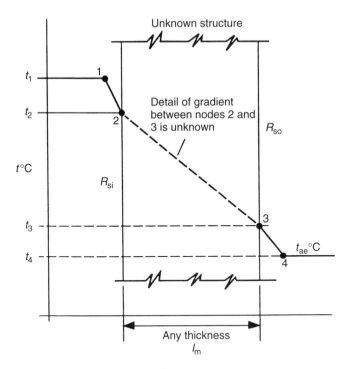

Unknown structure

Detail of gradient
between nodes 2 and
3 is unknown

R_{so}

t_1

t_2

$t°C$

R_{si}

t_3

t_4

$t_{ae}°C$

Any thickness
l_m

6.1 Temperature gradient through a structure.

that is made for calculating thermal transmittances. It is also true that the daily cyclic variation in outdoor air temperature and solar heat gains, plus the intermittent cooling effects of wind and rain, cause unsteadiness in the flow of heat from the building. The analysis of such heat transfers requires dedicated software, weather data and a computer model of the whole building. An awareness of the overall problem is helpful, however. There needs to be as large a temperature difference between indoors and outdoors as reasonably practical on the day of test. This is to minimize the effect of any errors in the measurement of the temperatures. When the indoor and external air temperatures are 20°C and 10°C, an error of 0.5°C in one of the temperatures will be $100 \times 0.5/(20 - 10)\%$, 5% of the difference. If overall inaccuracies can be kept within 5%, a reasonably reliable outcome can be obtained. The heating system and weather should also be functioning under steady conditions during the test period. Take the values of R_{si} and R_{se} to be 0.12 m² K/W and 0.06 m² K/W, as they would be for walls with normal exposure. Use other values if necessary. If, on a test, the temperatures t_1, t_2, t_3 and t_4 are 21°C, 17°C, 0°C and -2°C the rate of heat flow Q, thermal transmittance U and resistance R of the structure can be calculated by using the R_{si} or R_{se} equations:

$$Q = \frac{1}{R_{si}} \frac{W}{m^2 K} \times A m^2 \times (t_1 - t_2) K$$

The surface area $A\,m^2$ is taken as $1\,m^2$:

$$Q = \frac{1}{0.12} \frac{W}{m^2 K} \times (21 - 17) K$$

$$Q = 33.33\,W$$

The same answer results from the use of R_{se}:

$$Q = \frac{1}{R_{se}} \frac{W}{m^2 K} \times A m^2 \times (t_3 - t_4) K$$

$$Q = \frac{1}{0.06} \frac{W}{m^2 K} \times (0 - -2) K$$

$$Q = 33.33 \, W$$

Find the U value from:

$$Q = U \frac{W}{m^2 K} \times A m^2 \times (t_1 - t_4) K$$

$$33.33 = U \frac{W}{m^2 K} \times 1 m^2 \times (21 - -2) K$$

$$U = \frac{33.33}{(21 - -2)} \frac{W}{m^2 K}$$

$$U = 1.45 \frac{W}{m^2 K}$$

This is an elderly wall, which has a higher thermal transmittance than modern standards. Consideration can be given as to how much additional thermal insulation is possible. The thermal resistance of the existing structure, without the surface film resistances, can be found from:

$$Q = \frac{1}{R} \frac{W}{m^2 K} \times A m^2 \times (t_2 - t_3) K$$

$$33.33 = \frac{1}{R} \frac{W}{m^2 K} \times 1 m^2 \times (17 - 0) K$$

$$R = \frac{(17 - 0)}{33.33} \frac{m^2 K}{W}$$

$$R = 0.51 \frac{m^2 K}{W}$$

When the thermal transmittance is known from design calculations or in-situ measurements, the thickness of additional thermal insulation that is needed to reduce heat loss can be calculated. This may be desirable in order to align the building with current regulations and improve its energy-using efficiency. Outdated building designs will be less attractive to potential users than new or recently refurbished low-energy consumption residential, commercial and industrial alternative sites.

The wall U value that was considered here could be lowered from 1.45 W/m^2 K to, say, 0.4 W/m^2 K by the addition of thermal insulation. If the insulation can be injected into the wall cavity, no further constructional measures are needed. Where there is no cavity, or if rainwater penetration could result, then an additional internal or exterior layer of material is required. Thermal insulation may not be structurally rigid and it often does not provide a hard-wearing or weatherproof surface. Adding layers to either side of a wall necessitates architectural changes, particularly to fenestration and doorways. If polyurethane board and an internal surface finish of 10 mm plasterboard can be fitted to the interior surfaces, the necessary thickness of insulation can be calculated as follows.

Plasterboard λ 0.16 W/mK, polyurethane board λ 0.025 W/mK

$$New\ resistance\ of\ structure = \frac{1}{U}\frac{m^2K}{W}$$

$$New\ resistance\ of\ structure = \frac{1}{0.4}\frac{m^2K}{W}$$

$$New\ resistance\ of\ structure = 2.5\frac{m^2K}{W}$$

$$Resistance\ of\ plasterboard = \frac{0.01}{0.16}\frac{m^2K}{W}$$

$$Resistance\ of\ plasterboard = 0.0625\frac{m^2K}{W}$$

$$Resistance\ of\ existing\ wall = \frac{1}{1.45}\frac{m^2K}{W}$$

$$Resistance\ of\ existing\ wall = 0.69\frac{m^2K}{W}$$

The additional thermal insulation needed is found by subtracting the existing thermal resistance, and that for the new surface finish, from the target thermal resistance:

$$Resistance\ needed = (2.5 - 0.69 - 0.0625)\ m^2K/W$$

$$Resistance\ needed = 1.748\ m^2K/W$$

$$Insulation\ resistance = \frac{(d\,mm)}{\lambda}\frac{mK}{W} \times \frac{1\,m}{10^3\,mm}$$

$$1.748 = \frac{d\,mm}{0.025}\frac{mK}{W} \times \frac{1\,m}{10^3\,mm}$$

$$Insulation\ thickness\ d\ mm = 1.748\frac{m^2\,K}{W} \times 0.025\frac{W}{mK} \times \frac{10^3\,mm}{1\,m}$$

$$Insulation\ thickness\ d = 43.7\,mm$$

Materials are available in standard dimensions. The thickness to be used will be the next larger size, 50 mm. Check that the additional insulation calculations have been correctly made and find the real new U value:

$$R = \frac{1}{1.45} + \frac{0.01}{0.16} + \frac{0.05}{0.025}\frac{m^2K}{W}$$

$$R = 0.69 + 0.0625 + 2\frac{m^2K}{W}$$

$$R = 2.752\frac{m^2K}{W}$$

$$U = \frac{1}{2.752}\frac{W}{m^2K}$$

$$U = 0.36\frac{W}{m^2K}$$

The new thermal transmittance does not exceed the desired value of 0.4 W/m^2 K and is suitable. If the wall has an air cavity between the inner and outer surfaces, it may be possible to inject urea formaldehyde or phenolic foam, or blown rock wool, and achieve the desired result without architectural effects. Another possibility is the addition of insulation and a protective layer to the exterior surface.

EXAMPLE 6.7

A 20-year-old 150 mm thick ribbed concrete flat roof over an office is supported on a structural steel frame. A typical cross-section through the roof is shown in Figure 6.2. The concrete is waterproofed with 19 mm asphalt that is topped with 25 mm of white stone chippings. Beneath the concrete, there is a 400 mm deep unventilated air space for service cables and pipes. The ceiling tiles are 12 mm thick fibreboard supported on a lightweight galvanized steel frame. The ceiling tile frame is suspended from the structural steel by galvanized wires and self-tapping screws. All the lighting, electrical and other services that are within the ceiling space are supported by hangers from the roof structural steel frame. The roof has normal exposure and its thermal transmittance is to be reduced to 0.25 W/m^2 K. Thermocouple temperature sensors were used to assess the average thermal transmittance of the roof structure. On the day of test, the temperatures t_1, t_2 and t_5 were 19°C, 17.6°C and 5°C. The temperature at node 4 could not be measured owing to the roughness of the surface. The temperature at node 3 could be measured but it is not needed. Describe the features of two methods that could be used to insulate the roof. Decide which materials would be suitable and find the correct thickness for the insulation.

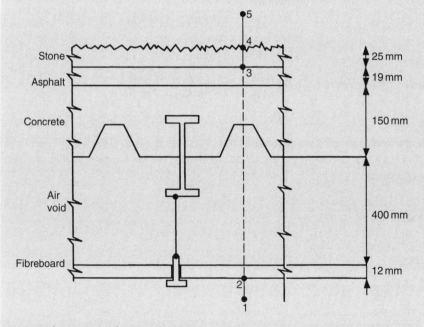

6.2 Roof construction for Example 6.7.

$R_{si} = 0.1\ m^2\ K/W$, $R_{se} = 0.05\ m^2\ K/W$ (low emissivity), $R_a = 017\ m^2\ K/W$ (high emissivity).

Cast concrete λ 0.38, asphalt 0.5, stone chippings 0.96, glass fibre 0.06, polyurethane board 0.025 W/mK.

The options to be tried are as follows.

Option (a): Remove some ceiling tiles and lay a lightweight blanket, such as glass fibre, on top of the tiles. This depends on whether the fibreboard tiles, support wires and screws are able to hold the additional weight. Extra support rods may be needed. There will be considerable disturbance to the room usage. This may preclude installation work during normal working hours. Removing the tiles will disturb dust and debris from the void and necessitate a cleaning operation in the room. Indoor scaffolding will be needed. Care must be taken not to lay the insulation on top of luminaires and electrical cables, to avoid the overheating of lamps and wiring.

$$Q = \frac{1}{R_{si}} \frac{W}{m^2 K} \times A m^2 \times (t_1 - t_2) K$$

$$Q = \frac{1}{0.1} \frac{W}{m^2 K} \times 1 m^2 \times (19 - 17.6) K$$

$$Q = 14\,W$$

$$14\,W = U \frac{W}{m^2 K} \times 1 m^2 \times (19 - 5) K$$

$$U = 1 \frac{W}{m^2 K}$$

$$New\ R = \frac{1}{0.25} \frac{m^2 K}{W}$$

$$New\ R = 4 \frac{m^2 K}{W}$$

$$Existing\ R = \frac{1}{1} \frac{m^2 K}{W}$$

$$Existing\ R = 1 \frac{m^2 K}{W}$$

$$New\ insulation\ R = (4 - 1) \frac{m^2 K}{W}$$

$$New\ insulation\ R = 3 \frac{m^2 K}{W}$$

$$Fibre\ thickness\ d\,mm = 3 \frac{m^2 K}{W} \times 0.04 \frac{W}{mK} \times \frac{10^3 mm}{1 m}$$

$$d = 120\,mm$$

Blanket thickness to be used is 150 mm.

$$New\ insulation\ R = \frac{1}{1} + \frac{0.15}{0.04} \frac{m^2 K}{W}$$

$$New\ insulation\ R = 4.75 \frac{m^2 K}{W}$$

$$New\ U = \frac{1}{4.75}\frac{W}{m^2 K}$$

$$New\ U = 0.21\frac{W}{m^2 K}$$

The new thermal transmittance is below the desired value of 0.25 W/m^2 K and is suitable. Check that the new U value is correct:

$$R = 0.1 + \frac{0.012}{0.06} + \frac{0.15}{0.04} + 0.17 + \frac{0.015}{0.38} + \frac{0.019}{0.5} + \frac{0.025}{0.96} + 0.05\frac{m^2 K}{W}$$

$$R = 4.73\frac{m^2 K}{W}$$

$$U = \frac{1}{4.73}\frac{W}{m^2 K}$$

$$New\ U = 0.21\frac{W}{m^2 K}$$

Option (b): Remove the stone chippings from the roof and lay sheets of rigid polyurethane or phenolic foam. An adhesive can be used to hold the sheets in place. The stone chippings are then placed on top of the foam. The foam is water-repellent and rot-resistant. Installation on the outer surface of the roof will not cause disturbance indoors. The roof will become a warm deck type and will gain the benefit of improved thermal storage capacity: that is, the building will remain warmed for longer periods. In summer, the concrete will be insulated from the solar heat gains and hot outdoor air, and will remain relatively cool.

$$Polyurethane\ thickness\ d = 3\frac{m^2 K}{W} \times 0.025\frac{W}{mK} \times \frac{10^3 mm}{1m}$$

$$d = 75\ mm$$

Polyurethane thickness to be used will be 100 mm and the new U value will be 0.2 W/m^2 K. The installation can be validated by measuring the three temperatures when steady-state conditions have been re-established. Calculate the ceiling surface temperature with the glass fibre insulation in place on a day when the indoor and outdoor air temperatures are 21°C and 3°C:

$$Q = 0.21\frac{W}{m^2 K} \times 1\,m^2 \times (21 - 3)K$$

$$Q = 3.78\,W$$

$$Q = \frac{1}{R_{si}}\frac{W}{m^2 K} \times 1\,m^2 \times (t_1 - t_2)K$$

$$3.78 = \frac{1}{0.1}\frac{W}{m^2 K} \times 1\,m^2 \times (21 - t_2)K$$

$$t_2 = 21 - 3.78 \times 0.1$$

$$t_2 = 20.6°C$$

Questions

1. State what is meant by the following terms:

    1. thermal resistance;
    2. thermal conductivity;
    3. thermal resistivity;
    4. specific heat capacity;
    5. thermal transmittance;
    6. orientation and exposure;
    7. surface resistance;
    8. cavity resistance;
    9. emissivity;
    10. admittance factor;
    11. heavyweight and lightweight structures.

2. The following materials are being considered for the internal skin of a cavity wall:

    1. 105 mm brickwork λ 0.62 W/mK;
    2. 200 mm heavyweight concrete block λ 1.63 W/mK;
    3. 150 mm lightweight concrete block λ 0.19 W/mK;
    4. 75 mm expanded polystyrene slab λ 0.035 W/mK;
    5. 100 mm mineral fibre slab λ 0.035 W/mK and 15 mm plasterboard λ 0.16 W/mK;
    6. 40 mm glass fibre slab λ 0.035 W/mK, 150 mm lightweight concrete block λ 0.19 W/mK and 15 mm lightweight plaster λ 0.16 W/Mk.

    Compare their thermal resistances and comment upon their suitability for a residence.
3. Calculate the thermal transmittances of the following:

    1. 6 mm single-glazed window λ 1 W/mK, R_{si} 0.13 m^{2}K/W, severe exposure R_{se} 0.04 m^{2}K/W.
    2. 6 mm double-glazed window, glass λ 1 W/mK, R_{si} 0.13 m^{2}K/W, R_a 0.18 m^{2}K/W, R_{se} 0.04 m^{2}K/W.
    3. 220 mm solid brick wall λ 0.84 W/mK and 13 mm lightweight plaster λ 0.16 W/mK, R_{si} 0.13 m^{2}K/W, R_{se} 0.04 m^{2}K/W.
    4. 220 mm solid brick wall λ 0.84 W/mK, 150 mm glass fibre quilt λ 0.035 W/mK and 10 mm plasterboard λ 0.16 W/mK, no cavity R_{si} 0.13 m^{2}K/W, R_{se} 0.04 m^{2}K/W.
    5. 105 mm brick wall λ 0.84 W/mK, 10 mm air space R_a 0.18 m^{2}K/W, 40 mm glass fibre slab λ 0.045 W/mK and 100 mm lightweight concrete block λ 0.19 W/mK, R_{si} 0.13 m^{2}K/W, R_{se} 0.04 m^{2}K/W.
    6. 40° pitched roof, 10 mm tile λ 0.84 W/mK, roofing felt λ 0.5 W/mK and 10 mm flat plaster ceiling λ 0.16 W/mK with 100 mm glass fibre quilt λ 0.04 W/mK laid between the joists, R_a 0.16 m^{2}K/W, R_{si} 0.1 m^{2}K/W, R_{se} 0.04 m^{2}K/W.
    7. 19 mm asphalt λ 0.5 W/mK flat roof, 13 mm fibreboard λ 0.06 W/mK, 25 mm air space R_a 0.16 m^{2}K/W, 100 mm mineral wool quilt λ 0.04 W/mK and 10 mm plasterboard λ 0.16 W/mK, R_{si} 0.1 m^{2}K/W, R_{se} 0.04 m^{2}K/W.

4. A lounge 7m long × 4m wide × 2.8 m high is maintained at a resultant temperature of 21°C and has 1.5 air changes per hour of outside air at −2°C. There are two double-glazed wood-framed windows of dimensions 2 m × 1.5 m U 3 W/m^{2}K, and aluminium-framed double-glazed door of dimensions 1 m × 2 m U 3.6 W/m^{2}K. Exposure is normal. One long and one short wall are external and constructed of 105 mm brick, 10 mm air space, 40 mm polyurethane board, 150 mm lightweight concrete block and 13 mm lightweight plaster. The internal walls are of 100 mm lightweight concrete block and are plastered. There is a solid ground floor with edge insulation U 0.34 W/m^2 K. Adjacent rooms are at

a resultant temperature of 18°C. Calculate the steady-state heat loss from the room for a convective heating system. Brick λ 0.84, polyurethane 0.025, lightweight concrete 0.2 and plaster 0.16 W/m. R_a 0.18 m^{2}K/W, R_{si} 0.13 m^{2}K/W, R_{se} 0.04 m^{2}K/W.

5. A single-storey community building of dimensions 20 m × 15 m × 3 m high has low-temperature hot water radiant panel heaters. There are ten windows of dimensions 2.5 m × 2 m. Natural infiltration amounts to one air change per hour. Internal and external design temperatures are 20°C and −1°C. Thermal transmittances are walls 0.6, windows 5.3; floor 0.5, roof 0.4 W/m^2 K. Calculate the steady-state heat loss.

6. A single-storey factory is allowed to have 35% of its wall area as single glazing U 5.7 W/m^2 K and 20% of its roof area as single-glazed roof-lights U 5.7 W/m^2 K as a design limitation, while wall and roof U values are not to exceed 0.6 W/m^2 K. An architect proposes a building to meet this standard of dimensions 50 m × 30 m × 4 m high with a wall U value of 0.4 W/m^2 K, a roof U value of 0.32 W/m^2 K, 20 double-glazed windows each of area 16 m^2 having a U value of 3.3 W/m^2 K and 35 roof-lights each of area 10 m^2 having a U value of 5.3 W/m^2 K. Does the proposal meet the design restriction and what is the rate of heat loss per m^2 floor area?

7. Calculate the boiler power required for a building with a heat loss of 50 kW and an indirect hot-water storage system for 20 people, each using 50 litre of hot-water at 65°C per day. The cylinder is to be heated from 10°C in 2.5 h. Add 10% for pipe and cylinder heat losses and 25% for rapid heating from a cold start.

8. A 30-year-old single-storey building has dimensions 40 m × 20 m × 4 m high with windows of area 80 m^2 and a door of area 9 m^2. It is to be maintained at a resultant temperature of 20°C when the outside is at −1°C and natural ventilation amounts to one air change per hour. Thermal transmittances are as follows: walls, 0.6 W/m^2 K; windows, 5.3 W/m^2 K; door, 5 W/m^2 K; floor, 0.6 W/m^2 K; roof, 0.8 W/m^2 K. A convective heating system is used. It is proposed to reduce the U values of the windows to 2.6, walls to 0.3 W/m^2 K and roof to 0.32 W/m^2 K. Calculate the percentage reduction in heater power that would be produced.

9. List the ways in which existing residential, commercial and industrial buildings can have their thermal insulation improved. Discuss the practical measures that are needed to protect the insulation from deterioration.

10. Review the published journals and find examples of buildings where the existing thermal insulation has been upgraded. Prepare an illustrated presentation or article on a comparison of the outcomes from the cases found.

11. Write a technical report on the argument in favour of adding thermal insulation to existing buildings. Support your case by referring to government encouragement, global energy resources, atmospheric pollution, legislation, cost to the building user and the profitability of the user's company.

12. A flat roof over a bedroom causes intermittent condensation during sub-zero outdoor air temperatures. The roof has normal exposure. The owners want to eliminate the condensation and reduce the thermal transmittance to 0.15 W/m^2 K. Thermocouple temperature sensors were used to assess the average thermal transmittance of the roof structure. On the day of test, the indoor air, ceiling surface and outdoor air temperatures were 16°C, 11°C and −2°C. Calculate the existing thermal transmittance of the roof and the thickness of expanded polystyrene slab that would be needed.

13. An external solid brick wall is to be insulated with phenolic foam slabs λ 0.025 W/mK held on to the exterior brickwork with UPVC hangers. Expanded metal is to be fixed onto the outside of the foam and then cement rendered to a thickness of 12 mm λ 0.5 W/mK. The wall has a sheltered exposure. The intention is to reduce the thermal transmittance to 0.3 W/m^2 K.

Thermocouple temperature sensors were used to assess the average thermal transmittance of the wall prior to the design work. On the day of test, the indoor air, interior wall surface and outdoor air temperatures were 15°C, 12.7°C and 6°C. Calculate the existing thermal transmittance of the wall and the thickness of phenolic foam that would be needed. If the foam is only available in multiple thicknesses of 10 mm, state the thermal transmittance that will be achieved for the wall. Calculate the internal surface temperature that should be found on the wall for a day when the indoor and outdoor air temperatures are 18°C and 0°C.

14. The roof over a car-manufacturing area consists of 4 mm profiled aluminium sheet λ 50 W/mK s on steel trusses. Wood wool slabs, 25 mm λ 0.1 W/mK, are fitted below the roof sheets. The roof trusses remain uninsulated as they protrude through the wood wool. The trusses cause condensation to precipitate onto the vehicle bodies during cold weather. The roof is to be insulated with polyurethane board λ 0.025 W/mK, which will be secured to the underside of the roof trusses. The roof has a normal exposure. The intention is to reduce the thermal transmittance to 0.25 W/m² K. Thermocouple temperature sensors were used to assess the average thermal transmittance of the roof prior to the insulation. On the day of test, the indoor air under the roof was 13°C, internal roof surface temperature was 11°C and the outdoor air temperature was 2°C. Calculate the existing thermal transmittance of the roof and the thickness of polyurethane that would be needed. The insulation is only available in multiple thicknesses of 10 mm. State the thermal transmittance that will be achieved for the roof. Calculate the internal surface temperature that should be found on the newly insulated roof for a day when the indoor and outdoor air temperatures are 16°C and −5°C.

15. Which of these buildings has a slow response, several hours, to variations in weather?

1. Concrete- and steel-framed 20-storey offices.
2. Traditional stone churches.
3. London Underground railway stations.
4. Large volume single-storey industrial buildings having lightweight thermal insulation to corrugated sheet steel wall and roof cladding, for example, aircraft hanger, car factory.
5. Small prefabricated building, transportable, temporary site accommodation, caravan, tent and marquee.

16. Which of these is correct?

1. Thermal resistivity is the fire resistance property of a material.
2. Thermal resistance is the total resistance to flow of water through a heating system circulation.
3. Thermal conductivity is used in calculating the resistance of an electrical heating wiring system.
4. Thermal resistance is a material component property and is measured in m² K/W.
5. Thermal resistance is how many hours electrical cable insulation resists fire in the building.

17. Which of these is correct?

1. The sheet of glass in a window provides a significant thermal resistance.
2. Thermal conductivity of window glass is around 1 W/m K.
3. Thermal conductivity of window glass is around 1 m K/W.
4. Window glass is only used to keep wind out of the building.
5. Windows create no thermal resistance to heat flow.

18. Which is the correct unit?

1. Thermal conductivity m² K/W.
2. Thermal transmittance W/m³ K.

3.   Thermal conductivity m K/kJ.
4.   Thermal resistivity W/m K.
5.   Thermal resistivity m K/W.

19. What does $\sum UA(t_1 - t_2)$ mean?

1.   Something in Greek.
2.   Universal ASHRAE temperature difference used for building heat gain calculation.
3.   Integration of U values and areas during a time interval.
4.   Summation of thermal transmittance, surface area and indoor-outdoor air temperature difference of each external element of the building.
5.   All the U values, surface area and temperature differences added together for the whole building.

20. What does $\sum 0.33NV(t_1 - t_2)$ mean?

1.   One-third of the volumetric air change rate multiplied by daily degree days above base temperature.
2.   33% of normal building volume per degree temperature difference to calculate energy usage cost.
3.   A design guide to the plant room floor area likely to be required for air handling units.
4.   A fraction of the nominal building volume multiplied by air temperature difference.
5.   Volumetric specific heat capacity of air, times number of air changes per hour, times room volume, times indoor-outdoor air temperature difference, calculates natural ventilation rate of heat loss for a heating system.

21. What are admittance values?

1.   Solar heat gain factors for windows and opaque structures.
2.   The opposite of resistance values.
3.   Number of people who can pass through the buildings' entry and transportation systems at peak flow periods.
4.   Always twice the thermal transmittance value.
5.   Thermal factors evaluating heat flows into thermal storage of the structure.

22. Which of these does not apply to admittance values?

1. Y W/m^2 K.
2.   Reciprocal of U value.
3.   Used instead of thermal transmittance in certain circumstances.
4.   Expresses heat flow inwards to a heavy mass structural component.
5.   Used for highly intermittently heated buildings.

23. Which explanation of thermal conductivity is correct?

1.   Ability of a material to conduct electricity.
2.   Property evaluating materials' ability to pass heat.
3.   Equal to resistivity multiplied by thickness.
4.   Units are W/m^3 K.
5.   Units are W/m^2 K.

24. Which is correct about thermal resistance?

1.   Calculated from data tables and computer programs.
2.   Calculated from material thickness divided by thermal conductivity.
3.   Calculated by dividing material thickness in metres by thermal resistivity in m K/W.
4.   Units are kJ/m^2 K.
5.   Units are W/m K.

25. Which of these calculated values of thermal resistance is not correct?

    1. 110 mm of brickwork is 0.13
    2. 150 mm of fibreglass roof insulation is 3.75
    3. 200 mm concrete having a thermal conductivity of 2.0 W/m K is 0.05
    4. A metal window frame 20 mm thickness has a thermal conductivity of 50.0 W/m K and has a thermal resistance of virtually zero.
    5. A low energy building wall has 1.0 m thickness of phenolic foam having a thermal conductivity of 0.04 W/m K making a thermal resistance of 25.0 m^2 K/W.

26. Which of these is correct?

    1. *U* value is the sum of all thermal resistivity in a structure.
    2. *R* value is the sum of all thermal resistivity in a structure.
    3. *Y* value is the sum of all thermal resistivity in a structure.
    4. *U* value is the sum of all thermal resistances in a structure.
    5. *R* value is the sum of all thermal resistances in a structure.

27. Which is correct about an existing structure's thermal transmittance?

    1. Can only be calculated from design information.
    2. Cannot be measured in situ.
    3. Measurement requires a thermal imaging camera.
    4. Measure structural temperatures to calculate *U* value.
    5. Thermocouple temperature sensors have to be buried into drilled holes through the structure.

7 Heating

Key terms and concepts

absorption heat pumps 155; building energy management system 152; ceiling heating 135; chemistry of combustion 146; chimney 148; combined heat and power 148; computer-based control 153; district heating 150; electrical power generation 148; embedded pipe system 135; energy management system 153; evacuated tube radiant system 136; flue gas constituents

Introduction

Terminal heat emitters, such as radiators, convectors and warm-air methods, pipework layouts, and pipe and pump sizing are discussed. Oil-firing equipment is described and the combustion process is analysed. Basic flue arrangements are shown. Electricity is generated at the expense of usable energy discharged to the atmosphere or the sea. The plant needed to convert this surplus into saleable heat for district heating is outlined. Interest in this subject will develop for various reasons, and the UK lags behind other European countries in the employment of combined heat and power stations. The control and operational monitoring of heating, air conditioning and other building services have been enhanced by the use of computer-based techniques known as building energy management systems (*BEMS*). These are explained and clear links with other services are shown.

Heating equipment

A wide variety of heating equipment is available that can heat the occupied space either directly by combustion of a fuel or indirectly by utilizing air, water or steam as a heat transfer fluid. The cost of electricity reflects the complexity of its production and distribution, but from the user's point of view it is a refined source of energy, which can be converted with 100% efficiency. Electrical energy purchased at night can be used to heat water, concrete or cast iron in insulated containers. This stored heat is released when needed.

An economic balance is sought between capital and running costs for each application, bearing in mind the building's use. Automatic controls can monitor water and air temperatures, operational times and weather conditions to minimize fuel and electricity consumption. In order to take maximum advantage of a building's thermal storage capacity, optimum-start controllers are used to vary start and stop times for systems that are used intermittently. Computer control is employed in large buildings, where the capital cost can be offset by reduced energy consumption and personnel savings. Heat emitters can be classified into the following types.

Radiators

Heat emitters providing radiation come into this group. A steel single-panel radiator emits about 15% of its total heat output by radiation and the remainder by convection. Radiant output from multiple panel and column types may be a lower percentage of the total. Electric, gas and coal appliances produce large amounts of convection and are partly radiators. Types of radiator are:

1. *hot water*: single, double or triple panel column radiators, skirting heaters, recessed panels, banks of pipes;
2. *electricity*: off-peak storage heaters, radiant appliances, convectors radiant ceiling systems;
3. *gas, coal and oil: radiant appliances*. A gas-fired domestic radiant and fan convective appliance often suitable for the elderly or infirm even when central heating provides air heating around the occupants. Air heating alone is not always warm enough for very sedentary residents.

The main advantage of radiant heaters is that they also provide convective heat output. Their characteristics are:

1. *Steel single panels*: neat appearance, narrow high heat output per square metre of surface area, easy to clean.
2. *Steel double panels*: greater heat output per square metre of wall area used, difficult to clean, protrude into the room, more costly. Anti-corrosion chemicals needed in heating water for all steel materials.
3. *Cast iron panels*: heavy and more obtrusive, low heat output, very long service period.
4. *Steel and cast iron columns*: high heat output per square metre of wall area used, bulky, heavy, often mounted on feet, difficult to clean except the hospital pattern which are smooth finished.
5. *Radiant panels*: flat cast iron or steel plates with water pipes bonded to their back. They are often mounted at high level in industrial workshops and require a large surface area.
6. *Banks of pipes*: bare steel or copper pipes fitted at skirting level in rooms or storage areas to provide an inexpensive heating surface. These can be installed in floor trenches beneath a decorative floor grille allowing indoor foot traffic to use the floor space unrestricted. Traditional churches often have these and modern buildings have them at the foot of floor-to-ceiling glazed areas to counteract downdraughts.
7. *Off-peak storage*: thermal storage heaters taking electricity at night during less expensive charging periods. The heat is stored at high temperature in cast iron or refractory bricks in an insulated casing. Heat is released continuously into the building unless the heater is fitted with a thermostatically controlled fan and a time switch that determines its operating period. The only other control is over the length of the charge period; this requires estimating the following day's weather pattern. Heaters are bulky and their weight requires attention to the floor structure to ensure sufficient strength.

Convectors

Natural convectors rely on gravity convection currents produced by the heater. Skirting heaters have a finned pipe inside a sheet metal casing. Their heat emission is about 480 W per metre run, they are light and easily handled and they are less obtrusive than taller equipment. Long lengths of unobstructed wall space are needed. Where they run behind furniture, the finned element is omitted and a plain pipe is installed to reduce heat output. They are always fitted onto two-pipe systems and the return pipe can be fitted inside the casing. Valves and air vents are enclosed in accessible boxes at the ends of continuous lengths. Natural convectors produce a uniformly rising current of warm air around the perimeter of the room and this is effective in producing a comfortable environment. There is negligible radiant heating. Other natural convectors are either 1 m high or extend up to room height. They create strong convection currents with little radiation and are particularly suitable for locations where elderly, very young or disabled people are being cared for as there are no hot surfaces that may cause skin burns or start fires.

Natural convectors have high heat outputs and can be built into walls, cupboards or adjacent rooms to improve their appearance. Electricity or low- or medium-temperature hot water can be used as the heating medium. The heating elements need periodic cleaning. Such heaters are used in locations where quiet operation and the lack of draughts or intense radiation are important design considerations, such as libraries, art galleries and antique furniture stores. Figures 7.1 and 7.2 show their principle.

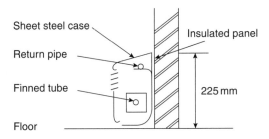

7.1 Skirting convector.

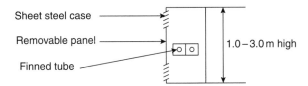

7.2 Natural convector.

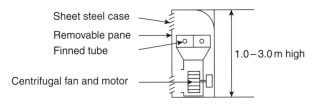

7.3 Fan convector.

Fan convectors have a similar construction to natural convectors with the addition of one or more centrifugal fans and an air filter. Heat output can be very high and fans may be operated at various fixed speeds or from variable-speed motors. Figure 7.3 shows a typical arrangement. Fan operation is controlled from built-in thermostats or remote temperature sensors. Installation can be at low or high level and the heated air stream is directed away from sedentary occupants. Fan convectors can be usefully sited at doorways to oppose incoming cold air and rapidly reheat entrance areas. A two-pipe circuit must be used, and fan convectors are installed on separate circuits from hot-water radiators as their control characteristics are different. Constant-temperature hot-water is supplied to them, whereas radiators may have variable water temperatures to reduce heat output in mild weather.

Low-temperature hot-water heating pipes or electric heating cables are buried in concrete walls, floors or ceilings to provide a large low-temperature surface that is maintained at a few degrees above room air temperature. Floor-to-ceiling air temperature gradients tend to be less than those obtained with more concentrated forms of heat emission and a uniform distribution of comfort is produced. An example is shown in Figure 7.4.

Soft copper pipes are laid in position on the concrete floor slab and held by clips, and the ends are connected to header pipes in service ducts. Joints are avoided for the underfloor sections. Steel or plastic pipes may be used in some situations. Thermal expansion and contraction of the pipework must be accommodated and the floor surface temperature is limited to avoid damage to the structure, surface finishes or occupants. This is done by enclosing the pipe in a hard sleeve

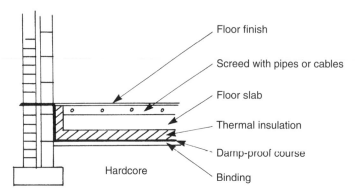

Floor finish

Screed with pipes or cables

Floor slab

Thermal insulation

Damp-proof course

Hardcore

Binding

7.4 Embedded panel heating.

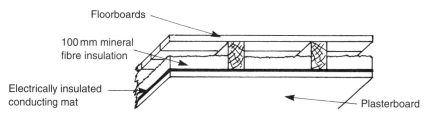

Floorboards

100 mm mineral
fibre insulation

Electrically insulated
conducting mat

Plasterboard

7.5 Radiant ceiling heating.

on water pipes operating at 85°C or by controlling water temperature to 45°C with a mixing valve system. Pipes are buried in the floor screed. Heating elements are evenly distributed to provide uniform radiation and convection to the occupants. Electric ceiling heating can consist of buried cables or a flexible conducting mat fixed between the ceiling joists and plasterboard, as shown in Figure 7.5. The mat is electrically insulated from the structure and connected to 240 V 50 Hz supplies to rise to a surface temperature of 40°C.

Radiant panel systems employ either a high- or a low-temperature surface to transmit heat by radiation directly to the occupants, and to other unheated surfaces, producing an elevated mean radiant temperature. Comfort conditions can be maintained with lower air temperatures than with convective systems. This should result in economical running costs. Convection heat output from the hot radiant source is minimized by placing thermal insulation over the reflector. High-temperature radiation is generated using gas combustion close to ceramic reflectors, which emit some heat in the visible part of the infrared region and consequently are seen to be contributing to a feeling of warmth. Domestic gas fires and industrial heaters are in this category. Covered pedestrian areas of shopping precincts can be warmed from recessed units in canopies. The effect of using high-temperature panels is to produce a series of localized 'sun spots' over a small floor area. Careful location is necessary to avoid overheating of people or objects. Low-temperature systems utilize hot water, air or flue gas to heat a metal sheet or pipe, which emits long-wave infrared radiation outside the visible band. They can be installed in factory or office environments and produce uniform overall warmth, assisted by re-radiation and convection from surfaces heated by the radiant source. Unlike convective systems, they are not adversely affected by room height. Complete systems can be suspended from the ceiling, leaving floors uncluttered. An evacuated tube system is shown in Figure 7.6. Flue products from the gas burners are drawn along steel pipes by a vacuum pump and discharged to the atmosphere.

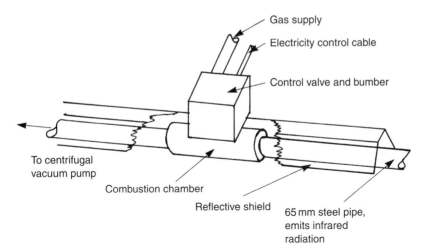

Gas supply

Electricity control cable

Control valve and bumber

To centrifugal
vacuum pump

Combustion chamber

Reflective shield

65 mm steel pipe,
emits infrared
radiation

7.6 Overhead industrial radiant tube heating system.

Warm air systems use recirculated room air, heated either directly or indirectly by the energy source. Direct firing of combustion gases into the air is permissible only in large well-ventilated factory premises. All other applications require a fuel-to-air heat exchanger where the combustion products are enclosed in a sheet metal passageway. Room air is passed over the outside of this heating surface. Heated air is passed through ducts to the occupied space. It is diffused into the room through a grille, which mixes it with room air convection currents and avoids draughts. Each grille has a damper to regulate the air flow. Extract grilles and ductwork return the air to the heater. Care is needed not to extract air directly from kitchens and bathrooms, as this would lead to odours and condensation in living areas. The main advantage of warm-air systems is quick heating up and response to thermostatic control. A source of radiant heat is needed in the sitting room to complement the otherwise purely convective heating. Domestic installation may be where the heater is fitted in a centrally located cupboard with short ducts connecting the heater with the supply and recirculation air grilles.

Hot-water heating systems

The basic arrangements of the various hot-water heating systems are shown in Figures 7.7–7.11. Hot-water heating systems are classified by the temperature and pressure at which they operate. Low temperature and pressure systems are open to atmospheric pressure and are limited to 80°C; medium temperature systems are pressurized up to 2 bars and 120°C, while high temperature may be used in excess of 3 bars and 120°C for the largest sites, district or industrial heating applications.

The pump position relative to the cold feed and vent pipe connections is important in systems with an open expansion tank. The water pressure rise across the pump can be considerable, and if the arrangement is incorrect, water can be pumped up the vent pipe and discharged into the open tank. The connection of the cold feed pipe to the circulation system is known as the neutral point (Figure 7.12). It is here that the water pressure is always equal to the static height of water above it, with the pump exerting no additional pressure. A satisfactory arrangement is shown in Figure 7.12. The hydraulic gradient shows the variation of total water pressure throughout

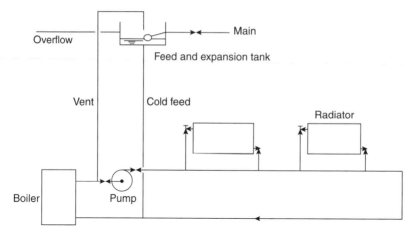

7.7 Low-temperature hot-water one-pipe heating system.

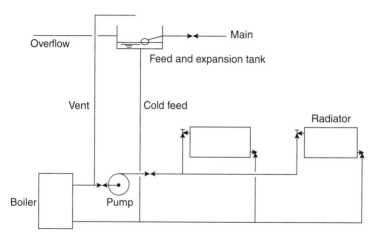

7.8 Low-temperature hot-water two-pipe heating system.

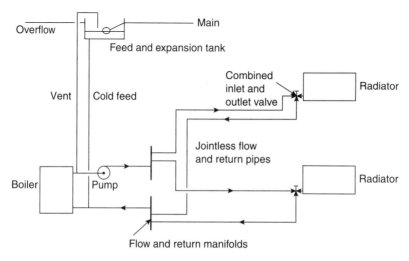

7.9 Low- or medium-temperature hot-water micro bore heating system.

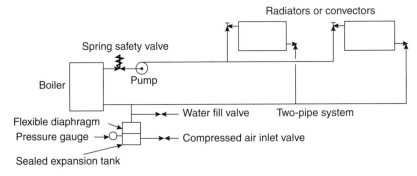

7.10 Low- or medium-temperature hot-water heating system using a sealed expansion tank.

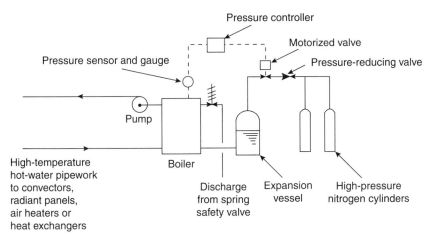

7.11 Pressurization equipment for a high-temperature hot-water heating system.

the circulation, i.e. the sum of the static head and the pump head, some of which generates suction pressure between the neutral point and the pump inlet.

The design of a pumped heating system is approached in the following way:

1. Calculate room heat losses.
2. Decide radiator and convector positions and then their sizes from manufacturers' literature.
3. Calculate the water flow rate for each heat emitter.
4. Design the pipework layout.
5. Mark the water flow rates on the pipework drawing and add them up all the way back to the boiler from the furthest heater, marking the drawing with each value.
6. Choose pipe sizes from a chart or data, using an estimate of pressure loss rate and maximum allowable water velocity.
7. Calculate the pump head and water flow rate; compare with the pump manufacturer's stated performance curves and choose a suitable pump.

The temperature of water flowing through a heat emitter or along a pipe will drop from its flow temperature t_f °C to its return temperature t_r °C and lose heat Q W. If there is one heater in a room, Q will be equal to the room steady-state heat loss. Heat losses from distribution pipework

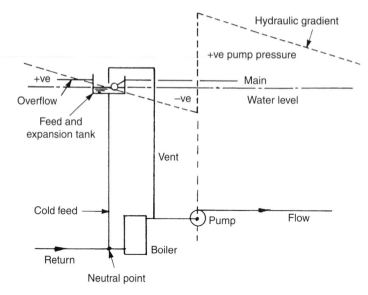

7.12 Neutral point in a heating system.

may initially be assumed to be 10% of the room heat loss, and the radiator water flow rate should be increased by this amount. The specific heat capacity (*SHC*) of water is 4.19 kJ/kg K. The heat balance equation is:

Heat lost by water = radiator heat output Q + pipe heat emission

Water flow rate q × SHC × temperature drop = Q + 10% × Q

$$\text{Water flow rate } q = \frac{1.1 \times Q}{SHC \times (t_f - t_r)} \quad \frac{kg}{s}$$

The radiator manufacturer's heat output data will be for a fixed temperature difference between the mean water temperature and the room air temperature. Usually a figure of 55 K is used. Comparison of design conditions and literature data can be made using the following equations:

$$\text{Radiator mean water temperature } MWT = \frac{(t_f + t_r)}{2}$$

$$\text{Radiator heat output at 55 K difference} = \frac{\text{room heat loss}}{\text{temperature correction factor}}$$

$$\text{Temperature correction factor} = \left[\frac{MWT - t_a}{55}\right]^{1.3}$$

t_a is the room air temperature °C.

EXAMPLE 7.1

A double-panel radiator is to be installed in a room where the air temperature is 22°C and the heat loss is 4250 W. Water flow and return temperatures are to be 82°C and 71°C respectively. An extract from a radiator manufacturer's catalogue for a temperature difference of 55 K is given in Table 7.1. Select a suitable radiator and find the water flow rate through it.

Table 7.1 Heat output from steel double-panel radiators.

Radiator length (m)	Heat output (kW) for 55 K difference		
	500 mm	700 mm	900 mm
1.720	2.00	2.60	2.90
1.920	2.30	2.90	3.30
2.200	2.60	3.25	3.75
2.400	2.85	3.60	4.25
2.600	3.10	3.90	4.80

Data in Table 7.1 is approximate and is only to be used in the examples within this book.

$$\text{Radiator mean water temperature } MWT = \frac{(82+71)}{2}$$

$$\text{Radiator mean water temperature } MWT = 76.5°C$$

$$\text{Temperature correction factor} = \left[\frac{76.5-22}{55}\right]^{1.3}$$

$$\text{Temperature correction factor} = 0.988$$

$$\text{Radiator heat output at } 55K \text{ difference} = \frac{4250\,W}{0.988}$$

$$\text{Radiator heat output at } 55K \text{ difference} = 4302\,W$$

The 2.6 m × 900 mm radiator is needed. Verify that the wall space is available and that there is not a clash with furniture and furnishings in the room. The water flow rate is:

$$\text{Water flow rate } q = \frac{1.1 \times 4.25}{4.19 \times (82-71)} \frac{kg}{s}$$

$$\text{Water flow rate } q = 0.1 \frac{kg}{s}$$

A first estimate of pipe sizes can be made using an average pressure loss rate $\Delta p/EL$ for the whole system (EL is the equivalent length). Either an arbitrary figure of, say, 300 N/m^3 is chosen, or a figure is evaluated from the expected pump head. The index circuit length is found from the total flow and return pipe length from the boiler to the furthest radiator. Pipe fittings increase frictional resistance and an initial 25% increase in measured pipe length is made in order to find the equivalent length of the system:

$$\textit{Pump pressure rise needed } \Delta p \frac{N}{m^2} = EL\,m \times \frac{\Delta p}{EL} \frac{N}{m^3}$$

This is often converted into head H m water gauge:

$$\textit{Pump head } H = \Delta p \frac{N}{m^2} \times \frac{m^3}{\rho kg} \times \frac{kg}{gN}$$

$$\textit{Pump head } H = \frac{\Delta p}{9.807 \times 1000}m$$

$$\textit{Pump head } H = \frac{\Delta p}{9807}m$$

Water flow rate through the pump is the sum of all the radiator water flow rates. Figure 7.13 shows typical pump performance curves. The allowed pressure loss rate can be assessed from the pump characteristic curves. For example, if pump B is to be used at a flow rate of 1 kg/s, the corresponding head developed is 3 m. This is equal to a pressure:

$$\Delta p = 9807 \times H$$

$$\Delta p = 9807 \times 3$$

$$\Delta p = 29421 \frac{N}{m^2}$$

$$29421 \frac{N}{m^2} = EL\,m \times \frac{\Delta p}{EL} \frac{N}{m^3}$$

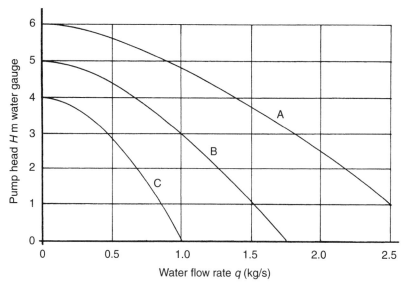

7.13 Pump performance curves.

Table 7.2 Flow of water in copper pipes of various diameters.

p/EL (N /m³)	Water flow rate q (kg/s) for diameters of					
	6 mm	15 mm	22 mm	28 mm	35 mm	42 mm
200	0.013	0.065	0.174	0.381	0.656	1.060 v 1.0
260	0.015	0.075	0.202	0.441	0.760	1.230
300	0.016	0.081	0.219	0.478	0.823	1.330
360	0.018	0.090	0.242	0.529	0.910	1.470
400	0.019	0.096	0.257	0.561	0.965	1.560
460	0.020	0.104	0.278	0.607	1.040	1.680 v 1.50
500 v 0.50	0.021	0.109	0.291	0.635	1.090	1.760
560	0.023	0.116	0.310	0.677	1.160	1.880
600	0.024	0.120	0.323	0.703	1.210	1.950
660	0.025	0.127	0.340	0.741	1.270	2.050
700	0.026	0.131	0.352	0.766	1.320	2.120
760	0.027	0.138	0.368	0.802	1.380	2.220
800	0.028	0.142	0.379	0.825	1.420	2.280

Note: v, water velocity (m/s); $\Delta p/EL$, rate of pressure loss due to friction.

Source: Reproduced from CIBSE Guide by permission of the Chartered Institution of Building Services Engineers.

Data is approximate and is only to be used for the examples in this book.

$$Allowed \ \frac{\Delta p}{EL} = \frac{29421}{EL} \ \frac{N}{m^3}$$

$$EL = 50\,m \times 1.25$$

$$EL = 62.5\,m$$

$$\frac{\Delta p}{EL} = \frac{29421}{62.5} \ \frac{N}{m^3}$$

$$\frac{\Delta p}{EL} = 471 \frac{N}{m^3}$$

This would be the maximum pressure loss rate, averaged over all the pipework, and pipe sizes would be read from the 460 N/m³ line in Table 7.2 to ensure that the available pump head was not exceeded.

EXAMPLE 7.2

Figure 7.14 shows the arrangement of a two-pipe low-temperature hot-water heating system serving three radiators. The pipe dimensions indicated apply to both flow and return pipes. The frictional resistance of the pipe fittings amounts to 25% of the measured pipe length. Flow and return water temperatures at the boiler are to be 85°C and 72°C respectively. Each radiator is situated in a room air temperature of 20°C. The heat outputs of radiators 1, 2 and 3 are 1 kW, 2 kW and 3 kW respectively. Pump C of Figure 7.13 is to be used. Find the pipe sizes for the system.

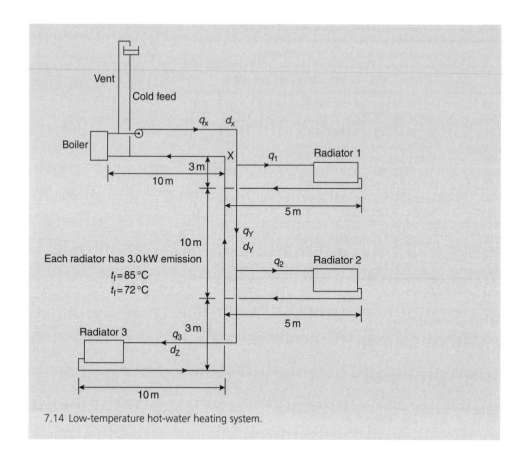

7.14 Low-temperature hot-water heating system.

For radiators 1 to 3:

$$q_1 = \frac{1.1 \times 1}{4.19 \times (85 - 72)}$$

$$q_1 = 0.02 \, kg/s$$

$$q_2 = 0.04 \, kg/s$$

$$q_3 = 0.06 \, kg/s$$

Water flow in the distribution pipework will be:

$$q_x = 0.12 \, kg/s$$

$$q_y = 0.1 \, kg/s$$

Water flow rate through the pump is 0.12 kg/s.
  Available pump head from Figure 7.13 is, $H = 6m$
  Pump pressure rise is:

$$\Delta p = 9807 \times 6 \frac{N}{m^2}$$

$$\Delta p = 58842 \frac{N}{m^2}$$

The index circuit is from the boiler to radiator 3; thus the measured pipe length is:

$$L = 2 \times (10 + 3 + 10 + 3 + 10)m$$

$$L = 72m$$

$$EL = 1.25 \times 72m$$

$$EL = 90m$$

$$58842\frac{N}{m^2} = 90m \times \frac{\Delta p}{EL}\frac{N}{m^3}$$

$$Maximum\ available\frac{\Delta p}{EL} = \frac{58842}{90}\frac{N}{m^3}$$

$$Maximum\ available\frac{\Delta p}{EL} = 654\frac{N}{m^3}$$

Table 7.2, using $\Delta p/EL = 600 \text{ N/m}^3$, the pipe sizes are:

$$d_x = 15mm$$

$$d_y = 15mm$$

$$d_z = 15mm$$

Note that the pressure loss rates in pipes Y and Z are 460 N/m^3 and less than 200 N/m^3 respectively. Not all the allowable pump head of 6 m will be absorbed in pipe friction losses. A gate valve beside the pump will be partially closed to increase the resistance of the system. Each radiator has two hand wheel valves, one for temporary adjustments by the occupant and the other for flow regulation by the commissioning engineer.

Oil-firing equipment

Fuel oil is graded in the Redwood no. 1 viscosity test according to its time of flow through a calibrated orifice at 38°C. Vaporizing and wall-flame burners in boilers of up to 35 kW heat output use 28 s oil, pressure jet burners use gas oil class D (35 s), and industrial boiler plant uses grade E (250 s), grade F (1000 s) and grade G (3500 s). Power stations may use 6000 s residual oil, heated to make it flow. This is the tar residue from crude oil distillation and can only be burnt economically on such a large scale. Figure 7.15 shows a typical domestic oil storage and pipeline installation. In the UK, domestic oils can be stored in outdoor tanks. Grades E, F and G require immersion heaters in the tank and pipeline heating to ensure flow.

Wall-flame burners have a rotating nozzle, which sprays oil onto peripheral plates around the inside of a water-cooled vertical cylindrical combustion chamber. An electric spark ignites the oil impinging on the plates, establishing a ring of flame around the walls of the boiler. Correct oil flow rate from the reservoir is controlled by a ball valve. Vaporizing burners consist of a vertical cylinder that is heated by the flame and evaporates further oil fed into its base from the reservoir flow control. Pressure jet burners are usually confined to boilers in plant rooms as they produce more noise. Oil is pumped at high pressure through a fine nozzle, forming a conical spray in the furnace. Combustion air is blown into this oil mist from a centrifugal fan. The turbulent interaction of oil and air causes further atomization of the oil droplets, and the mixture is ignited by an electric spark.

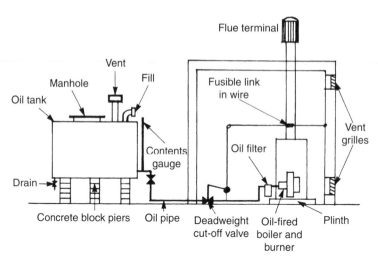

7.15 Oil storage tank installation.

Combustion

Combustion is an exothermic chemical reaction that liberates heat. Fuel must be intimately mixed with sufficient oxygen and raised to a temperature high enough for combustion to be maintained. All the carbon and hydrogen in the fuel are burnt into gaseous products that can be safely vented into the atmosphere. Hydrocarbon fuels are highly energy-intensive. They require little storage volume and their combustion is controllable. The constituents of dry air are 21% oxygen, 79% nitrogen and less than 1% other chemicals such as carbon dioxide, carbon monoxide, nitrous oxides and rare gases, measured by volume. Nitrogen is inert and takes no part in the chemistry of combustion, but it is heated in its passage through the furnace. Typical chemical compositions of fuels are given in Table 7.3.

The quantity of air required for complete combustion and the composition of the products can be evaluated from the fuel chemistry. For methane (CH_4), the complete volumetric analysis would be:

$$methane + air \rightarrow carbon\ dioxide + water\ vapour + nitrogen$$

Table 7.3 Fuel data.

Constituent %	Anthracite	Gas oil	Natural gas
Moisture	8.0	0.05	
Ash	8.0	0.02	
Carbon	78.0	86.0	
Hydrogen	2.4	13.3	
Nitrogen	0.9		2.7
Sulphur	1.0	0.75	
Oxygen	1.5		
Methane			90.0
Hydrocarbons			6.7
Carbon dioxide			0.6

Note: Data is approximate and is only to be used for the examples in this book.

The chemical symbols for these are oxygen O_2, nitrogen N_2, carbon dioxide CO_2, water vapour H_2O. Complete combustion:

$$CH_4 + 2O_2 + N_2 \rightarrow CO_2 + 2H_2O + N_2$$

This means that one volume of CH_4, when reacted with two volumes of O_2 during complete combustion, will produce one volume of CO_2 and two volumes of water vapour. All measurements are at the same temperature and pressure. It is assumed that the water vapour is not condensed. Some condensation is inevitable and when sulphur S is present in the fuel, it combines with some of the O_2 to form sulphur dioxide SO_2. If the gaseous SO_2 comes into contact with condensing water vapour and further O_2, weak sulphuric acid H_2SO_4 may be formed in the flue. Coagulation of liquid H_2SO_4 and carbon particles from chimney surfaces leads to the discharge of acid smuts into the atmosphere, causing local damage to washing, cars and stonework. Acidic corrosion of the boiler and chimney greatly reduce their service period. The flue gas temperature is kept above the acid dew-point of about 50°C to avoid such problems.

It can be seen from the methane combustion equation that $2\,m^3$ of O_2 are required to burn $1\,m^3$ of CH_4 completely. This O_2 is contained in $2/0.21 = 9.52\,m^3$ of air and this air contains $(9.52 - 2) = 7.52\,m^3\ N_2$. In order to ensure complete combustion under all operating conditions and to allow for deterioration of boiler efficiency between servicing, excess air is admitted. This ranges from 30% for a domestic pressure jet oil burner down to a few percentage in power station boilers where continuous monitoring and close control are essential. Excess O_2 from the excess air appears in the flue gas analyses. Measurement of O_2 and CO_2 levels reveals the quantity of excess air. The presence of carbon monoxide CO in the flue gas indicates that some of the carbon in the fuel has not been completely burnt into CO_2 and that more combustion air is needed. The theoretically correct air-to-fuel ratio is the stoichiometric ratio. Figure 7.16 shows the variation of flue gas constituents with the air-to-fuel ratio.

The CO_2 content of oil-fired boiler plant flues will be about 12% at 30% excess air, the combustion air volume required per kilogram of fuel burnt will be about $14.6\,m^3$ and the flue gas temperature leaving the boiler will be about 200°C. For domestic natural draught gas-fired boilers, excess air may be 60%, the flue gas temperature will be 165°C and the CO_2 content will be around 7.5%. Samples of flue gas taken during commissioning and routine servicing are tested for CO_2 and O_2 content by absorption into chemical solutions. The Orsat apparatus is typical.

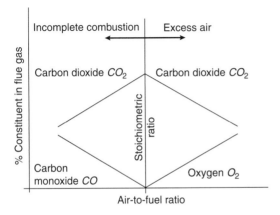

7.16 Variation of flue gas constituents with air-to-fuel ratio.

The smoke content is measured by drawing a sample through filter paper and comparing the discoloration with known values.

Flues

Flue systems for oil-fired boilers are of either the conventional brick chimney or free-standing pipe designs so as to discharge combustion products into the atmosphere and allow sufficient dilution so that fumes reaching ground level will not be noticeable. Chimneys from large boiler plant may be subject to minimum height specification by the local authority. Efflux velocity from the chimney can be increased by utilizing a venturi shape when fan assistance is used, effectively raising the chimney height. The flue pipe diameter will be equal to the boiler outlet connection. Each boiler in a multiple installation should have its own flue. Flues must be kept warm to prevent acidic condensation, and are therefore constructed within the building. Some useful heat is reclaimed in this way. Figure 7.17 shows the necessary separation of flue pipe from combustible materials at an intersection with a floor. External free-standing pipes are constructed of double-walled asbestos or stainless steel to reduce heat loss. Figure 7.18 shows two suitable arrangements.

Electrical power generation

Electricity is generated by alternators driven by steam turbines in power stations. The largest alternators produce 500 MW of electrical power at 33 kV. The steam is produced in a boiler heated by the combustion of coal or residual fuel oil, which could otherwise only be used for making tar. The oil is heated to make it flow through distribution pipework. Nuclear power stations produce heat by a fission reaction and the radioactive core is cooled in a pressurized reactor (PWR), carbon dioxide high temperature gas-cooled reactor (HTGR), liquid sodium fast breeder reactor or heavy water Canadian deuterium (CANDU) system. This fluid then transfers its heat to water, boiling it into steam to drive conventional turbines. Gas turbines directly drive alternators in gas-fired generators, avoid the use of steam, and allow the hot flue gas to be used to heat water for a combined heat and power plant. Smaller generation alternators are driven by methane combustion in gas turbine engines, gas reciprocating engines or by diesel engines. A large modern power station has four separate boiler-turbine-alternator sets, producing a total of 2000 MW at a maximum of 38% overall efficiency. Figure 7.19 shows the energy flows in a conventional power station.

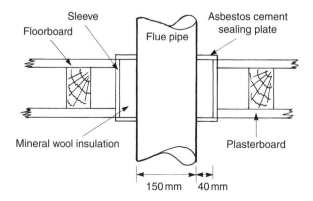

7.17 Separation of flue pipe from combustible materials in a floor.

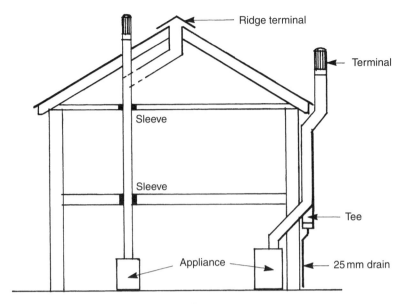

7.18 Internal and external free-standing flue pipes.

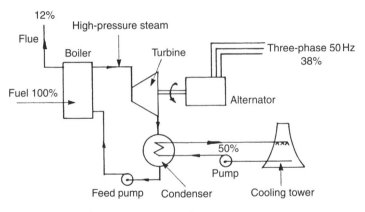

7.19 Conventional 2000 MW power station.

Approximately half the input fuel's energy is dissipated in natural-draught cooling towers or sea water, depending on the plant location. Steam leaves the turbine at the lowest attainable sub-atmospheric pressure so that as much power as possible is extracted from it as it passes through the turbines. The temperature of the cooling water may be as low as 35°C, which is of little practical use unless a mechanical heat pump is employed to generate a fluid at 60–90°C. The heat could then be pumped to dwellings. Power stations are normally sited away from centres of population and heat transport costs are high. During the next 25–100 years, the UK is going to have to make more efficient use of its indigenous hydrocarbon reserves, extend nuclear power generation capacity and develop alternative production methods such as photovoltaic panels, tidal, wave, solar concentrators, wind, geothermal and hydroelectric plants, as identified in the *HM Government Carbon Plan, 2011*. Existing power stations generate electricity only, at as high an efficiency as possible. Combined heat and power (CHP) stations produce less electricity

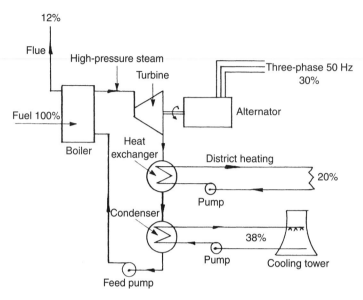

7.20 Combined heat and power plant.

and more heat but improve overall fuel efficiency to about 50%, as indicated in Figure 7.20. Future CHP plants will be smaller than the present electricity-only plants and will be sited close to centres of industry and population, reversing the trend of the twentieth century towards that of the nineteenth century when every town had its own city centre power station; look around and identify those power station sites that are now commercial or retail centres. The purpose of central, very large-scale generation was to minimize the cost of production; whether modern practice will improve on that principle remains to be proven. Coal-fired boilers will be used where practical. Fuel and ash will be mechanically handled and flue gases filtered to remove dust and impurities. Carbon capture and storage should evolve as a viable means of minimizing CO_2 emissions.

District heating

District medium- or high-pressure hot-water heating, employing two-, three- or four-pipe underground distribution systems, will provide heat primarily to the largest and most consistent users, such as hospitals, factory estates and city centres. Further custom will be won from existing buildings by straight price competition. The street distribution layout is indicated in Figure 7.21. Flow and return pipes will be well insulated and may be installed inside one large-diameter pipe, which will form the structural duct and moisture barrier.

CHP plant generates electricity for the locality and is connected into the national grid. It should also incinerate local refuse, utilizing the heat produced, and recycle materials such as metals and glass. It will provide hot water for sanitary appliances and air conditioning and, as these will be summer as well as winter heat loads, a method of separating them from the heating system will be used. This can be done with the three- and four-pipe arrangements shown in Figures 7.22 and 7.23 to economize on pump running costs and pipe heat losses during the summer.

The supply of heat to each dwelling will be controlled by an electric motorized valve, actuated by a temperature sensor in the heat exchanger, which will enable existing low-pressure hot-water

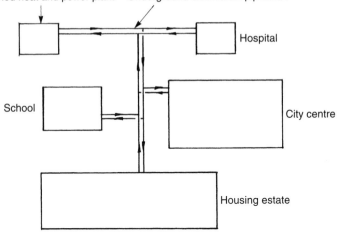

7.21 Medium- or high-pressure hot-water district heating system.

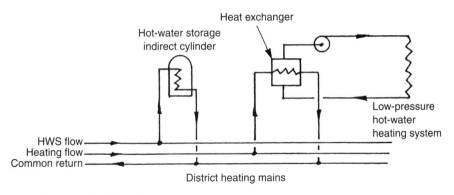

7.22 Three-pipe district heating system.

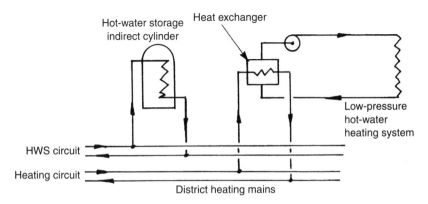

7.23 Four-pipe district heating system.

systems to be connected. A heat meter, consisting of a water flow meter and flow and return temperature recorders, will continuously integrate the energy used, and quarterly bills could be issued through a directly linked computer. Medium- and high-temperature hot-water heating systems are sealed from the atmosphere. Pressurization methods involve restraining thermal expansion, charging with air or nitrogen, or making use of the static head of tall buildings. As the boiling point of water increases with increasing pressure, high flow temperatures can be used. This permits a large drop in temperature from flow to return of 50 K or more, reducing water flow rates compared with low-pressure hot-water open systems. Pipe sizes are smaller and the system is more economical to install when used on a large scale. District cooling from a central refrigeration plant serving air-conditioning units in commercial buildings can be developed alongside a CHP scheme. Underground chilled water pipework will be separate from the heat network, and space, cost and acoustic advantages could be gained in comparison with individual systems. A higher standard of service should be available from centralized services, with fewer breakdowns and closer control of pollution.

Building energy management systems

Computer-based remote control and continuous monitoring of energy-using systems, such as heating, air conditioning, electrical power, lighting and transportation, provide a higher standard of service than can be achieved manually (Figure 7.24). The following types of control can be used.

1. The programmable logic controller (PLC) is a dedicated microprocessor programmed to operate a particular plant item such as a boiler, refrigeration compressor or passenger lift.
2. The energy management system (EMS) is a dedicated microprocessor that is linked to all the energy- and power-using systems such as heating, air conditioning, electrical power, lighting, lifts, diesel generators and air compressors. It may appear as a metal box on a wall of the plant room, having numbered buttons and a single line of screen display for maintenance staff to use for carrying out a limited range of routine changes. Such a unit may serve as an outstation that is either intelligent, having its own microprocessor, or dumb, merely passing data elsewhere.
3. A building energy management system (BEMS) is a supervisory computer that is networked to microprocessor outstations, which control particular plant such as heating and refrigeration equipment. All the energy-using systems within one building are accessed from the supervisor computer, which has hard disk data storage, a display unit, a keyboard, a printer and mimic diagrams of all the services. Additional buildings on the same site can be wired into the same BEMS by means of a low-cost cable. Remote buildings or sites are linked to the supervisor through a telephone modem. A modem is a modulator-demodulator box, which converts the digital signals used by the computer into telecommunication signals suitable for transmission by the telephone network to anywhere in the world.
4. The plant management system (PMS) is used to control a large plant room such as an electrical power generator or district heating station. A PMS can be anything from a small dedicated PLC on a water chiller to an extensive supervisory computer system.
5. The building management system (BMS) is used for all the functions carried out by the building including the energy services, security monitoring, fire and smoke detection, alarms, maintenance scheduling, status reporting and communications. Types of BMS range from systems serving one small office, shop or factory to systems serving government departments and international shopping chains. Systems may carry out financial audits, stocktaking and

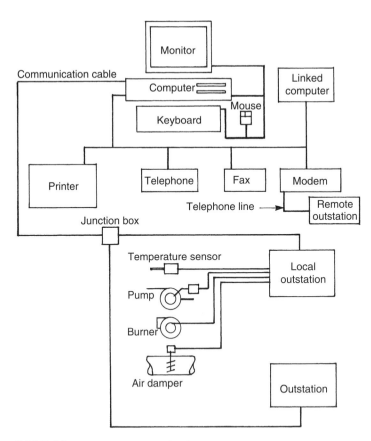

7.24 Building energy management system.

ordering of supplies each night utilizing telecommunications. Suppliers of, say, refrigeration equipment, maintain links with all their installations in clients' buildings and are informed of faults as they occur and often prior to clients being aware of the problem.

An outstation is a microprocessor located close to the plant that it is controlling and is a channel of communication with the supervisor. An intelligent outstation has a memory and processing capability that enables it to make decisions on control and to store status information. A dumb outstation is a convenient point for collecting local data such as the room air temperature and whether a boiler is running or not, which is then packaged into signals for transmission to the supervisor in digital code. Each outstation has its own numbered address so that the supervisor can read the data from that source only at a discrete time. The supervisor is the main computer, which oversees all the outstations, PLCs and modems, contacting them through a dedicated wiring system using up to 10 V and handling only digital code. Such communication can be made every few seconds, and accessing the data can take seconds or minutes depending on the quantity of data and the complexity of the whole system, i.e. the number of measurements and control signals transmitted. The engineer in charge of the BEMS receives displayed and printed reports from the supervisor and has mimic diagrams (Figure 7.25) of the plant, which enable identification of each pump, fan, valve and sensor together with the set points of the controllers that should be maintained. Alarm or warning status is indicated by means of flashing symbols

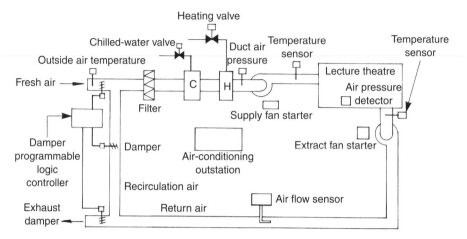

7.25 Mimic diagram on a BEMS screen.

and buzzers, indicating that corrective action by the engineer is required. Plant status, such as the percentage opening of a control valve and whether a fan is running, is recorded, but only the engineer can ascertain whether such information is correct, as some other component may have been manually switched off in the plant room by maintenance staff, or may have failed through fan belt breakage. Therefore a telephone line between the supervising engineer and the plant room staff is desirable to aid quick checking of facts. Connections are made to an outstation by means of a pair of low-voltage cables.

The energy management engineer has seven main functions, which are accomplished by physical work, calculation and word processing:

1. supervision of the plant;
2. choosing the energy supply tariff;
3. reading energy consumption meters and calculating consumption;
4. organizing maintenance work and keeping records;
5. liaising with all levels of staff and management;
6. producing energy management reports;
7. monitoring fire and security systems.

BEMS can automate functions:

1. Supervise the plant continually by means of analogue sensors, analogue-to-digital interfaces and communication cables.
2. Perform remote reading of energy meters and of electricity, gas, oil and heat flow, integrate these with time to calculate total energy consumption, compare with desired values and control the plant to minimize consumption.
3. Compare electricity and gas consumption rates with published tariffs and advise which would be most economical. It may be possible to switch off electrical loads automatically to keep to the lower-cost tariff.
4. Continuously monitor fire detectors and security systems, such as door entry control and video camera operation, detect faults, inform about status and start alarm signals.

5. Access BEMS control and monitor information at outstations by means of passwords, which disseminate information to, and allow restricted use by, different types of employee.
6. Detect faults and alarm conditions as soon as they occur and display the warnings at the supervisory computer where they will be noticed without delay.
7. Carry out normal control functions for energy-using systems and supervise what is going on.

Data which are sent to an outstation include:

1. measurement sensor data such as temperature, humidity, pressure, flow rate and boiler flue gas oxygen content;
2. control signals to or from valve or damper actuators;
3. plant operating status, which can be determined from the position of electrical switches which are open or closed, for example, to check whether a pump is operational.

Geothermal heating

The Southampton geothermal heating system removes heat from salt water which is pumped up from 1.7 km depth by a unique down-hole pump. The well has been operational since 1981 and has had pumped circulation since 1988. The well-head water temperature is 74°C, which is sufficient to provide up to 2 MW of heating and hot-water services within the city centre when working in conjunction with an absorption heat pump. This renewable energy resource is the core of the district heating system. The source of the warm water is a 16 m thick layer of porous Triassic Sherwood sandstone, which is maintained at 76°C by the natural circulation of groundwater and heat flow from greater depths. Several other UK sites have been identified as potential sources; geothermal heat energy greatly reduces greenhouse gas emissions as part of a sustainability initiative and is an example to us all of what can be done.

Questions

1. Sketch and describe two different types of heating system for each of the following applications: house, office, commercial garage, shop, warehouse and heavy engineering factory.
2. Why might the water in large heating systems be pressurized? Explain how pressurization systems work.
3. How do heating systems alter the mean radiant temperature of a room? Give examples.
4. What factors are included in the decision on the siting of a heat emitter? Give examples and illustrate your answer. What safety precautions are taken in buildings occupied by the very young, elderly, infirm or disabled people?
5. How can radiant heating minimize fuel costs while providing comfortable conditions? Give examples.
6. Sketch the installation of a ducted warm-air heating system in a house and describe its operation.
7. List the characteristics of electrical heating systems and compare them with other fuel-based systems.
8. Outline the parameters considered when deciding whether to use a one- or two-pipe distribution arrangement for a radiator and convector-low pressure hot-water heating system.

9. Three rooms have heat losses of 2 kW, 4 kW and 5 kW respectively. Double-panel steel radiators are to be used on a two-pipe low-pressure hot-water system having flow and return temperatures of 85°C and 72°C respectively. Room air temperatures are to be 20°C. Choose suitable radiators and calculate the water flow rate for each.

10. Sketch and describe a micro bore heating installation serving hot-water radiators. State its advantages over alternative pipework systems.

11. A medium-pressure hot-water heating system is designed to provide a heat output of 100 kW with flow and return temperatures of 110°C and 85°C respectively. Calculate the pump water flow rate required in litres per second.

12. Find the dimensions of a double-panel steel radiator suitable for a room having an air temperature of 15°C when the water flow and return temperatures are 86°C and 72°C respectively and the room heat loss is 4.25 kW.

13. The two-pipe heating system shown in Figure 7.14 is to be installed in an office block where radiators 1, 2 and 3 represent areas with heat losses of 12 kW, 20 kW and 24 kW respectively. Water flow and return temperatures are to be 90°C and 75°C respectively. The pipe lengths shown are to be multiplied by 1.5. Pump A (Figure 7.13) is to be used. Pipe heat losses amount to 10% of room heat losses. The friction loss in the pipes is equivalent to 25% of the measured length. Find the pipe sizes.

14. A hot-water radiator central heating system is commissioned and tested while the average outdoor air is 3°C and there is intermittent sunshine and a moderate wind. The building is sparsely occupied. Water flow and return temperatures at the boiler are 90°C and 80°C respectively and the room average temperature is 27°C. The heating system was designed to maintain the internal air at 22°C at an external air temperature of −1°C with flow and return temperatures of 85°C and 73°C respectively. State whether the heating system met its design specification and what factors influenced the test results.

15. List and discuss the merits of the methods used to generate electrical power. What should the UK policy be for the next 100 years in relation to the *HM Government Carbon Plan, 2011*? Should any country rely upon another country for its power supply in the long term? Will one European country become the dominant supplier of power? If so, what would be the advantages and disadvantages of such a policy?

16. Discuss the application of CHP systems in relation to density of heat usage, local and national government policy, possible plant sites, complexity of existing underground services, ground conditions, costs of competing fuels, type and age of buildings, traffic disruption during installation and better control of pollution. (The term 'density of heat usage' refers to the actual use of heat in megajoules per unit ground plan area m^2, including all floors of buildings and appropriate industrial processes requiring the sort of heat to be sold.)

17. How often does the building management system communicate data with sensors and actuators?

    1. Continuously.
    2. Once per hour.
    3. Daily.
    4. Every few seconds.
    5. Annual reports.

18. Which of these comments are factually correct about a building management system and are not just an opinion?

    1. Physical security protection is now out of date.
    2. Allows one person to control and monitor a large facility.

3. Digital recording cameras stop illegal break-ins and escapes.
4. Turn off the power source and it is useless.
5. RS232 and RS484 are types of automatic control system.

19. What does the mechanical services switchboard (MSSB) do?

1. Is the router for all telephone calls between Property Services staff.
2. Automatically controls all air conditioning and transportation systems on the campus.
3. It is the manually operated switchboard for all mechanical services systems within the building.
4. Switches all the electrical sub-circuits for the whole building.
5. Only needed in buildings that do not have a computer-based building management System.

20. Commissioning of a building management system is carried out:

1. With a screwdriver.
2. At the server computer.
3. Remotely through the internet.
4. By calibrating room air temperature sensors with a thermometer.
5. With a laptop computer communicating directly with each control box.

21. What does TCP/IP stand for?

1. Television control programming, internet post.
2. Transmission control protocol, internet protocol.
3. Telephone control program, internet protocol.
4. Transmission control program, internet packages.
5. Telephone communication package, internal protocol.

22. Which statement relating to combustion is correct?

1. CO_2 means two molecules of carbon monoxide.
2. $2O_2$ means two molecules of oxygen.
3. $2H_2O$ means two atoms of hydrogen plus two atoms of ozone.
4. CO is carbon oxide.
5. SO_2 means sodium dioxide.

23. What does stoichiometric ratio mean?

1. Optimum efficiency.
2. Maximum oxygen in flue gas.
3. Maximum carbon dioxide in flue gas.
4. Poor combustion.
5. 100% excess air provided.

24. Which is correct about nuclear-sourced conventional power generation?

1. Nuclear power stations never create any greenhouse gases.
2. They will become the sole means of generating electricity.
3. They are too dangerous to build.
4. Spent nuclear fuel rods are safe to handle.
5. Spent nuclear fuel rods remain radioactive for thousands of years.

25. Which is correct about nuclear-sourced conventional power generation?

1. Uranium is combusted to produce steam.
2. Uranium fusion releases heat.

3. Fission of Uranium releases heat.
4. Uranium corrodes into lead in releasing heat.
5. Radiation from Uranium releases heat.

26. Which is correct about the use of carbon dioxide sensors?

1. Detect ingress of pollution from road traffic.
2. Used to vary the supply of outdoor air into rooms having VAV systems.
3. Control the intake of outdoor air into an air handling unit.
4. Warns the fire and smoke detection systems of a fire source.
5. Used to control underground car park mechanical ventilation systems.

8   Water services

Learning objectives

Study of this chapter will enable the reader to:

1. use pH value;
2. explain water hardness;
3. identify and apply appropriate water treatment methods;
4. discuss storm water harvesting;
5. decide appropriate applications for mains pressure and storage tank cold- and hot-water systems;
6. understand pressure-boosted systems for tall buildings;
7. understand primary and secondary pipe circulation systems;
8. calculate the heater power for hot-water devices;
9. understand how cold- and hot-water pipe systems are designed;
10. have a basic understanding of how solar energy can be utilized in the provision of hot water in buildings;
11. understand the design principles for underground drainage pipework;
12. understand sewage-lifting requirements;
13. know testing procedures;
14. carry out a design assignment;
15. explain the principles of below-ground drain layout;
16. know the location and types of access fitting;
17. define the parts of waste and drain systems;
18. understand the type of fluid flow in a waste pipe;
19. explain the ventilation requirements of drainage pipes;
20. list and explain the ways in which the water seal can be lost from traps;
21. know the permitted suction pressure in a waste system;
22. know the diameters, slopes and maximum permitted lengths of waste and drain pipes for above-ground systems;

23. understand how drain systems are tested;
24. explain how drain systems are maintained;
25. calculate rainfall run-off into surface-water drains;
26. calculate gutter water flow capacity;
27. choose appropriate gutter and rainwater down pipe combinations to create an economical design;
28. assess methods for the disposal of surface-water;
29. calculate soakaway pit design.

Key terms and concepts

access 191; acidic 161; alkaline 161; blockage 191; corrosion 161; drain 179; filtration 160; flow 168; gulley 193; gutter 188; hardness 161; harvesting 163; hot water 164; inspection 191; maintenance 163; manhole 191; pH 161; pipe sizing 168; pressure boosting 164; reverse osmosis 162; sewer 193; soak away 189; solar collector 177; syphonage 180; testing 184; trap 179; urinal 184; waste pipe 179; water treatment 160; zeolite 161.

Introduction

Safe and hygienic water supplies are of paramount importance and a considerable amount of engineering is involved. The basics of water treatment are outlined, and then the ways in which water is distributed throughout buildings are discussed. Storm water harvesting is discussed. Flow of water to sanitary appliances depends upon their frequency of use which is not accurately predictable. The concept of pipe sizing is shown. Water system materials and the application of solar heating are explained. Terminology of drainage systems is outlined and then the characteristic flow within the pipework is explained. Understanding how fluid flows through waste and drain pipework is fundamental to correct design. The potential for water seal loss in traps beneath sanitary fittings and the deposition of solids in long sloping drains is examined. Various standard pipework layouts for above-ground systems are shown. Fluid flow through drain pipes is subject to diversity in timing and duration as are the hot- and cold-water supplies to the same appliances. Materials and jointing methods used for pipe systems are mentioned, as are the testing and maintenance procedures. Reference should be made for *CIBSE Guide G Public Health Engineering, 2004* for detailed information and data for design as only approximate data is provided in this sixth edition for illustration purposes.

Water treatment

Global water resources never vary, only water distribution creates problems of over- or under-supply. Rainfall collection, river, lake and below ground aquifers plus desalination of sea water are all used. Evaporation into the atmosphere, drained storm water, grey, and black water from buildings return H_2O to nature for later re-use. Planet earth never loses water; it has a perfect natural recycling system. Storage in reservoirs allows sedimentation of particulate matter; water is filtered through sand and injected with chlorine for sterilization in treatment plant. Slow sand filtration consists of a large horizontal bed of sand or a sand and granulated activated carbon sandwich. Carbon comes from coal and acts as a very efficient filter trap for microscopic traces of pesticides and herbicides. Water percolates down through the bed by gravity. Rapid sand filters have raw water pumped through a pressurized cylinder that contains the filter medium. This filter material is either crushed silica, quartz or anthracite coal. Filtering removes metallic salts, bacteria and turbidity, muddiness. It also removes colouring effects, odours and particles, which affect the taste of the water. The naturally occurring pH value and the total dissolved salt concentration are virtually unaltered by the water supply authority.

Water quality varies with the local geology and can be classified as hard, soft, acidic or alkaline. Mineral salts of calcium and magnesium have soap-destroying properties and are considered in the evaluation of water hardness. Temporary hardness is due to the presence of calcium carbonate, calcium bicarbonate and magnesium bicarbonate, which dissolve in water as it passes through chalky soil. These salts are deposited as scale on heat transfer surfaces during boiling, causing serious reduction in plant efficiency. They are known as carbonate hardness. Permanent hardness is due to the presence of the non-carbonate salts calcium sulphate, calcium chloride, magnesium chloride and other sulphates and chlorides. Neutralization of these is achieved by means of chemical reactions. Soft water contains up to 100 mg/l of hardness salts, as in Cornwall, and hard water contains as much as 600 mg/l, as in parts of Leicestershire. Acidic water is produced by contact with decomposing organic matter in peaty localities and normally occurs in soft-water regions. This water is very corrosive to steel, is plumbo-solvent and can cause dezincification of gunmetal pipe fittings.

Acidity or alkalinity pH value is due to the presence of free hydrogen ions in the water: acidic water, pH < 7; neutral water, pH = 7; alkaline water, pH > 7. Copper and plastic pipes and fittings can be used in acidic water regions. Hard-water areas are generally alkaline. Water treatment for large boiler plants includes chemical injection to reduce corrosion from dissolved oxygen, and the pH value is raised to 11. Galvanized metal can be used where the pH value is 7.4 if the carbonate hardness is greater than 150 mg/l. Users of large amounts of water may have treatment plant that removes or converts hardness salts to less harmful salts.

Base exchange

Raw water from the mains passes through a tank of zeolite chemicals where a base exchange takes place:

calcium carbonate (in water) + sodium zeolite (in tank) → sodium carbonate (in water)

+ calcium zeolite (left in tank)

Similar base exchanges occur between the zeolite and other hardness salts in the raw water, turning them into non-scale-forming salts. On complete conversion of all the sodium zeolite, the filter bed is backwashed with brine (sodium chloride solution) which undergoes exchange with the calcium zeolite.

calcium zeolite (in tank) + sodium chloride (in flush water) → sodium zeolite (in tank)

+ calcium chloride (flushed to drain)

The normal flow direction can then be resumed. The running cost of the system is limited to consumption of common salt, a small pump, periodic replacement of the zeolite and a small amount of maintenance work.

Steam boilers accumulate the salts passing through the treatment plant, and if they were allowed to become too numerous, they would be carried over into the steam pipes and clog safety valves and pressure controllers. Either continuous or intermittent blow-down of boiler water to the drain is designed to control salt concentration. The high-pressure blow-down water is cooled before being discharged into the drains, and the heat is recycled.

Demineralization

Complete removal of mineral salts is very expensive, but it is essential for power station steam boilers, high-performance marine boilers and some manufacturing processes, where the presence of impurities is unacceptable. Raw water is passed through chemical filters in several stages to complete the cycle.

Reverse osmosis desalination is a filtration technique in which ground water and sea water are pumped alongside a semi-permeable membrane in a pipe system. Clean water passes through the membrane. This method is used to produce drinking water in desert regions where rainfall is unreliable. There is a demand for electricity, for example, 120 MW for the Australian Victorian sea water desalination plant to produce 150 Gigalitres per year costing A$4bn; if amortized over three years, water cost would be under 1p per litre and energy requirement 0.8 W/kl, both of these numbers per unit are very small; compare them to what you pay for a bottle of water in a supermarket. It has a compact plant and is used to produce high purity water for industry and public water supply. If a renewable source such as solar, wind or nuclear energy is used to provide power, it may be claimed to be less resource-intensive than for hundreds of square kilometres of rainfall catchment land, reservoirs and dams.

Cold-water services

Mains water is used in two ways: direct from the main and as low-pressure supplies from cold-water storage tanks.

Mains supplies

At least one tap per dwelling and taps at suitable locations throughout large buildings are connected to the main for drinking water. The main also supplies ball-valves on cold water storage tanks and machines requiring a high-pressure inlet. Economical use of water is important for safety, environmental and cost control reasons. The manual flush control of WCs and the tap operation of other appliances allow responsible usage. Urinals present a particular hygiene and water consumption contradiction. The user has no control over the flushing of water through the trough or bowl. The absence of flushing water leaves the urinal unpleasantly odorous and discoloured. Cleaning staff may counteract this by the excess dumping of deodorant blocks into the urinal. Perfumed toilet blocks are up to 100% paradichlorobenzene. Toilet-cleaning fluid contains phosphoric acid. These toxic chemicals are passed to the sewage treatment plant through the drain system. Uncontrolled flushing when the urinals remain unused, particularly overnight, results in wasteful water consumption and no benefit to the user. In the UK the supply of potable water plus the removal of waste water from consumers may cost around £3 per m^3, kl, from a meter on the supply inlet pipe. An uncontrolled urinal cistern of 9 litres would flush, say, four times per hour, 24 h per day for 365 days in a year and consume 315 m^3 of water costing up to £900. The installation of a water inlet flow control valve to a range of urinals will only allow flushing when appliances have been used, saving consumption. The valve may be operated from a passive infrared presence detector, discharge water temperature sensor or a variation in the water pressure within the same room. A short-term water flow to a WC or basin causes the stored water pressure within bellows to exceed the pressure in the pipeline. A diaphragm opens and allows water to flow to the urinal cistern until the accumulator pressure again equals the pipeline pressure; water flow can be adjusted to avoid wastage.

Purchased water usage can be reduced by collecting rain water from roofs and storing it in above or below ground steel or concrete tanks. Those living in arid climates rely on harvested storm water, lakes, private dams, reservoirs and below ground aquifers. For many, in remote farm homesteads it is not an option. Sites in such as the UK might reduce purchased water where a regular availability of storm water is hardly in doubt. Finding ground space for a large enough tank capacity, polyethylene 2650 litres 1.5 m diameter and 1.5 m high in a small house rear garden and small commercial sites will often be problematical. What to use the harvested water for is another consideration. Clean rain water from a food grade tank can be filtered for consumption but it will not be dosed as is the public supply. It can be used for toilet flushing, watering gardens and washing cars, but drinking may be seen as a health risk. Large commercial, health care and industrial users might find it worthwhile to invest in filtering and dosing equipment or it could be used in cooling tower systems. Ongoing maintenance costs are for pumping, filter changing and tank cleaning. A three-bedroom household might consume $70 \, m^3$ of mains water in a year, costing £3 per m^3 (kl) or £210 for the water supplied and disposed into the public sewer. Whether saving that by harvesting is financially justifiable, say on a 3–5-year capital payback, depends on each site usage, costs and potential benefits.

Low-pressure supplies

Static water pressures in tall buildings are reduced by storing water at various levels. Sealed storage tanks are used for drinking water. Open water tanks become contaminated with airborne bacteria and are only used for sanitary purposes. Cold-water services are taken to taps, WC ball valves, hot-water storage cylinders and equipment needing low-pressure supplies. A separate cold feed is taken to a shower or group of showers to avoid the possibility of scalding. Tanks are sized to store the total cold-water requirement for a 24-h period. Minimum mains water pressure available in the street is 100 kPa (1 bar), which is 1 atmosphere gauge or 10 m height of water. The water supplier may be able to provide 300 kPa, or enough pressure to lift water to the top of a building 30 m high; however, allowance has to be made for friction losses in pipelines and discharge velocity, which effectively limit the vertical distance to between two and six storeys. Separation of the contaminated water being used within the building for washing, flushing sanitary appliances, circulating within heating and air-conditioning cooling systems, evaporative cooling towers, ornamental fountains, agricultural irrigation or manufacturing processes from potable mains water is achieved by using a storage tank with ball valve (break tank); a permanent air gap between the tap discharge and the contaminated water level (wash basin); a single-seat non-return valve (check valve); and a double-seat check valve.

Potential contamination from the building services reaching upstream into the water main fall into three groups: serious danger to life from sanitary appliances; causation of illness from washing machines for clothes and dishes; odour and discoloration from mixer taps and water softeners. Protective measures are by creating a permanent air gap or break tank with the two most serious risks and with a backflow preventer elsewhere.

Cold-water storage tanks are expected to contain water of similar quality to that supplied from the main and so must be covered to exclude foreign matter, insects and light as well as being thermally insulated and not contaminating the stored water themselves. Tanks are generally not larger than 2 m long by 1 m wide by 1 m high, and pipe connections must ensure that water flushes through all of them to eliminate stagnation. Servicing or isolating valves are located on the inlet to all ball valves on storage tanks and WC cisterns to facilitate maintenance without unnecessary water loss or inconvenience to the occupier. A servicing valve is required on all outlets from tanks of more than 15 litres, i.e. larger than a WC cistern. The drinking and food-rinsing water tap at a kitchen sink must be connected to the water main before any water softener enters and a check valve is required between this tap and the softener.

Service entry into a building is via an underground pipe passing through a drain pipe sleeve through the foundations and rising in a location away from possible frost damage. An external stop tap near the boundary of the property is accessible from a brick or concrete pit. A ground cover of 760 mm is maintained over the pipe. A stop valve and drain tap are fitted to the mains on entry to the building to enable the system to be emptied if the building is to be unoccupied during cold weather. A water meter is the next pipe fitting. This has a rotary flow sensor, which is used to integrate the quantity of water that has passed. The cubic metres of water that are supplied, and charged for, are assumed to be discharged into the sewer. A separate charge is levied for the supply of potable water and for the acceptance of the contaminated discharge of foul water. The consumer normally has no choice but to pay both the charges.

In tall buildings the pressure required to reach the upper floors can be greater than the available head, or pressure, in the mains. A pneumatic water pressure boosting system is used, as shown in Figure 8.1. Float switches in the storage tanks operate the pump to refill the system and minimize running times to reduce power consumption. A delayed-action ball valve on the cold-water storage tanks can be used; this delays the opening of the ball valve until the stored water has fallen to its low-level limit. System pressure is maintained by a small air compressor and pneumatic cylinder. The controller relieves excess pressure and switches on the compressor when the air pressure falls. During much of the day, water is lifted pneumatically at much lower cost than if it were pumped. Cold-water storage to cover a 24-hour interruption of supply ranges from 45 litre/person for offices to 90 litre/person for dwellings and 135 litre/person for hotels.

Hot-water services

Hot water can either be generated by the central boiler plant and stored, or produced close to the point of use by a more expensive energy. Central hot-water storage uses low-cost fuel as for central heating. Located within the main boiler house, a large-volume storage cylinder is employed. A small power input boiler is run almost continuously, winter and summer, under thermostatic control from the stored hot water. Primary circulation pipes are kept short and well insulated. This system meets sudden large demands for hot water. Secondary circulation pipes distribute hot water to sanitary appliances. A pump is fitted in the secondary return; its function is to circulate hot water when the taps are shut and it does not appreciably assist draw-off rates from taps. Connections from the secondary flow to the tap are known as dead-legs and are limited to 5 m of 15 mm diameter pipe. This minimizes wastage of cold water in the non-circulating pipework when running a tap and waiting for hot water to arrive.

A decentralized system is mainly for small hot-water service loads distributed over a large building or site where it would be uneconomic to use a central storage cylinder and extensive secondary pipework. Electricity or gas can be used in small storage or instantaneous water heaters

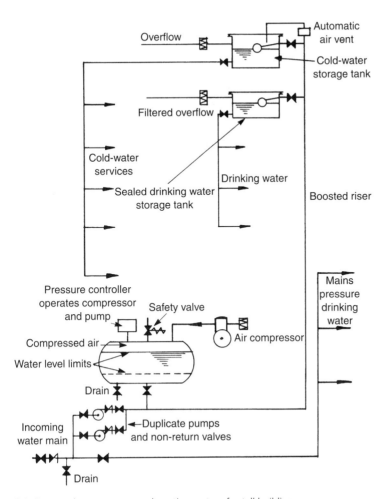

8.1 Pneumatic water-pressure-boosting system for tall buildings.

located at the point of use connected directly to the water main. Figures 8.2 and 8.3 show the operational features of gas-fired instantaneous and storage water heaters.

Mains-connected storage water heaters are protected from excess pressure and water temperature by a combined safety valve, such as that shown in Figure 8.4. On the rise of mains water pressure, an internal spring relieves water to outdoors through the female screwed pipe connection. On the rise of water temperature within the stored volume, the wax thermal probe expands to open the relief valve. Manual testing of the spring valve during routine inspections is done by raising the lever to discharge water. When testing like this, collected debris from corrosion of the water storage tank or salt encrustation often causes the spring-controlled valve to remain cracked open and leak water, resulting in a valve replacement task.

Electric instantaneous heaters have power consumptions of up to 6 kW and produce water at 40°C and up to 3 litre/min at 100% efficiency. Small electric storage heaters of 7 litres are fitted over basins or sinks. Capacities of up to 540 litres operate on the off-peak storage principle. Immersion heaters are controlled by time switches and thermostats and are connected in 3 kW stages. Air to water heat pumps can be used, preferably where outside air temperatures remain mild in winter to maximize efficiency; some have the outdoor air evaporator as a solar collector

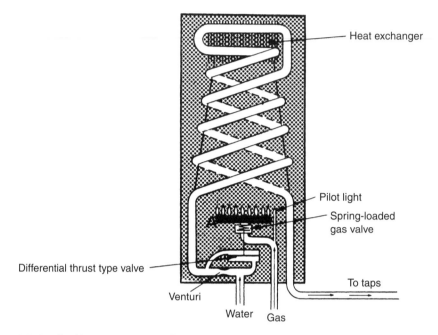

8.2 Gas-fired instantaneous water heater.

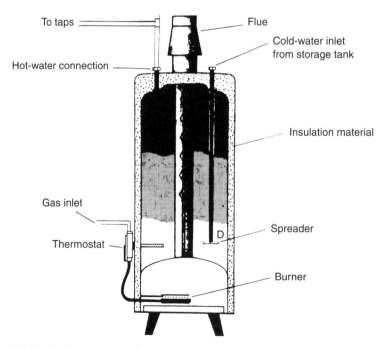

8.3 Gas-fired storage water heater.

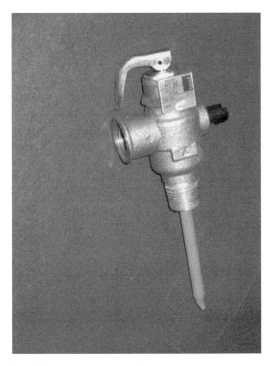

8.4 Mains water pressure safety valve.

to increase heat gain. Solar water heating parabolic concentrators or evacuated tubes can heat domestic water to 65°C. Flat glass insulated panels generally achieve a water temperature of around 40°C and act as a pre-heater to gas or electric heating.

Combined central heating and indirect hot-water service system is shown in Figure 8.5. The cylinder is insulated with 75 mm fibreglass and should have a thermostat attached to its surface at the level of the primary return. Water is stored at 65°C, and when fully charged the thermostat closes the motorized valve on the primary return. This 'off' signal may also be linked into the pump and boiler control scheme to complete the shut-down when the central heating controls are satisfied.

Hot-water pipes are insulated with a minimum of 25 mm of insulation, as are tanks exposed to frost. The primary system feed and expansion tank has a minimum capacity of 50 litres and the cold-water storage tank has a capacity of at least 230 litres. Hot-water storage requirements at 65°C are: office, 5 litres per person; dwelling, 30 litre per person; hotel and sports pavilion, 35 litre per person.

EXAMPLE 8.1

A dwelling has a 135-litre hot-water storage indirect cylinder. The stored cold-water temperature is 10°C and the hot water is to be at 65°C. Calculate the necessary heat input rate to provide a 3-hour recovery period from cold.

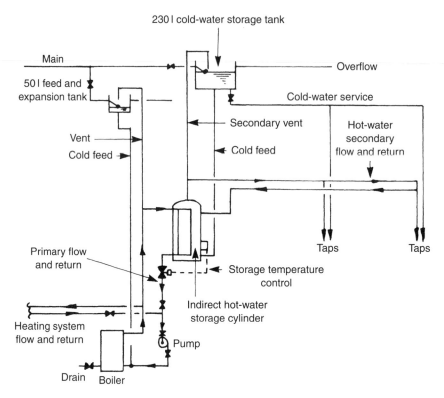

8.5 Indirect hot-water storage system.

$$Heat\ input = \frac{mass\ kg}{time\ h} \times SHC \frac{kJ}{kg\ K} \times temperature\ rise\ K$$

$$Heat\ input = \frac{135\,l}{3\,h} \times \frac{1\,kg}{1\,l} \times 4.19\frac{kJ}{kg\ K} \times (65-10)K \times \frac{1\,h}{3600\,s} \times \frac{kWs}{kJ}$$

$$Heat\ input = 2.881\,kW$$

This can be rounded to 3 kW to allow for heat losses and is adequate for most dwellings with two to four occupants.

Pipe sizing

Demand for water at sanitary appliances is intermittent and mainly random but has distinct peaks at fairly regular times. Pipe sizes for maximum possible peak flows would be uneconomic. Few appliances are filled or flushed simultaneously. To enable designers to produce pipe systems that adequately match likely simultaneous water flows, demand units, DU, are used. DU are dimensionless numbers relating to fluid flow rate, tap discharge time and the time interval between usage. They are based on a domestic basin cold tap water flow rate of 0.15 l/s for duration of 30 s and an interval of 300 s. This application is given a theoretical DU of 1.0 and other appliances are given relative values. Table 8.1 lists practical DUs.

Table 8.1 Practical demand units (DU).

Fitting	Application		
	Congested	Public	Private
Basin	10	5	3
Bath	47	25	12
Sink	43	22	11
WC (13.5-litre cistern)	35	15	8

Note: This data is approximate and only to be used in the examples in this book.

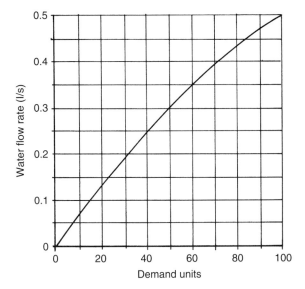

8.6 Simultaneous flow data for water draw-off points.
   Note: Only for use with examples in this book as data is approximate

   The use of spray taps and small shower nozzles greatly reduces water consumption. Design water flow rates of 0.05 l/s for a spray tap, 0.1 l/s for a shower spray nozzle over a bath and 0.003 1/s per urinal stall can be used in place of DU. Figure 8.6 is used to convert DU into water flow rates.
   Design procedure for water service pipe sizing:

1. Draw the pipe layout on the building plans.
2. Mark the DU appropriate to each sanitary fitting on the drawing.
3. Sum DU along the pipes to the water source, which will be the storage tank or incoming water main.
4. Convert DU to water flow rates using Figure 8.6.
5. Find the head of water H m causing the flow to each floor level.
6. Estimate the equivalent length EL of the pipe run to each floor in metres; this can be taken as the measured length plus 30% for the frictional resistance of bends, tees and the tap.

Table 8.2 Flow of water in copper tube of various diameters.

$\Delta p/EL$ (N/m³)	Water flow rate (kg/s)				
	15 mm	22 mm	28 mm	35 mm	42 mm
1000	0.160	0.429	0.933	1.60	2.58
1500 v = 1.5	0.201	0.537	1.170	2.00	3.22 v = 3
2000	0.236	0.630	1.370	2.34	3.77
2500 v = 2	0.268	0.712	1.540	2.65	4.26 v = 4
3000	0.296	0.787	1.710	2.92	4.70

Notes: v = water velocity (m/s).
Data in this table is approximate and is only to be used in the examples
within this book.

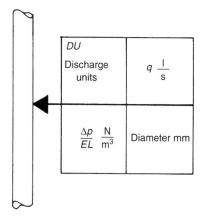

8.7 Notation for pipe-sizing data on drawings.

7. Find the index circuit; the circuit with the lowest ratio of H to EL.
8. Choose pipe sizes from Table 8.2 for the index circuit.
9. Determine the other pipe sizes from the H/EL figure appropriate to each branch of the index circuit.
10. Determine the water flow rate and head for a bronze-body hot-water service secondary pump, if one is required.
11. Notation system can be adopted to gather pipe-sizing data together on drawings, as shown in Figure 8.7.

EXAMPLE 8.2

A domestic cold-water service system is shown in Figure 8.8. The water main in the street is 50 m from the entry point shown and the supply authority provides a minimum static pressure of 20 m water gauge. The velocity energy and the frictional resistance at the ball valve amount to 2 m water gauge. Determine the pipe sizes.

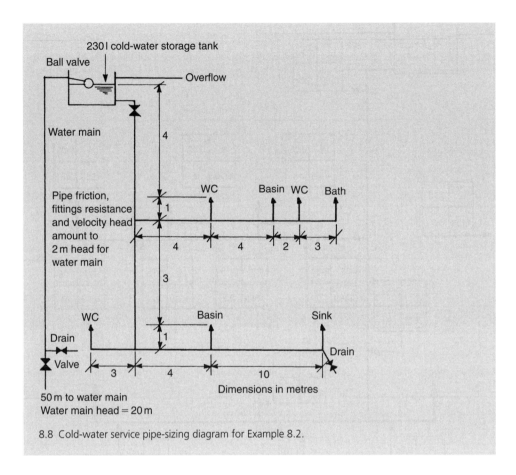

8.8 Cold-water service pipe-sizing diagram for Example 8.2.

Pipe-sizing data are shown in Figure 8.9. Taps and ball valves will be 15 mm on domestic sanitary appliances and 22 mm on baths. DU values are taken from Table 8.1: WC, 8; basin, 3; sink, 11; bath, 12. These are entered into the appropriate boxes on the working drawing and then totalled back to the water main. Great accuracy is not needed in reading the water flow rates appropriate to each DU.

The heads of water causing flow are 4 m for the upper floor and 8 m for the lower floor. The measured length of pipe from the storage tank to the furthest fitting on the upper floor, the bath:

$$L_1 = (4+1+4+4+2+3+1)m$$
$$L_1 = 19m$$

Equivalent length of the circuit to the bath:

$$EL_1 = 1.3 \times 19m$$
$$EL_1 = 24.7m$$

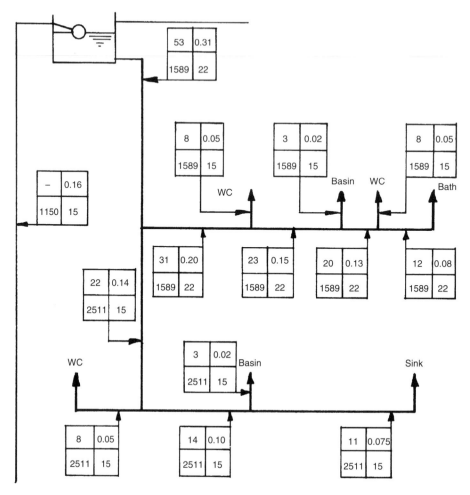

8.9 Pipe-sizing working drawing for Example 8.2.

Similarly, the equivalent length for the lower floor circuit to the sink is:

$$EL_2 = 1.3 \times (4+1+3+1+4+10+1)m$$

$$EL_2 = 31.2m$$

Head loss rate to the bath:

$$\frac{H_1}{EL_1} = \frac{4m\ head}{24.7m}$$

$$\frac{H_1}{EL_1} = 0.162\frac{m\ head}{m}$$

Head loss rate to the sink:

$$\frac{H_2}{EL_2} = \frac{8m\ head}{31.2m}$$

$$\frac{H_1}{EL_1} = 0.256\frac{m\ head}{m}$$

Pipe circuit to the bath has the lowest H/EL figure; this is the index circuit. H_1/EL_1 is the available pressure loss rate, which drives water through the upper part of the system. Branches to other fittings on the same floor level can be sized from the same figure. All pipes below the upper floor are sized using the value of H_2/EL_2 that is appropriate to that circuit. Relationship between head and pressure:

$$H = \frac{\Delta p}{9807}mH_2O$$

Δp is the pressure exerted by a water column of height H m.

$$\frac{\Delta p_1}{EL_1} = \frac{H_1}{EL_1} \times 9807\frac{N}{m^3}$$

$$\frac{\Delta p_1}{EL_1} = 0.162 \times 9807\frac{N}{m^3}$$

$$\frac{\Delta p_1}{EL_1} = 1589\frac{N}{m^3}$$

$$\frac{\Delta p_2}{EL_2} = 0.256 \times 9807\frac{N}{m^3}$$

$$\frac{\Delta p_2}{EL_2} = 2514\frac{N}{m^3}$$

These pressure loss rates are rounded and entered on Figure 8.8. A different pressure loss rate could be calculated for each pipe but sizing them all on the index values is sufficiently accurate at this stage. Suitable pipe diameters are chosen. Notice that the bath connection size has been used rather than the 15 mm pipe, which would satisfy the design data for much of the upper floor branch. The lower water velocities produced will minimize noise generation in the pipes. Pressure loss rate causing flow in the water main:

$$\frac{H_3}{EL_3} = \frac{head\ available\ for\ overcoming\ pipeline\ friction}{equivalent\ pipe\ length}$$

$H_3 = water\ main\ pressure - vertical\ lift - ball\ valve\ resistance$

$\quad\quad - water\ velocity\ pressure\ head\ at\ ball\ valve\ eiftk.\ The\ p$

$H_3 = 20 - (4 + 1 + 3 + 1) - 2m$

$H_3 = 9m\ water\ gauge$

$EL_3 = 1.3 \times (50 + 4 + 1 + 3 + 1)m$

$EL_3 = 76.7m$

$$\frac{H_3}{EL_3} = \frac{9m\ water}{76.7m}$$

$$\frac{H_3}{EL_3} = 0.1173\frac{m\ water}{m}$$

$$\frac{\Delta p_3}{EL_3} = 0.1173 \times 9807\,\frac{N}{m^3}$$

$$\frac{\Delta p_3}{EL_3} = 1150\,\frac{N}{m^3}$$

While this pressure loss rate is available, a water main 15 mm in diameter would provide a flow of a little over 0.16 kg/s. Then, while the taps are shut, the storage tank would refill in:

$$Refill\ time = 230\,kg \times \frac{s}{0.16\,kg} \times \frac{1\,h}{3600\,s}$$

$$Refill\ time = 0.4\,h$$

EXAMPLE 8.3

A hot-water service secondary system is shown in Figure 8.10. Estimate the sizes of the pipes and specify the pump size.

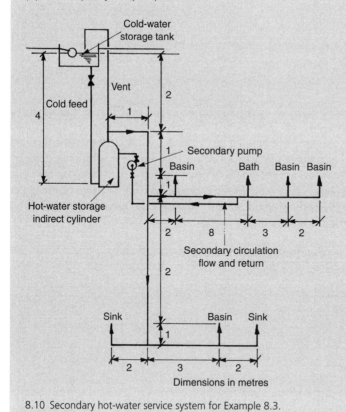

8.10 Secondary hot-water service system for Example 8.3.

Taps on the lower floor are within the limit for dead-leg non-circulating pipes and a secondary return is shown for the group of appliances on the upper floor. Figure 8.11 is the working drawing. Water flow through the cold-feed pipe into the indirect cylinder is at the same rate as

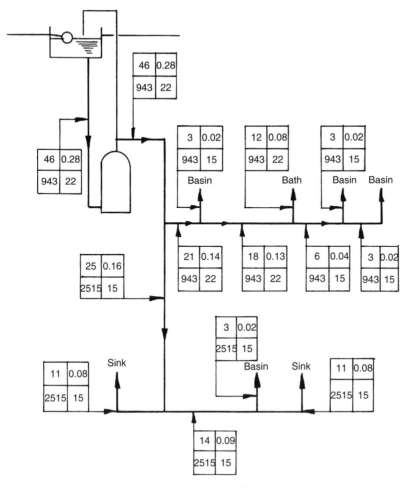

8.11 Hot-water service pipe-sizing working drawing for Example 8.3.

the expected outflow. Pressure loss rate to X from the cold-water storage tank:

$$\frac{H_1}{EL_1} = \frac{3\,m\;water}{(1.3 \times 24)\,m}$$

$$\frac{H_1}{EL_1} = 0.0962 \frac{m\;water}{m}$$

$$\frac{\Delta p_1}{EL_1} = 0.0962 \times 9807 \frac{N}{m^3}$$

$$\frac{\Delta p_1}{EL_1} = 943 \frac{N}{m^3}$$

The pressure loss rate to Y:

$$\frac{\Delta p_2}{EL_2} = \frac{6}{1.3 \times 18} \times 9807 \frac{N}{m^3}$$

Table 8.3 Insulated pipe heat emission.

Pipe diameter (mm)	15	22	28	35	42
Heat emission (W/m K)	0.19	0.23	0.25	0.29	0.32

Note: Data in this table is approximate and is only to be used in the examples within this book.

$$\frac{\Delta p_2}{EL_2} = 2514 \frac{N}{m^3}$$

The secondary return pipe is not intended to play an active part in water discharge from the taps but only to pass an amount of water relating to circulation pipe heat losses. Pipes insulated with 25 mm glass fibre have heat emissions as shown in Table 8.3.

Heat loss from the secondary circulation is,

$$Heat\ loss = 13\,m \times 0.23 \frac{W}{mK} \times (65 - 20)K + 13\,m \times 0.19 \frac{W}{mK} \times (65 - 20)K$$

$$Heat\ loss = 0.25\,kW$$

Water flow rate necessary to offset this pipe heat loss while losing, say, 5°C between the outlet and inlet connections at the cylinder:

$$q = \frac{0.25}{4.19 \times 5} \frac{kg}{s}$$

$$q = 0.0117 \frac{kg}{s}$$

By reference to Table 8.2, a pipe 15 mm or 22 mm in diameter carrying 0.0117 kg/s has a pressure loss rate of much less than 200 N/m^3. A gross overestimate of the pump head for this circuit is:

$$H = 26\,m \times 1.3 \times 200 \times \frac{1}{9807} m$$

$$H = 0.7\,m\,H_2O$$

A pump that delivers 0.0117 kg/s at a head of 0.7 m would meet the requirement. Pump C in Chapter 7 (Figure 7.19) would provide a far higher flow rate than this, and either a smaller pump would be used or the control valves would be partially closed to avoid the production of noise due to high water velocities.

Materials for water services

Hot- and cold-water systems are normally semi-hard copper tube with compression, soldered or push-fit polybutylene connections. Galvanized and stainless steel pipes and fittings are used in large installations such as industrial, hospital and campus sites. Lead pipes may still be found in old buildings or as underground water mains and these should be replaced. Galvanized steel pipes corrode and eventually block with rust deposits, discolouring and stopping water flow. Corrosion protection is provided by ensuring that incompatible materials such as copper and zinc galvanizing are not mixed in the same pipe system. Hot- and cold-water service systems are continually flushed with fresh water, making it necessary to use galvanized metal, copper or stainless steel as the water cannot be treated with chemicals. Copper and galvanized steel should not be used in the same system because electrolytic action will remove the internal zinc coating and cause pipe failure. A galvanized metal cold-water storage tank can be successfully used with copper pipes as the low temperature in this region limits electrolytic action. Heat accelerates all corrosion activity.

Black mild steel is used in recirculatory heating systems. An initial layer of mill scale, which is metal oxide scale formed during the high-temperature working of the steel during its manufacture, helps to slow further corrosion. Discoloration of the central heating water from rust to black during use shows steady corrosion. A black metallic sludge forms at low points after some years. Large hot-water and steam systems have the mill scale chemically removed during commissioning and corrosion-inhibiting chemicals are mixed with the water to maintain cleanliness and avoid further deterioration. Methane gas forms in closed heating systems during the first year of use, due to early rapid corrosion and radiators need frequent venting to maintain water levels. Proprietary inhibitors should be added to all central heating systems. These control methods of corrosion are anti-bacterial. Without them, steel boilers and radiators can rust through in 10 years.

Solar heating

Solar heating can be employed to assist the generation of hot water in secondary storage systems with a consequent reduction in energy costs. The highly variable nature of solar radiation in the cool climates makes a financial return on the capital invested in equipment of around 10% per annum when electrical water heating is used. Solar energy is used to provide:

1. space heating through architectural design in a passive system;
2. space heating using collectors with air as the heat transfer fluid in an active system;
3. heating and hot-water using collectors with water as the heat transfer fluid in an active system.

Thermal storage of the collected energy is needed to balance the supply of solar heat with its times of use, usually when the supply has greatly diminished. Methods of thermal storage used are: thick concrete or brick construction for passive systems combined with large areas of south-oriented glazing; rock heat storage, heated by air recirculation; water tanks; phase change salts. Flat-plate solar collectors are the most popular as they can heat water to 30–40°C without the danger of boiling. They can be incorporated into the architectural design of a sloping roof and are reasonably cheap. They are used to preheat the water supplied to the hot-water service storage cylinder, as shown in Figure 8.12. Only the simplest arrangement is indicated, where the cold-water storage tank supplies only the hot-water cylinder. A wide variety of system designs are in use, depending upon requirements. The solar pump switches off when the temperature sensors find no measurable water temperature rise. All the water then drains back into the storage tank to avoid frost damage. The system will be shut down during the coldest weather.

Concentrating solar collectors are used to generate water temperatures of 80°C and over for hot-water service heating or air-conditioning systems. They are usually driven by electric motors, through reduction gears, to track the sun's movement during each day so that the collecting pipe is kept in the focal plane of the parabolic mirror or polished aluminium reflector. Typical flat-plate and concentrating solar collectors are shown in Figure 8.13.

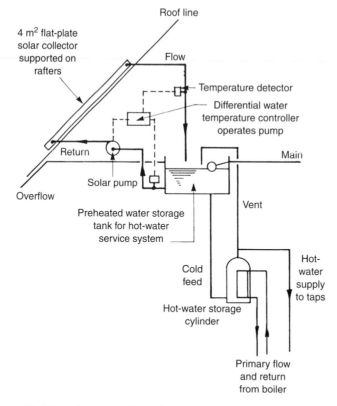

8.12 Solar collector to preheat a hot water-service system.

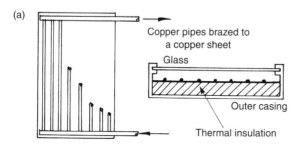

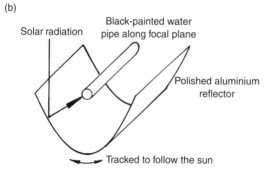

8.13 (a) Flat-plate solar collector; (b) parabolic trough concentrating solar collector.

Drain systems

Correctly functioning sanitation is essential to healthy habitation. In the UK, we tend to consider well-maintained plumbing and drainage functions as the normal arrangement but not everyone in the world is so fortunate; try camping for a week to find out the alternatives. Terminology of drain systems includes: bedding material around a buried pipeline assisting in resisting imposed loads from ground and traffic; benching of curved smooth surfaces at the base of manholes which assist the smooth flow of fluids; combined system in which foul and surface-water are conveyed in the same pipe; crown is the highest point on the internal surface of a pipe; discharge stack is a vertical pipe conveying foul fluid/solid; foul drain conveying black waste water material; sewer pipe system provided by the local drainage authority; invert is the lowest point on the internal surface of a pipe; manhole access chamber to a drain or sewer; separate system in which foul and surface-water are discharged into separate sewers or places of disposal; subsoil drains of underground porous or un-jointed pipes to collect groundwater and convey it to its discharge point; surface- or storm-water drain conveying rain water away from roofs or paved areas; waste pipe from a sanitary appliance to a stack.

Discharge from a sanitary appliance into drain pipes is a random occurrence of surges of fluid. The pipe flows full at some time and a partially evacuated space appears between the remaining trap seal and the surging fluid. Separation between the water attempting to remain in the P-trap and the plug falling into the soil stack causes an air pocket to form. The static pressure of this air will be sub-atmospheric. Air from the room and the ventilated soil stack bubbles through the water to equalize the pressures and a noisy appliance operation results. The inertia of the discharge may be sufficient to syphon most of the water away from the trap, leaving an inadequate or non-existent seal. The problem is avoided by using 32 mm basin waste pipes when the length is restricted to 1.7 m at a slope of 20 mm/m run, about 1° (Figure 8.14).

The sloping waste pipe can be up to 3 m long if its diameter is raised to 40 mm after the first 50 mm of run. This allows aeration from the stack along the top of the sloping section. Longer waste pipes with bends and steeper or even vertical parts have a 25 mm open vent pipe as shown in Figure 8.15.

Vertical soil and vent stacks are open to the atmosphere 900 mm above the top of any window or roof-light within 3 m. Underground foul sewers are atmospherically ventilated. Water discharged into the stack from an appliance entrains air downwards and establishes air flow rates of up to a hundred times the water volume flow rate. Air flow rates of 10 to 150 litre/s have been measured. The action of water sucking air into the pipe lowers the air static pressure, which is further reduced by friction losses. Water enters the stack as a full-bore jet, shooting across to the opposite wall, falling and establishing a downward helical layer attached to the pipe surface. Restricted air passageways at such junctions further lower the air static pressure

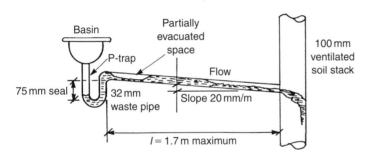

8.14 Design of a basin waste pipe to avoid self-syphonage.

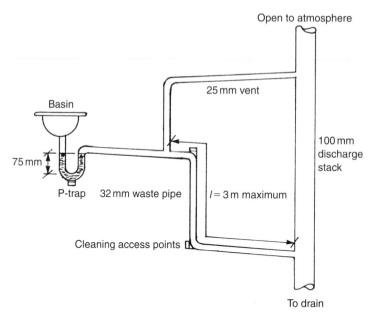

8.15 Vented basin waste pipe.

by their resistance to flow. Atmospheric pressure will be re-established at the base of the stack because of the flow of air into the low-pressure region. The falling fluid tends to fill the pipe near the base and positive air static pressures can be generated. Appliances connected to such a region may have their water seals intermittently forced out. Figure 8.16 shows the probable air static pressures during the simultaneous discharge of three appliances.

Maximum permitted suction pressure is -375 Pa as this is equivalent to the recommended trap depth of 75 mm water gauge for single-stack drain installations. When suction of this magnitude is applied to a 75 mm water seal, some of the water is sucked from the trap, leaving about 56 mm. This is sufficient to stop fumes entering the building. Loss of water seal from a trap can occur in the following cases:

1. *Self-syphonage*: this can be avoided by placing restrictions on lengths and gradients and venting long or steep gradients.
2. *Induced syphonage*: water flow past a waste pipe junction in a stack or along a sloping horizontal range of appliances can suck out the water seal; overcome this by suitable design of the pipe diameters, junction layouts and venting arrangements.
3. *Blow-out*: a positive pressure surge near the base of a stack could push out the water seals of traps connected in that region; waste pipes are not connected to the lower 450 mm of vertical stacks, measured from the bottom of the horizontal drain.
4. *Cross-flow*: flow across the vertical stack from one appliance to another; pipes are not connected to soil and vent pipes where cross-flow, particularly from WC branches, could be caused, as shown in Figure 8.17.
5. *Evaporation*: amounts to about 2.5 mm of seal loss per week while appliances are unused.
6. *Wind effects*: wind-induced pressure fluctuations in the stack may cause the water seal to waver out; vent terminal should be sited away from areas subject to troublesome effects.

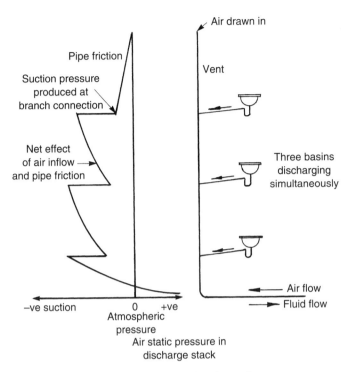

8.16 Air static pressure distribution in soil and vent pipes.

7. *Bends and offsets*: sharp bends in a stack can cause partial or complete filling of the pipe, leading to large pressure fluctuations; foaming of detergents through highly turbulent fluid flow will aggravate pressure fluctuations; connections to the vent stack before and after an offset equalize air pressures; a bend of minimum radius 200 mm is used at the base of a soil stack to ensure constant ventilation.
8. *Surcharging*: an underground drain that is allowed to run full causes large pressure fluctuations; additional stack ventilation is required.
9. *Intercepting traps*: where a single-stack system is connected into a drain with an interceptor trap nearby, fluid flow is restricted; additional stack ventilation is used.
10. *Admission of rainwater into soil stacks*: when a combined foul and surface-water sewer is available, it is possible to admit rainwater into the discharge stack; continuous small rainwater flows can cause excessive pressure fluctuations in buildings of about 30 storeys; flooding of the stack during a blockage would cause severe damage.
11. *Pumped or pneumatically ejected sewage lifting*: the discharge stack is gravity-drained into a sump from where it is pumped into a street sewer at a higher level; a separate vent is used for the sump chamber and pumped sewer pipe to avoid causing pressure surges.
12. *Capillary*: lint or hair remaining in a trap may block flow or capillary action may empty it; additional maintenance is carried out in high-risk locations.
13. *Leakage*: occurs through mechanical failure of the joints or the use of a material not suited to the water conditions.

Figure 8.18 shows the principle of operation of an anti-syphon trap. When excessive suction pressure occurs in the waste pipe, some of the water in the trap is syphoned out. When the

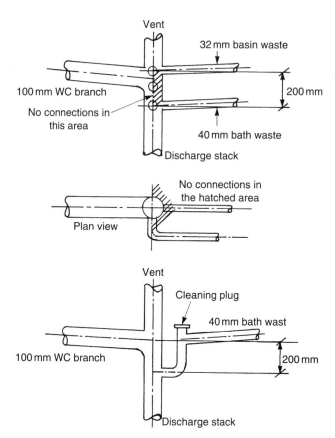

8.17 Permitted stack connections avoiding cross-flow.

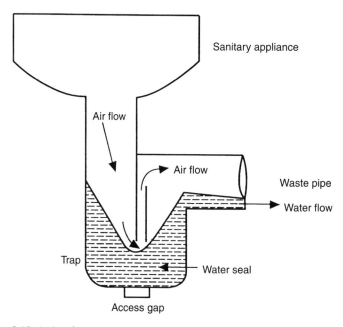

8.18 Anti-syphon trap.

central ventilation passage becomes uncovered, it connects the inlet and outlet static air pressures. This returns the waste pipe to atmospheric pressure and the syphonage ceases. Sufficient water remains in the trap to maintain a hygienic seal.

Drainage installations should remove effluent quickly and quietly, be free from blockage and be durable and economic. They are normally expected to last as long as the building and be replaced only because of changed requirements or new technology. Blockages occur when the system is overloaded with solids, becomes frozen, suffers restricted flow at poorly constructed bends or joints, or has building material left inside pipe runs. Each section of discharge pipe must be accessible for inspection and internal cleaning.

Transport of solids from WCs is the controlling problem in the design, installation and maintenance of sloping drains. Pipe bends produce rapid deceleration of solids downstream, followed by velocity regain as the remaining flush water catches up with and accelerates the solids with minimal loss of inertia. When minimum gradients are used, solid deposition could occur at a bend. Solid deposition can also occur at a top entry into a sewer. Branch connections should be at 45° to the horizontal. The arrangement of pipes for various sanitary appliances is shown in Figures 8.19–8.23.

Groups of appliances for dwellings are depicted in Figures 8.24 and 8.25. A pumped WC discharge unit, as shown in Figure 8.26, enables the use of a 22 mm diameter copper pipe to run long distances, and upwards, to connect into the soil and vent stack at a convenient location.

Materials used for above-ground drain systems range from cast iron socket and spigot lead-caulked joints; soldered lead pipes, both in property built before the 1950s; screwed and socketted joints in galvanized steel pipes, silver soldered, push-fit or compression jointed copper pipes; and various forms of plastic with O-ring push fit seals or solvent welded joints. Secure bracketing to the structure is essential and also allowance for thermal expansion. Pipes passing through walls

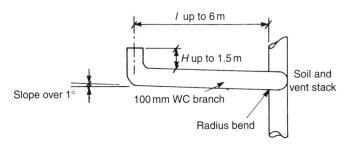

8.19 Branch pipe to a WC.

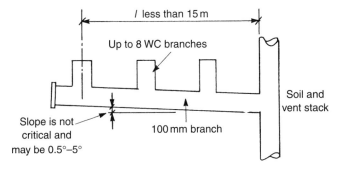

8.20 Branch for a range of WCs.

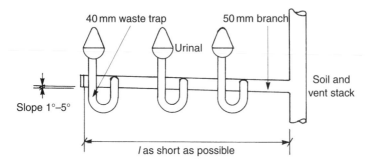

8.21 Branch for a range of urinals.

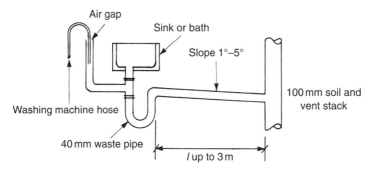

8.22 Branch from a sink or bath.

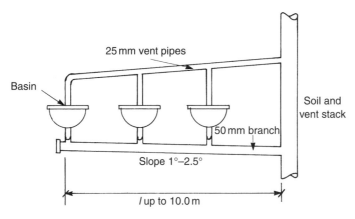

8.23 Branch discharge pipe for a range of up to ten basins.

or floors are sleeved with a layer of inert material to prevent the ingress of moisture into the building and provide the elasticity required for thermal movement.

Inspection and commissioning tests on drainage installations involve the following:

1.   Inspections are made to check compliance with specifications and codes; particular attention is given to quality of jointing, security of brackets and removal of swarf, cement or rubble from inside pipe runs.

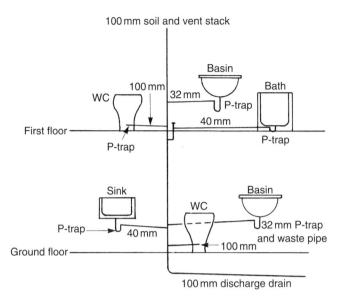

8.24 Soil and vent stack in housing.

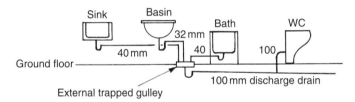

8.25 Drainage pipes for a bungalow.

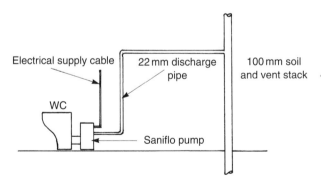

8.26 Pumped WC discharge system.

2. Prefabricated waste pipe systems are factory tested before delivery to the site; complete system tested on completion by filling the water seals and inserting air bags, expandable bungs, into the ends of stack pipes; a rubber hose is inserted into the vent stack through a WC water trap; the air pressure in the stack is hand-pumped up to 38 mm water gauge, measured on a U-tube manometer; this pressure must remain constant for 3 minutes without further pumping; a soap solution wiped onto joints will reveal the leak locations.

3. Syphonage by simultaneous discharge of several appliances should reveal a minimum remaining water seal of 25 mm in all traps; discharge should take place quietly and smoothly.

Periodic inspection, testing, trap clearance, removal of rust and repainting should be a feature of an overall service maintenance schedule. Washers on access covers require occasional replacement. The use of chemical de-scaling agents, hand or machine-operated rodding and high-pressure blockage removal must be carefully related to the drainage materials and the skill of the operator.

Lime scale is found in hard-water areas. A dilute corrosion-inhibited acid-based de-scaling fluid is applied directly to scale visible on sanitary appliances and is then thoroughly flushed with clean water. The fluid is a mixture of 15% inhibited hydrochloric acid and 20% orthophosphoric acid. For removal of grease and soap residues, a strong solution of 1 kg of soda crystals and 9 litres of hot water is flushed through the system. The soda crystals are mixed with the hot water in a basin. When the soda is fully dissolved, the plug is released. This may be necessary frequently in commercially used appliances.

A hand plunger may be sufficient but repeated blockage should be investigated. Hand rodding from the nearest access point can be performed using various tools as appropriate. A kinetic ram gun can be used for blockage in branch pipes. The impact of compressed air from the gun creates a shock wave in the water, which dislodges the solids. However, a blow-back from a stubborn blockage may injure the operator and damage the pipework and therefore the ram gun must be limited to the removal of soft materials. Coring and scraping mechanical tools can be used to remove hard lime scale in 100 mm pipes, provided that the materials will withstand the maintenance operation. A steel cutter is turned by a flexible drive fed through the drain pipes.

EXAMPLE 8.4

Draw a suitable arrangement of a sanitary pipe system for a 22-storey block of flats with two groups of appliances on each floor. Each group consists of a WC, a bath, a sink and a basin, all sited close to the stack.

A 100 mm soil and vent stack can be used. Typical floor layout is shown in Figure 8.27.

EXAMPLE 8.5

A 14-storey office block is to have ranges of three WCs, three basins and three urinals on each floor. The WCs and urinals are situated on each side of the stack, but the common waste pipe from the basins is to be 5 m long and have four bends. Draw a suitable discharge pipe arrangement and state the pipe sizes to be used.

A 100 mm discharge and vent stack can be used in the arrangement shown in Figure 8.28 with an additional ventilating stack of diameter 40 mm.

Surface water drainage

Ground surface-water systems may be designed on the basis of a rainfall intensity of 50 mm/h and 75 mm/h for roofs. Quantity of water entering a drain depends on the amount of evaporation

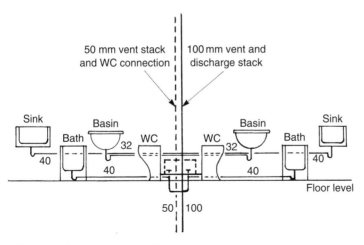

8.27 Typical floor layout for two flats in Example 8.4.

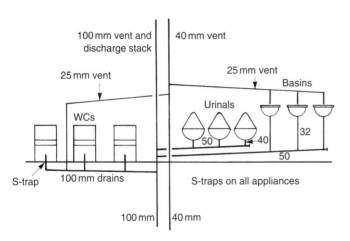

8.28 Typical floor layout for Example 8.5.

into the atmosphere and natural drainage into the ground. Drain flow load is represented by the impermeability factor; typically 0.95 roof, 0.9 roads, 0.75 paths, 0.25 gardens and 0.2 woodland; rain water flow rate Q is given by:

$$Q = area\ m^2 \times rainfall\ \frac{mm}{h} \times impermeability$$

EXAMPLE 8.6

Footpaths, roadways and gardens on a commercial estate cover an area of 75000 m², of which 20% is garden and grassed areas. Estimate the surface-water drain flow load in litres per second. How many 15 litre/s surface water drain gulleys are needed?

Impermeability factors are 0.9 for the roads and paths and 0.25 for gardens and grass.

Average impermeability $= 0.2 \times 0.25 + 0.8 \times 0.9$

Average impermeability $= 0.77$

$$Q = 75000 \ m^2 \times 50\frac{mm}{h} \times \frac{1\,h}{3600\,s} \times \frac{1\,m}{10^3\,mm} \times 0.77 \times \frac{10^3\,l}{1\,m^3}$$

$$Q = 802\frac{l}{s}$$

$$\textit{Drain gulleys} = 802\frac{l}{s} \times \frac{\textit{gulley}\,s}{15\,l}$$

$$\textit{rain gulleys} = 54$$

Cast iron covers over a drainage gulley in roadways pass 20 litre/s or more, depending upon surface-water speed, degree of flooding and blockage from debris. Roof drainage rainfall intensity of 75 mm/h occurs for about 5 minutes once in 4 years. An intensity of 150 mm/h may occur for 3 minutes once in 50 years and where overflow cannot be tolerated, this value is used for design. The flow load Q for a roof is calculated from:

$$Q = \textit{area } m^2 \times 75\frac{mm}{h} \times \frac{1\,h}{3600\,s} \times \frac{1\,m}{10^3\,mm} \times \frac{10^3\,l}{1\,m^3}$$

$$Q = (0.021 \times \textit{area})\frac{l}{s}$$

Depth of flow in a level gutter discharging freely increases from the outlet up to a maximum at the still end. Depth of flow at the outlet is about half that at the still end. A fall of 1 in 600 increases flow capacity. Frictional resistance and bends of a sloping gutter reduce water flow. Water flow in down pipes is much faster than in the gutter and they will never flow full. Their diameter is usually taken as being 66% of the gutter width. Typical gutter flow capacities may be 1 l/s for 100 mm, 1.5 for 125 mm and 3 for 150 mm diameter when the outlet is at one end. A centre outlet doubles flow capacity for the same diameter. A roof is divided into areas served by a gutter with end-outlet rainwater down pipe. If the whole roof is drained into a gutter with an outlet at one end, then the gutter carries water from the entire roof area. However, when a centre outlet is used, a smaller gutter size might be possible as it only carries half the flow load. The number and disposition of the down pipes are considered in relation to the gutter size, architectural appearance, cost and complexity of the underground drainage system.

EXAMPLE 8.7

A sports centre roof of dimensions 15 m × 8 m is laid to fall to a PVC half-round gutter along each long side. Find an appropriate gutter size when the gutter is to slope at 1 in 600. Each gutter can have a centre or end outlet.

$$Q = (0.021 \times \textit{area})\frac{l}{s}$$

$$Q = (0.021 \times 15 \times 8)\frac{l}{s}$$

$$Q = 2.52\frac{l}{s}$$

For a centre outlet, the gutter will carry half of Q, 1.26 l/s and a 125 mm gutter is correct. An outlet at one end means 150 mm gutter is used.

EXAMPLE 8.8

A sloping roof has a level box gutter 125 mm wide and 50 mm deep. The roof is 25 m long and 5 m up the slope. Calculate whether the gutter will adequately convey rainwater when the rainfall intensity is 75 mm/h and recommend the outlet location.

$$Q = (0.021 \times area)\frac{l}{s}$$

$$Q = (0.021 \times 25 \times 5)\frac{l}{s}$$

$$Q = 2.65\frac{l}{s}$$

The rectangular gutter is smaller than a 125 mm semi-circular shape that has a flow capacity of 1.5 l/s; use a centre outlet or two outlets to avoid overflow.

Surface-water can be removed from a site by one or more of the following methods:

1. Sewer where the local authority agrees that there is adequate capacity; surface-water is drained into either a combined sewer or a separate surface-water sewer. Surface-water from garage forecourts and car parks is run in open gullies to an interceptor chamber. Ventilation of explosive and poisonous petrol vapour is essential, as a concentration of 2.4% in air is fatal. It is illegal to discharge petrol, oil or explosive vapour into public sewers. The interceptor chamber is an underground storage tank of concrete and engineering bricks, which allows separation of the clean water from the oily scum remaining on its surface. It is intermittently pumped out and cleaned. The discharge drain to the sewer is turned downwards to near the bottom of the interceptor and three separate chambers are used in series.
2. Soak away after ground permeability is established from borehole tests to measure the rate of natural drainage within a site. If running underground water is found, a simple rock-filled pit can be used. Slow absorption is overcome by constructing a perforated precast concrete, dry stone or brick pit, which stores the rainfall quantity. The stored volume is found from an assumed steady rainfall of 15 mm/h over a period of 2 h. This is exceeded around once in 10 years, so there may be occasional flooding for short periods. A soak away pit is circular with its depth equal to its diameter.
3. Storage in an artificial pond or lake, or even an underground storage tank, will be necessary if the expected run-off from the site is at a greater rate than could be accommodated by a sewer or watercourse.
4. The local authorities may allow the disposal of surface-water into watercourses. Expected flow rates at both normal and flood water levels must be established.

EXAMPLE 8.9

Storage soak-away pits 2.25 m deep are to be employed for a tarmac-covered car park of dimensions 100 m × 30 m. Determine the number and size of the pits needed. Draw a suitable drainage layout for the car park.

$$Volume\ to\ be\ stored = 15\frac{mm}{h} \times 2h \times \frac{1m}{10^3 mm} \times 100m \times 30m \times 0.9$$

$$Volume\ to\ be\ stored = 81\,m^3$$

Pits 2.25 m deep and diameter of 2 m to be used.

$$Pit\ volume = \frac{\pi \times 2^2}{4} \times 2.25\,m^3$$

$$Pit\ volume = 7.07\,m^3$$

$$Number\ of\ pits = 81m^3 \times \frac{pit}{7.07\,m^3}$$

$$Number\ of\ pits = 12$$

Figure 8.29 shows the suggested layout.

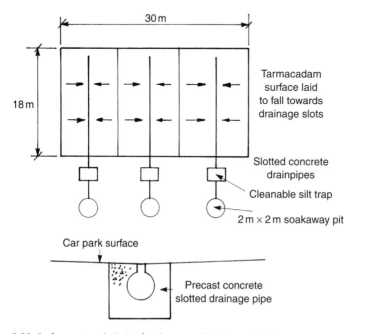

8.29 Surface-water drainage for the car park in Example 8.9.

Below-ground drainage

Below-ground drainage systems are designed to operate without the input of energy, wherever possible, to be reliable and to require little, if any, maintenance. Their layout has to be such that drains are not subject to undue stress from foundations or traffic and are fully accessible for occasional clearance. Sanitary discharge services operate by gravity flow and require no energy input. Parts of buildings or sites that are below the sewer invert require a pump to raise the fluid. These operate intermittently to minimize electrical power consumption. Drains are laid to fall at an even gradient, which produces a self-cleaning water velocity so that potential deposits are accelerated and floated downstream. Large drops in drain level are accommodated in a back-drop manhole, rather than a lengthy steep slope, in order to minimize excavation.

Pipes are laid in a series of straight lines between access points used for inspection, testing and cleaning. Branch connections are made obliquely in the direction of flow. There is a preference against running drains beneath buildings owing to possible settlement and the potential cost of later excavation to make repairs, and because the drain invert is lower than the floor damp-proof membrane and rising drains need a waterproof seal against groundwater. Selected trench bedding and backfill material are used to provide continuous pipe support, to spread imposed ground loads due to the weight of soil and passing traffic, to protect drains from sharp objects and other services, as well as to divert stresses imposed by building foundations. Temporary boards are used to protect exposed drain trenches during construction work. Figure 8.30 shows a typical foul drain system from a single stack.

Blockages may happen as drain systems are likely to be in place for a hundred years or more and demands upon them continue to increase. Ability to clean drains is an essential feature of good design. Good health depends upon satisfactory drainage. Domestic drains are likely to be located less than 1.5 m below ground level, at a maximum gradient of 1:40 and 100 mm in diameter, possibly increasing to 150 mm in diameter at the downstream end of an estate or large building.

Vitrified clay or PVC pipe and fittings are often used with flexible joints to accommodate ground movement and thermal expansion due to variations in fluid temperature. Brick, concrete, PVC or glass-reinforced plastic (GRP) access chambers are used. All changes in direction are either through 135° or large radius swept bends. Access points are provided for removing compacted material and for using rigid rods to clear blockages in the direction of flow, even though flexible water-jetting techniques are currently available and it is possible to clear obstructions

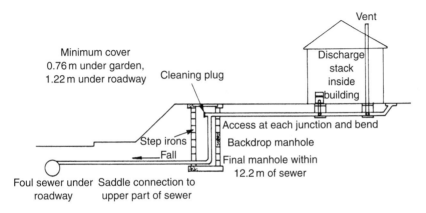

8.30 Foul drainage installation.

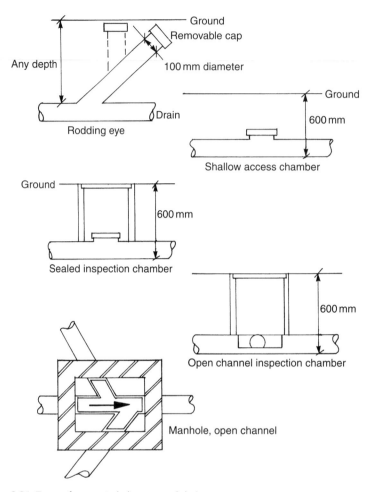

8.31 Types of access to below ground drainage.

from either direction. Airtight covers are desirable to avoid access points allowing a health hazard or flooding. Figure 8.31 shows types of access.

A rodding eye is a 100 mm diameter drain pipe extended from any depth to ground level to allow rodding in the downstream direction A shallow access chamber has a removable threaded cap on a branch fitting to allow access in either direction located such that the distance from ground level to drain invert is less than 600 mm to facilitate reaching into the drain. A sealed inspection chamber is a 600 mm deep, 500 mm diameter chamber for access to screwed caps on drain junctions. An open-channel inspection chamber is a 600 mm deep, 500 mm diameter access chamber with benched smooth surfaces for drain junctions. Manholes are the main access point for an operative wearing breathing apparatus to climb down steps to any depth; a 1 m deep manhole is 450 mm square, and a 1.5 m deep manhole has dimensions of 1200 mm × 750 mm or 1050 mm diameter, and a cover 600 mm square. Gulleys are ground-level connection points for various waste pipes into the below-ground 100 mm diameter drain, providing a water trap against sewer gas and allowing debris removal and rodding access; it may have a sealed lid or open grating.

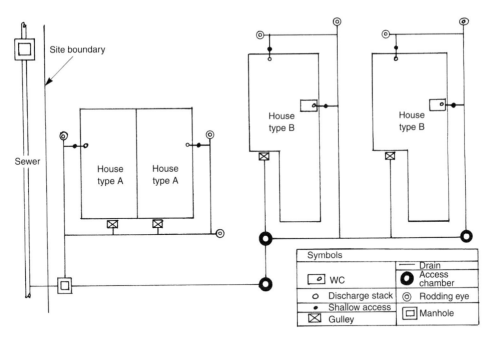

8.32 Typical site layout showing access.

The first access point close to the building is a gulley, a removable WC or a shallow access chamber just after the base of the internal drainage stack. It is not necessary to fit access points at every change in drain direction, but pipe junctions are made with access chambers. The maximum spacing between access points is 12 m from the start of the drain to the first access, 22 m from a rodding eye to a shallow access chamber, 45 m from a rodding eye to an access chamber or manhole and 90 m between manholes. Figure 8.32 demonstrates a typical housing estate drain layout. Careful integration with the surface-water drainage system is necessary as falls to the sewers are preconditioned by the sewer inverts, and the two drains may run within the same trenches and cross each other where branch connections are made.

Traditionally, glazed vitrified clay pipes were used because they represent an efficient use of UK national resources. The finished internal surface of GVC pipes offers less frictional resistance to flow than that of concrete pipes and is resistant to chemical attack and abrasion. Rigid joints consist of a socket and spigot cemented together. The brittle nature of such pipe runs has led to the introduction of flexible joints, which can withstand ground movement due to thermal and moisture variations and settlement of buildings. Plastic and rubber sealing ring joints allow up to 5° of bending and longitudinal expansion and contraction. Pipe sizes range from 75 mm to 750 mm in diameter.

Spun concrete drain pipes of diameter up to 1.83 m with oval cross-sections, which maintain flow velocity at periods of low discharge, are used. Plastic sleeves with rubber sealing rings give joints flexibility and a telescopic action.

Plastics have increasingly replaced naturally occurring materials owing to their low weight and high degree of prefabrication. Complete systems from the sanitary appliance to the sewer, using one supplier and material, are common. Such materials are derived from crude oil and their higher cost needs to be compensated by reduced site time. Smooth bore drain systems can be assembled with minimum skill and they are highly resistant to corrosion. Thermal expansion is greater, and

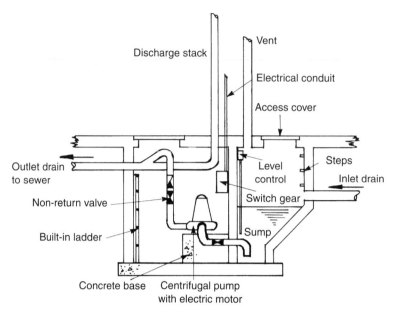

8.33 Sewage-lifting pump.

telescopic joints are used. Short-term discharges from some appliances, for example, some types of washing machine, can be at temperatures of 80°C or higher. Polypropylene and acrylonitrile butadiene styrene (ABS) pipes are suitable for the high-temperature applications.

Where sanitary appliances discharge into drain pipes below the foul sewer invert, a collecting sump, pump and fluid level controller are used in the manner shown in Figure 8.33. Either a large clearance centrifugal pump, driven by a 440 V three-phase electric motor, or a pneumatic ejector, is used. The storage chamber is sized to accommodate several hours of normal discharge so that the pump only runs for short periods and total electrical power consumption need not be high. Duplicate pump sets ensure a continuity of service during breakdowns and maintenance.

The pneumatic ejector collects the discharge in a steel tank containing a float. At the upper fluid level, the float operates a change-over valve, which admits air from a compressor and storage vessel. The incoming compressed air drives the sewage into the outlet drain at the higher level. Non-return valves are fitted to the inlet and outlet pipes to stop the possibility of reverse flow. Both types of sewage-lifting equipment have open vent pipes to ensure that back pressures are not imposed upon the soil stack.

Water and air pressure testing is done on below ground drains. Completed sections are tested before, and sometimes after, backfilling. Drain runs are tested between manholes. The lower end is sealed with an expandable plug. A temporary up stand plastic or aluminium pipe is connected at the higher manhole. Water is admitted to produce a static head of 1.4–2.4 m and maintained for an hour or more. Some drainage materials will absorb water, and the initial water level is replenished from a measuring cylinder or jug. The maximum allowable water loss is 1.0 litre/h for 10 m of 100 mm diameter pipe or pro rata for other diameters and lengths. An air test can be conducted in a similar manner: a static pressure of 75 mm water gauge on a U-tube manometer should be maintained for a period of 5 minutes without further pumping.

Questions

1. Sketch and describe the Earth's natural water cycle.
2. List and describe the sources of water and the methods used for its storage and treatment.
3. What pollutants are present in naturally occurring water and where do they come from?
4. Explain the terms 'temporary' and 'permanent' hardness, and list their characteristics.
5. State the use of service reservoirs and describe public mains water distribution methods.
6. State the design parameters for cold-water mains and storage systems within buildings, giving particular information on protection from frost damage, suitability for drinking and protection of the mains against contamination from the building.
7. List the design parameters for hot-water service systems, giving typical data.
8. Sketch the layout of a water services system in a house, showing typical sizes of equipment and methods of control. Show how wastage of water is minimized.
9. Sketch and describe the methods used to generate hot water, noting their applications, economy and thermal performance.
10. Sketch and describe a suitable cold-water services installation for a 20-storey hotel where the mains water pressure is only sufficient to reach the fifth floor.
11. Draw the layout of an indirect hot-water system employing a central-heating boiler and secondary circulation. Show all the pipe work and control arrangements.
12. A small hotel is designed for 20 residents and four staff. Hot water to be stored at 65°C is taken from a cold-water supply at 10°C and heated during a 4 h period. Calculate the heat input rate required.
13. A cold-water storage tank in a house with five occupants is to have a capacity of 100 litres/per person and be fed from a water main able to pass 0.25 l/s. How long will it take to fill the tank?
14. A secondary hot-water service system has 55 m of 28 mm circulation pipes and 40 m of 15 mm circulation pipes. Water leaves the cylinder at 65°C and returns at 60°C. Air temperature around the pipes is 15°C. Pressure loss rate in the pipes is 260 N/m^3. The frictional resistance of the pipe fittings is equivalent to 25% of measured pipe length. Specify the head and flow rate required for the secondary circulation pump and choose a suitable pump.
15. Draw cross-sections through four different types of pipe joint used for water services, showing the method of producing a water seal in each case.
16. List the factors involved in the provision of a pipe work system for the conveyance of drinking water within a curtilage. Comment on the suitability, or otherwise, of jointing materials, lead pipes and storage tanks in this context.
17. Describe the corrosion processes that take place within water systems and the measures taken to protect equipment.
18. Sketch and describe the ways in which solar energy can be used within buildings for the benefit of the thermal environment and to reduce primary energy use for hot-water production. Comment on the economic balance between capital cost and expected benefits in assessing the viability of such equipment.
19. List the types of sanitary appliance available and describe their operating principles, using appropriate illustrations. Comment on their maintenance requirements, water consumption, long-term reliability and materials used for manufacture.
20. Why is domestic hot water stored at high temperature?

    1.   Maximizes the thermal efficiency of the water heater.
    2.   Minimizes pipe sizes for flow rates.
    3.   Minimizes storage quantity.

4. It should not be, as it can scald users.
5. Kills Legionella bacteria in water.

21. Which water description is correct about soft water?

1. Occurs in granite rocky soil parts of the country.
2. High pH value.
3. Low suspended salt concentration.
4. Total dissolved salt concentration of over 200 ppm.
5. High soap-destroying capability.

22. Which water description is correct about hard water?

1. Occurs in chalk soil parts of the country.
2. Very low pH value.
3. High suspended salt concentration and suspended vegetable matter.
4. Total dissolved salt concentration below 50 ppm.
5. Laundries consume minimal soap.

23. Which causes temporary hardness in water?

1. Suspended solids.
2. Acidic ground water sources.
3. Precipitation of salts during storage in the building.
4. Sulphates and chloride salts in the water.
5. Dissolved carbonate salts.

24. Which is correct about permanent water hardness salts?

1. Cannot be removed from water.
2. Only removed by steam generation and condensation.
3. Salts deposited on heat transfer surfaces during water boiling.
4. Only removable by dosing ground water with acid.
5. Due to presence of non-carbonate dissolved salts.

25. Which is not correct about water pH value?

1. A measure of free hydrogen ions in water.
2. pH below 7 is always found in hard water.
3. Steam boiler water is treated to a pH of 11.
4. Acidic water from granite ground has a low pH.
5. Low pH is preferable for closed black steel pipe heating systems.

26. Which is correct about water treatment for closed circulation water systems?

1. Not needed.
2. Water purifies itself due to release of dissolved oxygen and salts during commissioning.
3. Absence of fresh oxygen in closed system avoids need for corrosion treatment.
4. There is no difference between the effects of pH value, acidity, alkalinity and corrosive-ness of water.
5. Treatment provided to combat electrolytic corrosion.

27. Which applies to the base exchange water treatment system?

1. Calcium carbonate in public mains water chemically reacts with sodium zeolite in treatment tank.
2. Calcium carbonate in public mains water is absorbed by zeolite salt in the treatment tank and remains there.

3. Zeolite salts are consumed by incoming hardness salts and residue has to be removed for disposal.
4. Zeolite salts filter out calcium carbonate and other hardness salts and must be disposed of to waste water when fully clogged.
5. Zeolite salts destroy hardness salts in the public mains water supplied to steam boilers.

28. What is the meaning of low pressure water supply main?

1. Building has a pressure reduction valve on the incoming public water main.
2. Street water main is at very low pressure.
3. There is no such thing as all public water supply systems function at well above atmospheric pressure.
4. All water services in buildings below 4-storey height.
5. Mains water pressure supplies storage tanks that service the building's systems.

29. Which technical feature ensures a consumer building does not contaminate the public water supply pipe system?

1. Non-return valve on the supply pipe to each building.
2. Manual isolating valve at each building entry pipe.
3. Maintaining the building's piped water systems at a lower pressure with a pressure reducing valve at entry from the public main.
4. Dosing all pipe systems in a building with biocide.
5. An air gap.

30. What is the health risk, if any, from stored domestic hot water?

1. No health risk when public mains water is heated and stored as it always remains dosed with drinkable chemicals.
2. Heating water does not raise its health risk.
3. Only mains water that has come into contact with the atmosphere could become contaminated.
4. Mains water bacteria always remain dormant.
5. Legionella bacteria present in water grow rapidly between 20°C and 40°C.

31. How do we protect ourselves from bacteria in domestic hot water systems?

1. Heat and store hot water at 42°C.
2. Only use hot water at less than 45°C.
3. Water stored at 65°C so all bacteria killed.
4. Heat stored water to 65°C to kill bacteria then wait for it to cool to 40°C prior to use.
5. Heat stored water to 100°C to sterilize bacteria, wait for it to cool to 40°C prior to use.

32. A 1000 litre domestic hot water storage calorifier has a heating water rate of input of 50 kW. How long will it take to raise the water content from 10°C to 65°C if heat loss from the cylinder is negligible?
33. List the ways in which an above-ground drainage installation satisfies its functional requirements.
34. Describe, with the aid of sketches, the ways in which the water seal can be lost from a trap and the precautions taken to avoid this happening.
35. Describe the types of fluid flow encountered in drainage pipes.
36. State the meaning of the following terms: bedding; combined system; drain; sewer; manhole; separate system; stack; discharge pipe; vent, and explain how they interact.

37. Sketch the pipe layout for a typical group of sanitary appliances in a dwelling, where they are all connected into a stack. Show suitable pipe sizes, slopes and details of the connections at the stack.
38. Sketch and describe the methods of testing drainage installations.
39. What type of water flow occurs in the waste pipe from a basin?

    1. Turbulent.
    2. Laminar.
    3. Steady continuous stream along lower half of sloping pipes.
    4. Water swirls clockwise down vertical and along sloping pipes, adhering to the walls of the pipe due to the Coanda effect.
    5. Full bore surge followed by dribbling.

40. Which is a cause of induced syphonage in a waste pipe?

    1. Positive back-pressure from downstream pipes.
    2. Low atmospheric air pressure.
    3. Waste pipes running full.
    4. High atmospheric air pressure.
    5. Insufficient ventilation of vertical stack.

41. How are waste pipe system syphonage risks reduced?

    1. Adequate ventilation to open air.
    2. Installing larger diameter waste pipes than normally recommended.
    3. Connecting every waste trap individually to the vertical stack.
    4. Maintaining air tightness of the drainage system.
    5. Regular internal cleaning of all waste and drain systems.

42. List the principal requirements for an underground drainage installation.
43. Sketch and describe the operation of a sewage pumping installation. Draw the details of the construction of the below ground chambers.
44. Design a below-ground drainage system for the Pascal Sports Club shown in Figure 8.34. The foul sewer is at an invert of 2 m and 25 m to the right of the east wall of the club. Only one connection is allowed to be made to the 300 mm diameter sewer, and this is to its upper half. It will be necessary to design the above-ground waste pipes from all sanitary appliances in order to optimize the gulley positions, the 100 mm diameter pipe routes and the location of the one ventilation stack at the high point of the whole system. Minimize the use of underfloor pipes, all of which must be 100 mm in diameter and fully accessible. Modifications can be made to the building to construct above-ground service ducts to accommodate hot- and cold-water pipes as well as wastes and drains. A 100 mm diameter rainwater down pipe is located 500 mm from each external corner of the building on the north and south sides. These connect to 100 mm diameter below-ground drains, which run to the surface-water sewer alongside the foul sewer. Ensure that both drain systems are fully integrated and separated by a bedding of at least 100 mm thick shingle or broken stones of maximum size 5–10 mm. Access to the surface-water pipes is of the same standard as that to the foul pipes. The last access prior to the sewer for both drains should be a manhole. There is no manhole at the junction of the drain and sewer. The shower rooms will have trapped floor gulleys that connect to the foul drain. No model solution is provided as the design should be discussed with tutor and colleagues, and reference should be made to manufacturers' guides.
45. How are below-ground drains tested?

    1. Filling with water and pumping the pressure up to 30 m water gauge for an hour.

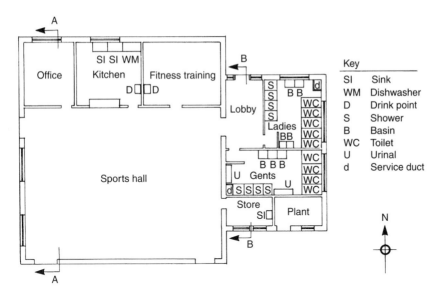

8.34 Pascal Sports Club.

2. Sealing ends of completed system, hand pumping air pressure up to 150 mm water gauge for an hour without further pumping.
3. Prior to backfilling trench, subject completed drain system to a static water height of 2.4 m for an hour.
4. Internal camera survey.
5. Watching for leaks prior to backfilling trench.

46. A housing estate has footpaths and roads covering an area of 4000 m². Calculate the rainwater flow load and the number of drain gullies required.
47. The flat roof of a school is to be of dimensions 40 m × 20 m with a rectangular gutter on each long side and sloping at 1 in 600. Design a suitable gutter and down pipe system.
48. Storage soak away pits 2 m deep are to be used to dispose of rainwater from a roof of dimensions 10 m × 8 m. Determine a suitable size and number of pits.
49. Describe the features and maintenance requirements of surface-water drainage systems for car parking, garage forecourts and large paved areas in shopping centres.

9 Electrical installations

Learning objectives

Study of this chapter will enable the reader to:

1. understand how electricity is generated and distributed;
2. know the difference between single- and three-phase electricity;
3. distinguish line, neutral and earth conductors;
4. calculate the resistance of conductors;
5. understand the temperature effect of a current;
6. calculate current and power in electrical circuits;
7. know how to measure current and voltage;
8. use power factor;
9. calculate series and parallel circuit resistances;
10. find the current capacity of cables;
11. choose cable sizes;
12. calculate permissible cable lengths;
13. understand temporary electrical installations for construction sites;
14. calculate the total electrical loading in kilovolt-amperes and amperes for installation;
15. estimate the total cost of electricity likely to be consumed in an installation during normal use;
16. choose the correct fuse rating;
17. understand the operation of fuses and circuit-breakers;
18. know the distribution of electricity within buildings;
19. identify the use of isolating switches, distribution boards and meters;
20. understand earth bonding of services;
21. know the types of cable conduits and their applications;
22. understand the principles of ring circuits;
23. understand how electrical systems are tested;
24. design lightning conductor systems;

25. have an understanding of data installations and wireless communications;
26. discuss on-site generation.

Key terms and concepts

cables 216; capacitor 205; cartridge fuse 213; conduits 218; data systems 220; distribution board 216; electric shock 213; electrical measurements 204; fault current 214; fuses and circuit-breakers 213; kilowatts (kW) and kilovolt-amperes (kVA) 205; lightning conductor systems 220; line current 207; miniature circuit-breaker 214; Ohm's law 207; on-site generation 222; power factor 204; residual current device 213; ring circuit 217; series and parallel circuits 204; single- and three-phase 205; specific resistance 203; telecommunications 220; temporary electrical installations on sites 209; testing 219; triple pole and neutral switch 217; voltage, current, power and resistance 203.

Introduction

Safe and economical use of electricity is of paramount importance to the building user and the world as it is the most highly refined form of energy available. Electricity production consumes up to three times its own energy value in fossil fuel, nuclear energy and renewable sources. Atmospheric emissions accrue from every lamp, wire and electric motor. Electricity use defines us as developed societies; how long can you cope without it? A couple of hours are an inconvenience due to a temporary system breakdown; weeks would be a major disaster to our lifestyle. Yet, some societies live without it. Electricity in its distributed form is potentially lethal. In this chapter the handling methods and safety precautions for utilizing electricity are explained and a range of calculations, which can easily be performed by the services designer or constructor prior to employing specialist help, is introduced.

Electricity distribution

Electrical power generating companies supply electrical power into the national 400 kV grid system of overhead bare wire conductors. This very high voltage is used to minimize the current carried by the cables over long distances. Step-down transformers reduce the voltage in steps down to 33 kV, when it can be supplied to industrial consumers and to other transformer stations on commercial and housing estates (CIBSE, 2005).

An electricity-generating alternator rotates at 50 Hz (3000 rev/mm) and has three coils in its stator. The output voltages and currents from each coil are identical but are spaced in time by one-third of a revolution, 120°. Each coil generates a sine wave or phase voltage that has the same heating effect as a 240 V continuous direct current supply. This is its root mean square (RMS) value. The RMS value of the three phases operating together is 415 V. Figure 9.1 shows the connections to a three-wire three-phase 415 V, 50 Hz alternating current supply entering a non-domestic building. Various circuits of different voltages are supplied from the incoming mains. Equal amounts of power are fed into each phase, and so it is important that power consumption within a building is equally shared by each line. The neutral wire is a live conductor in that it is the return path to the alternator for the current which has been distributed.

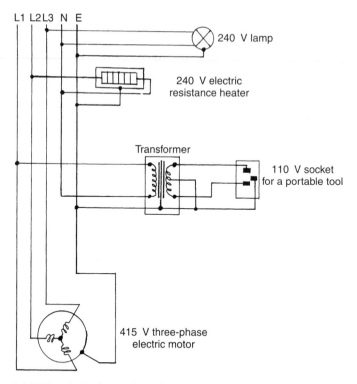

L1 L2L3 N E

240 V lamp

240 V electric
resistance heater

Transformer

110 V socket
for a portable tool

415 V three-phase
electric motor

9.1 Wiring circuits from a three-phase 415 V incoming supply.

Figure 9.2 shows a single phase from one of the generator coils where the effective voltage is the RMS value:

$$RMS = \sqrt{\frac{\sum V_n^2}{n}}$$

Where V is the voltage at each of n measurements, say at $15°$ intervals.

A balanced load, such as a three-phase electrical motor driving an air-conditioning fan, water pump or lift motor, does not produce a current in the neutral wire. This is because an alternating current flows alternately in the forward and backward directions along the line wire. The overall effect of three driving coils in the motor is a balance in the quantity and direction of the current taken from the line conductors. There is no net return current in the neutral wire from such a balanced load. Single-phase electrical loads, which are not in balance, produce a net current in the neutral conductor.

The casings of all electrical appliances are connected to earth by a protective conductor, the earth wire, connected to the earthed incoming service cable of the electricity supply authorities or an earth electrode in the ground outside the building. Gas and water service pipes are bonded to the earth by a protective conductor.

Circuit design

The resistance R ohms (Ω) of an electrical conductor depends on its specific resistance ρ Ω m, its length l m and its cross-sectional area A m^2. The specific resistance of annealed copper is 0.0172

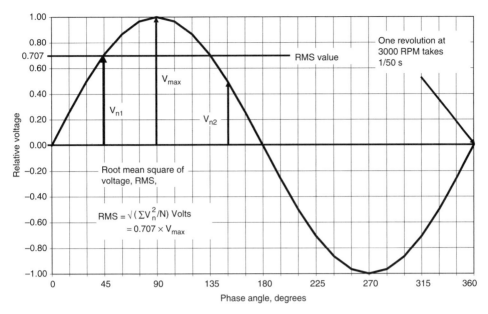

9.2 Single-phase RMS.

$\mu\Omega\, m$ (μ, micro stands for 10^{-6}) at 20°C.

$$R = \rho\frac{l}{A}\,\Omega$$

EXAMPLE 9.1

Calculate the electrical resistance per metre length at 20°C of a copper conductor of 2.5 mm² cross-sectional area.

$$R = \frac{0.0172}{10^6}\,\Omega m \times \frac{1\,m}{2.5\,mm^2} \times \frac{10^6\,mm^2}{1\,m^2}$$

$$= 0.0069\,\Omega$$

The resistance of a cable increases with increase in temperature and the temperature coefficient of resistance (α) of copper is 0.00428 Ω/Ω °C at 0°C. If the resistance of the conductor is R_o at 0°C, then its resistance at another temperature R_t can be found from:

$$R_t = R_o(1 + \alpha t)\,\Omega$$

where t is the conductor temperature (°C).

EXAMPLE 9.2

Find the resistance of a 2.5 mm^2 copper conductor at 40°C.

R_O is not known but the resistance of this conductor at 20°C was found in Example 9.1 and t can represent the increase in temperature above this value. A graph of resistance versus temperature would reveal a straight line of slope α.

$$R_{40} = R_{20}(1 + \alpha \times 20)\,\Omega$$

$$= 0.0069 \times (1 + 0.00428 \times 20)\,\Omega$$

$$= 0.0075\,\Omega$$

The relation between applied voltage, electric current and resistance is given by Ohm's law:

$$I \text{ amps} = \frac{V \text{ volts}}{R \text{ ohms}}$$

Figure 9.3 shows how an ammeter and a voltmeter are connected into a circuit to measure power consumption. The load may be an electrical resistance heater or tungsten filament lamp, in which case the power consumption in watts is found from:

$$\text{power in watts} = V \text{ volts} \times A \text{ amps} \times \cos\phi$$

for single phase and

$$\text{power in watts} = V \text{ volts} \times A \text{ amps} \times \sqrt{3} \times \cos\phi$$

for three phase, where $\cos\phi$ is the power factor (zero to unity). An electrical resistance heater and tungsten lamps are purely resistive loads whose power factor, $\cos\phi$, is unity, 1.

Loads such as electric motors and fluorescent lamps have a property known as inductance, which causes the current to lag behind the voltage that is producing that current. This is due to an electromotive force (e.m.f.), i.e. a voltage, which opposes the incoming e.m.f. The densely packed electromagnetic windings of a motor have a high inductance and thus the 50 Hz cyclic variation of voltage and current is opposed by the 'inertia' of the equipment. The opposing e.m.f. comes from the expanding and collapsing magnetic fields of the input power around the conductors. The lag angle $\phi°$ between peak voltage and peak current means that the instantaneous available

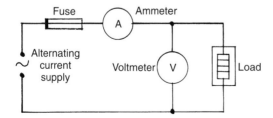

9.3 Measurements of power consumption in an electrical circuit.

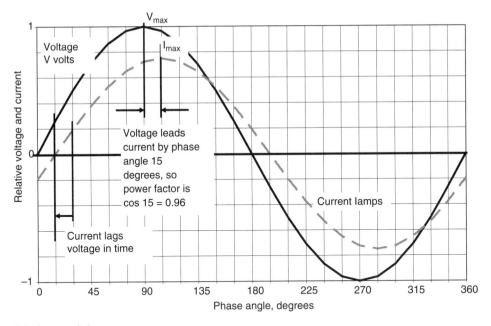

9.4 Phase angle lag.

power is less than the product of the two peaks. Figure 9.4 shows voltage applied across a circuit creating a current that lags in real time, peak current occurs after the peak voltage.

Power factor is the term used to differentiate between useful output power in watts and the input instantaneous product of voltage and current to the load:

$$\text{power factor } PF = \frac{\text{useful power in watts}}{\text{input volts} \times \text{amps}} = \frac{\text{watts}}{\text{volt} \times \text{amps}}$$

$$= \frac{\text{real power (do to work)}}{\text{apparent power}}$$

$$\text{power factor } PF = \frac{\text{kW}}{\text{kVA}}$$

$$= \frac{\text{kilowatts}}{\text{kilovolt-amps}} = \frac{\text{kW real}}{\text{kVA (apparent)}}$$

Figure 9.5 shows all three phases as they occur in real time, separated by 120° phase angle. Because of the three voltage sine waves within each cycle, the overall power generated is higher and smoother than with only a single phase motor. Capacitors have an electrical storage capability, which is used to overcome the effects of inductance. Power factors of electrical equipment are commonly 0.85 and these can be improved to 0.95 by the addition of power-factor-correcting capacitors.

Several loads to a circuit may be connected either in series or in parallel with each other. For series-connected resistances:

$$R = R_1 + R_2 + R_3 + \cdots$$

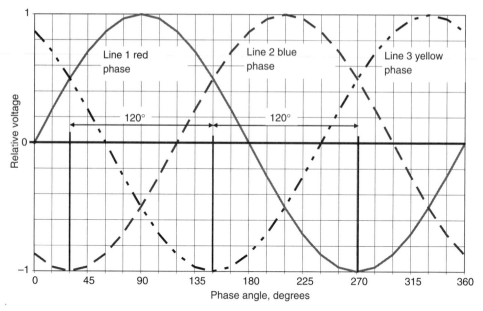

9.5 Three phases.

For parallel-connected resistances:

$$\frac{1}{R} = \frac{1}{R_1} + \frac{1}{R_2} + \frac{1}{R_3} + \cdots$$

The resistance of electrical cables must be sufficiently low that the cables do not become significant sources of heat and run at temperatures that could be a fire hazard or could damage their electrical insulation. Such heat generation would generally be wasted energy. The maximum permissible voltage drop in a cable is 4% of the nominal supply voltage from the consumer's intake terminal to any point in the installation at full load current.

Cables that are grouped together run in conduits or are covered with thermal insulation, say, in a roof space, and can operate at a temperature above the 30°C ambient condition assumed in the selection of their size and their electrical insulation. Where their temperature is likely to rise above this value, their current-carrying capacity is reduced by appropriate rating factors during design of the system. Care should be taken to allow natural cooling of all cable routes. The current-carrying capacity has to be 1.33 times the design current for cables partly surrounded by thermal insulation and twice the design current if they are wholly surrounded. This will generally mean an increase by one or two cable sizes.

EXAMPLE 9.3

Calculate the power consumption and resistance of a 240 V filament lamp if it has 1.5 A passing through it.

power in watts $=$ volts $\times$ amps

$$= 240\,V \times 1.5\,A$$

$$= 360\,W$$

From Ohm's law:

$$I = \frac{V}{R}$$

$$R = \frac{V}{I}$$

$$= \frac{240}{1.5}$$

$$= 160\,\Omega$$

EXAMPLE 9.4

Insulation on a conductor carrying 415 V has an electrical resistance to earth of 500 MΩ. What leakage current could flow through the PVC when the cable is laid on an earthed metal support (1 MΩ = 10^6 Ω)?

The difference between line and earth is 240 V. From Ohm's law:

$$I = \frac{V}{R}$$

$$= \frac{240\,V}{500 \times 10^6\,\Omega}$$

$$= 0.48 \times 10^{-6}\,A$$

$$= 0.48\,\mu A$$

EXAMPLE 9.5

Compare the currents carried by an overhead line at 400 kV and 33 kV for the transmission of 10 MW of power for unity power factor, $\cos\phi = 1$.

For a three-phase system, using:

$$\text{watts} = \text{volts} \times \text{amps} \times \sqrt{3} \times \cos\phi$$

$$\text{current in amps} = \frac{\text{watts}}{\text{volts}} \times \frac{1}{\sqrt{3}} \times \frac{1}{\cos\phi}$$

For 400 kV:

$$\text{current} = \frac{10 \times 10^6 \ W}{400 \times 10^3 \ V} \times \frac{1}{\sqrt{3}} = 14.4 \ A$$

For 33 kV:

$$\text{current} = \frac{10 \times 10^6 \ W}{33 \times 10^3 \ V} \times \frac{1}{\sqrt{3}} = 175 \ A$$

This demonstrates the advantages of high-voltage transmission of electrical power as smaller cable sizes can be used for the long distances involved.

Cable capacity and voltage drop

The maximum current-carrying capacities and actual voltage drops for unenclosed copper cables thermoplastic sheathed cross-linked polyethylene (XLPE), or older PVC, clipped to the surface of the building, are given approximate values in Table 9.1 for use in this book's examples only. Flexible connections to appliances may use 0.5 mm² conductors for 3 A and 0.75 mm² conductors for 6 A loads. The maximum voltage drop allowed is 4% of the 240 V nominal supply.

EXAMPLE 9.6

Find the maximum lengths of 1 mm², 1.5 mm² and 2.5 mm² copper cable which can be used on a 240 V circuit to a 3 kW immersion heater.

$$\text{current} = \frac{3000 \ W}{240 \ V} = 12.5 \ A$$

$$\text{allowed voltage drop} = \frac{4}{100} \times 240 = 9.6 \ V$$

$$\text{maximum length or run} = \frac{\text{maximum voltage drop allowed mV}}{\text{load current A} \times \text{voltage drop mV/A m}}$$

Table 9.1 Electrical cable capacities.

Nominal cross-sectional area of conductor (mm²)	Maximum current rating (A)	Voltage drop in cable (mV/Am)
1	15	44
1.5	19.5	29
2.5	27	18
4	36	11
6	46	7.3
10	63	4.4
16	85	2.8

Note: Data is approximate and only to be used within the examples in this book.

For 1 mm² cable:

$$I = \frac{9.6 \times 10^3}{12.5 \times 44} \, m \ = 17.5 \, m$$

For 1.5 mm² cable:

$$I = \frac{9.6 \times 10^3}{12.5 \times 29} \, m \ = 26.5 \, m$$

For 2.5 mm² cable:

$$I = \frac{9.6 \times 10^3}{12.8 \times 18} \, m \ = 42.7 \, m$$

Construction site distribution

A list is made of all electrical equipment to be used on site in order to assess the maximum demand kilovolt-amperes and cable current rating required. An estimate of the cost of electricity for running the site may also be made.

EXAMPLE 9.7

A building site is to have the following electrical equipment available for use:

(a)   tower crane, electric motors totalling 75 kW at 415 V;
(b)   sump pump, 5 kW at 240 V;
(c)   60 tungsten lamps of 100 W each at 240 V;
(d)   12 flood lamps of 400 W each at 240 V;
(e)   20 hand tools of 400 W at 110 V.

1.   Find the total kilovolt-amperes to be supplied to the site if the power factor of all rotary equipment is 0.8.
2.   Find the electrical current rating for the incoming supply cable to the site.
3.   Estimate the cost of electricity consumed on the site during a 12-month contract.

For rotating machinery:

$$\text{power VA} = \frac{\text{useful power W}}{PF} = \frac{W}{0.8} \text{ and, kVA} = \frac{kW}{PF}$$

For single-phase current:

$$\text{line current} = \frac{VA}{V} A$$

$$= \frac{kVA}{kV} A$$

Table 9.2 Building site plant schedule.

Equipment	Power (kW)	Number	kW	kVA	V	A
Tower crane	75	1	75	93.75	415	130.4
Sump pump	5	1	5	6.25	240	26
Lamps	0.1	60	6	6	240	25
Flood lamps	0.4	12	4.8	4.8	240	20
Hand tools	0.4	20	8	19	110	90.9
Totals			98.8	120.8	n/a	n/a

For three-phase current:

$$\text{line current} = \frac{VA}{V \times \sqrt{3}} A$$

$$= \frac{kVA}{kV \times \sqrt{3}} A$$

Results of the power calculations are given in Table 9.2.
  The answers required are as follows.

1. The total input power kilovolt-amperes required for site is 120.8 kVA.
2. The incoming supply cable capacity at 415 volt, 3 phase, 50 Hz required is:

$$\text{Current} = \frac{120.8 \times 1000}{415 \times \sqrt{3}}$$

$$\text{Current} = 168\,A$$

This is the input current to the site at the voltage of that cable. This is not the same as the total of the currents calculated in Table 9.2 as these larger numbers only appear at their reduced voltages in the relevant sub-circuits.

3. Assume that the crane, pump and tools are running for 25% of an 8 h working day, 5 days per week for 48 weeks, 20 of the tungsten lamps are for security lighting 16 h every night, and the remaining 40 tungsten lamps and the flood lamps are used for 3 h per day, 5 days per week for the winter period of 20 weeks. The crane, pump and tools are working for:

$$0.25 \times 8\frac{h}{day} \times 5\frac{days}{week} \times 48\,weeks = 480\,h$$

The security lamps are working for:

$$16\frac{h}{day} \times 7\frac{days}{week} \times 52\,weeks = 5840\,h$$

The other lamps are working for:

$$3\frac{h}{day} \times 5\frac{days}{week} \times 20\,weeks = 300\,h$$

Table 9.3 Building site energy use.

Equipment	Power (kW)	Number	Time (h)	Energy (kWh)
Tower crane	75	1	480	36000
Sump pump	5	1	480	2400
Tungsten lamps, security	0.1	20	5824	11648
Lamps	0.1	40	300	1200
Flood lamps	0.4	12	300	1440
Tools	0.4	20	480	3840
Total power used				56528

The total energy used by the systems is found from:

kWh = number of appliances × kW per appliance × operation hours as shown in Table 9.3.

If electricity costs 8 p per unit (kWh) then the estimated cost for the 1-year contract will be:

$$cost = 8\frac{P}{kWh} \times 56528\,kWh \times \frac{£1}{100\,p}$$

$$= £4522$$

Construction site safety

Adequate safety in the use of electricity on site is essential and is a legal obligation upon employers. The area electricity supply authority must be contacted before any site work, to establish the locations of overhead and underground power cables. Assume that all lines are live. Overhead lines are not insulated except at their suspension points. Roadways for site vehicles should be made underneath overhead cables by the erection of clearly marked goalposts, on each side of the cable route, through which traffic must pass. These goalposts form the entrance and exit from the danger area and are spaced at 1.25 jib lengths of the mobile crane to be used on site, or at a minimum of 6 m either side of the cables. Entry to the roadway other than through the goalposts is barred with wooden fencing or tensioned ropes with red and white bunting at high and low levels.

Underground cables that become exposed during excavations must remain untouched until the electricity authority has given advice. Safe working clearances will be ascertained at this time. Hand lamps and tools are operated from a transformer at 110 V to reduce the damage caused by an electric shock. For work within tunnels, chimneys, tanks or drains, 25 V lamps, or battery lamps, are advised. Each portable appliance should be checked by the operator before use and also inspected and tested by a competent person at intervals not exceeding 7 days. Records of maintenance and safety checks should be kept.

A weatherproof cubicle is provided at the edge of the site by the main contractor for the electricity authority's temporary fuses and main switch. Site distribution cables are supported from hangers on an independent wire suspended between poles around the edge of the site, with spur branches to site accommodation and work areas. The minimum clear heights under cables should be 4.6 m in positions inaccessible to vehicles, 5.2 m in positions accessible to vehicles and 5.8 m across roads.

The site programme for the main contractor is as follows.

1. Arrange a pre-contract meeting between the executives responsible for the work.
2. List electrical requirements for all temporary plant.
3. Prepare layout drawings showing equipment siting and electrical loads, including site offices, stores, canteen, sanitary accommodation and illumination. Carefully site equipment to minimize interference with the construction work.
4. Apply to the electricity supply authority for a temporary supply to the site, stating maximum kilovolt-ampere demand and voltage and current requirements.
5. Provide the electrical distribution equipment: 415 V, three-phase, 50 Hz for fixed plant and movable plant fed by trailing cables; 240 V, single-phase, 50 Hz for site accommodation and site illumination; 110 V, single-phase, 50 Hz for portable lighting and tools; 50 V or 25 V, single-phase, 50 Hz for portable lamps to be used in confined spaces and damp areas.
6. Once site accommodation is in place, ensure that a satisfactory semi-permanent electrical system is provided.
7. A competent electrician is to carry out all site work. His name, designation and location are to be prominently displayed on site, so that faults, accidents and alterations can be expedited. All plant and cables are regularly inspected and tested.
8. Display the electricity regulations placard and the first aid instruction card.
9. Appraise the use of electrical equipment and distribution arrangements weekly to ensure that the most efficient use is made of the system. Idle equipment is returned to the supervised store.

Distribution equipment for site use is housed in weatherproof rugged steel boxes on skids. Built-in lifting lugs facilitate crane or manual transportation. The different equipment required is as follows:

- *Supply incoming unit (SIU)*: a unit to house the electricity supply authority's incoming cable, service fuses, neutral link, current transformers and meters. Outgoing circuits of 100 A or more are controlled by triple pole and neutral (TPN) switches for three phase and either cartridge fuses or residual current devices to break the circuits in the event of a current flowing to earth from a fault in a wire or item of plant.
- *Main distribution unit (MDU)*: a cubicle, which may be bolted to the SIU, providing a number of single- and three-phase outlets through weatherproof plugs and sockets. Each outlet is protected by a residual current device.
- *Transformer unit (TU)*: a unit providing 110 V single- and three-phase supplies with 65 V between any line and earth for safety in the use of hand tools. It may be coupled to an SIU. Each outlet circuit is protected by a residual current device.
- *Outlet unit (OU)*: easily portable distribution box fed from an MDU or TU for final connection of sub-circuits to tools, lighting or motors. Each circuit is protected by a residual current device and clearly labelled with its voltage, phases and maximum current capacity.
- *Extension outlet unit (EOU)*: similar to an OU but fed from a 16 A supply, and circuit protection is by a cartridge fuse. It may have up to four 16 A outlets on a cubicle metal box.
- *Earth monitor unit (EMU)*: flexible cables supplying electricity to movable plant; may incorporate a separate pilot conductor in addition to the protective conductor to earth. A small current passes through the portable plant and the EMU via the pilot and protective conductors. If the earth conductor is broken, the EMU current is interrupted and the circuit is automatically isolated at its circuit breaker.

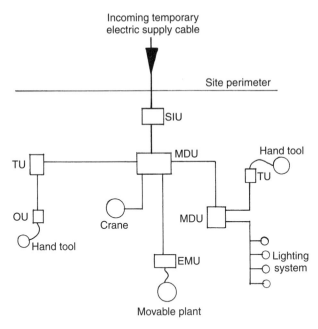

9.6 Distribution of electricity during site construction.

Semi-permanent installations are run in metal-sheathed or armoured flexible cable. The metal sheath is permanently earthed as is the earth wire. An over-sheath or an oil-resisting and flame-retardant compound is provided.

Connections from outlet units to hand-operated tools and lighting systems are made in tough rubber-sheathed (TRS) flexible cable. Walkways and ladders must be kept clear of cables and the cables must be kept 150 mm from piped services. Cables under site roadways are installed in a temporary service duct, such as drain pipes, at a depth of 600 mm and with markers at each end of the crossing. Figure 9.6 shows an arrangement of a site's electrical distribution.

Safety cut-outs

An electric shock is sustained when part of the human body establishes contact between a current-carrying conductor and earth. It is also likely from contact with two conductors of different phase. Voltages of less than 100 V have proved fatal under certain circumstances. The size of the current depends on the applied voltage and the body's resistance to earth. Rubber shoes or flooring greatly increase the resistance of the shock circuit. Body resistance with damp skin is around 1100 Ω at 240 V; thus a current of 218 mA could flow. At 55 V, body resistance is 1600 Ω and this could produce 34 mA. A current of 1–3 mA is generally not dangerous and can just be perceived. At 10–15 mA, acute discomfort and muscle spasm occur, making release from the conductor difficult. A current of 25–50 mA causes severe muscle spasm and heart fibrillation, and will probably be fatal. Prolonged exposure to a shock current causes burns from the heating effect of the supply. If electric shock occurs, switch off the supply without contacting any metal component. If necessary, begin resuscitation and summon qualified medical assistance immediately.

During normal operation, current flows from the 240 V (or other nominal value) line, through the appliance and along the neutral conductor back to the power station alternator. 240 V is the nominal drop of voltage across the appliance. Should either the line or the neutral conductors come into contact with a conducting material that is earthed, owing to a wiring fault, the current will choose the lower-resistance path to earth on its return journey to the earthed alternator. Immediately, a higher current will flow and the appliance has become a shock hazard. The increased heating effect of the fault current can be used to melt a rewirable or cartridge fuse at the appliance, its fault current being 60% above the stated continuous rating. High rupturing capacity (HRC) cartridge fuses have silver elements in a ceramic tube, which is packed with granulated silica. They allow for the high starting currents required for electric motors. The correct fuse rating must be used for each appliance to avoid damage to cables and buildings from overheating through the use of too high a fuse capacity. Fuse ratings are quoted in Table 9.4 and are found from:

$$\text{fuse current rating} = 1.6 \times \frac{\text{appliance input VA}}{\text{circuit voltage}}$$

A faster-acting protection, with greater reliability, whose operation can be tested, is provided by a miniature circuit-breaker (MCB), which opens switch contacts upon detection of an excess current. Circuit faults that cause a leakage to earth are detected by a residual current circuit-breaker (Figure 9.7).

During normal operation, the line and neutral coils around the electromagnetic core generate equal and opposite magnetic fluxes, which cancel each other out. A current leakage to earth at the appliance reduces the current in the neutral conductor and a residual current is generated in the core by the line coil. This residual current generates an e.m.f. and current in the detector

Table 9.4 Fuse ratings for 240 V single-phase and unity power factor.

Power consumption (W)	Fuse required (A)
120	0.8
240	1.6
720	4.8
1200	8
3120	20.8
3600	24
7200	48

Note: Data is approximate and only to be used within the examples in this book.

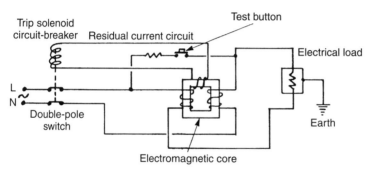

9.7 Residual current circuit breaker.

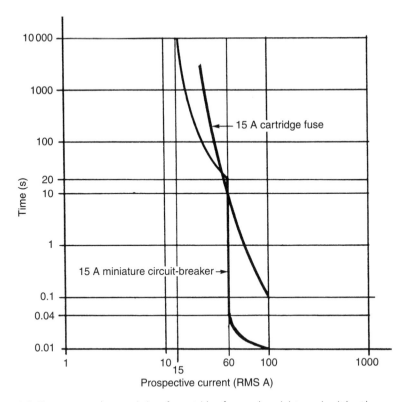

9.8 Time-current characteristics of a cartridge fuse and a miniature circuit breaker.

circuit, which in turn energizes the trip solenoid, which opens the double-pole switch, or TPN switch in a three-phase circuit, and isolates the appliance. Residual currents of 30 mA are set for sensitive applications and outdoor equipment. They are frequently used in addition to fuses. Pressing the test switch short-circuits the line coil and a residual current is generated in the core, tripping the residual current device for test purposes.

Fuses and circuit-breakers are selected for their time-current characteristics in relation to the risk that is being protected against. Figure 9.8 shows the performance curve for a cartridge fuse having a 15 A continuous rating for domestic installations and a comparative miniature circuit-breaker The horizontal and vertical scales of the graph are logarithmic. It can be seen that both devices will pass the design maximum 15 A current without opening the circuit. The cartridge fuse is designed to melt if a current of 46 A were to flow for a period of 5 s, or 97 A for 0.1 s. Other combinations of heating effect are in proportion. These points lie to the left of the fuse curve and do not cause it to break.

The MCB is designed to open the circuit when 60 A flows for between 0.1 and 5 s. A lower current value will take in excess of 20 s to open the contacts. A fault in the protected circuit that causes the current to rise above 60 A will open the MCB in less than 0.1 s. A miniature circuit-breaker and a residual current device (RCD) may be combined in one moulded casing to protect against excess current, short circuit and a leakage to earth. The MCB has a bimetallic strip thermal and magnetic trip mechanism. The speed of operation of an RCD is typically 20 ms.

Electrical distribution within a building

The incoming cable, residual current device and meter are the property of the electricity supply authority. Underground cables are at a depth of 760 mm under roads, and enter the building through a large radius service duct of 100 mm internal diameter. A drain pipe can be used for this purpose, laid through the foundations and rising directly to the meter compartment. External meter compartments can be used. The meter should not be exposed to damp or hot conditions and the electricity supply authority's advice should be sought. Figure 9.9 shows a distribution system for a dwelling.

Each circuit has a fuse or circuit-breaker and the fused distribution board connects the neutral and earth protective conductors to the supply cable. Appliances have a cartridge fuse at their connecting plug. Three-phase distribution in a large building is shown in Figure 9.10. Switches are used to enable separation of individual circuits as well as appliances. A fuse or circuit-breaker is always fitted on the live line so that the incoming current is disconnected.

Power socket outlets are fitted into a ring circuit as shown in Figure 9.11. Care must be taken not to overload the circuit by connecting appliances whose total current consumption would exceed the 30 A limit, particularly in kitchens.

Types of cable used in distribution systems are as follows.

* *Thermo plastic sheathed (TPS)*: TPS, XLPE insulated and sheathed cables consist of copper conductors of multi-stranded or solid wire having sizes from 1 mm^2 to 16 mm^2 cross-sectional area. Single-, twin- or three-core cables, with or without earth wires, are used. They are among the cheapest cables available and can be pulled through conduits or holes bored in floor joists. Such holes are drilled 50 mm below the floorboards. Ambient temperature limits for the cable are 0–65°C and the cable must not exceed 70°C in use. Colour coding of the

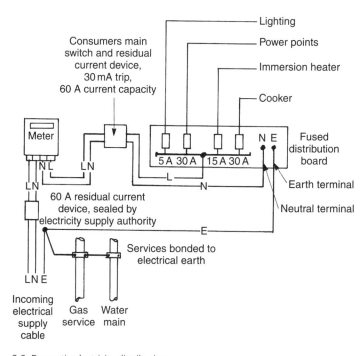

9.9 Domestic electricity distribution.

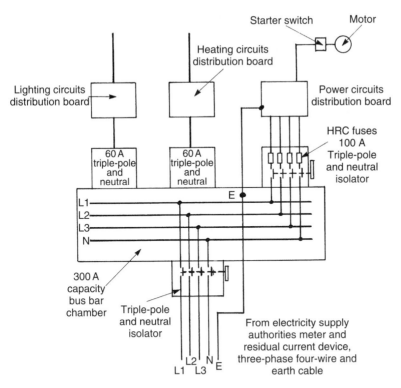

9.10 Three-phase electricity distribution.

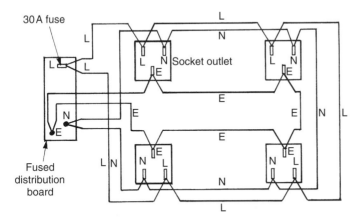

9.11 Ring main to socket outlets.

insulation ensures correct polarity at terminals. The earth, or protective conductor, is always green, neutral is always black and the line conductor is red on single phase. Three-phase line conductors are one red, one yellow and one blue. TPS is a flexible, non-hygroscopic, tough, durable, corrosion-resistant and chemically inert material, which is used for both electrical insulation and conduits. Installations are tested for earth leakage through their insulation at regular intervals and cables are replaced after about 20 years because of ageing of the material.

- *Mineral insulated copper conduit (MICC)*: Solid copper conductors surrounded by compressed magnesium powder are factory-fitted inside a copper tube or sheath. A TPS over a sheath gives extra protection for cables that are to be buried in building materials. These cables are used for both internal and external wiring and withstand severe conditions, even continuing to operate during a fire. They can be operated continuously at temperatures of up to 250°C, compared with only 70°C for *TPS* covered cables. The soft copper cable is supplied in rolls and is run continuously from the distribution board to the switch or power point. Screwed gland joints are designed to exclude dampness from the hygroscopic insulant. One sheath may encapsulate up to 19 1.5 mm^2 conductors. They are non-ageing and unlikely to require replacement during the building's period of service. Particular applications are fire alarm systems, in petrol filling stations and within boiler plant rooms. The copper sheath is used as the earth protective conductor and also withstands severe abuse, flattening or twisting without short-circuiting the conductors. The cables can be bent by hand or machine, and conduit fittings are not needed. The overall diameters of mineral-insulated cables are much smaller than those of other types of comparably rated cable system. Only a thin plaster covering is needed if the cables are not to be surface mounted.
- *Armoured TPS insulated and sheathed*: Copper or aluminium conductors in TPS insulation, a TPS bedding, galvanized steel wire armour and a TPS sheath are used for heavy-current cables to large machinery and mobile plant on site. The cable can be run on the building surface, laid on the ground or put in a trench. Screwed gland nuts are used to bond the armour to the appliance casing.

Busbar

Bare copper or aluminium rectangular bar conductors are supported on insulators within a sheet metal duct or conduit. Vertical service shafts within buildings may have a rising busbar system with tap-off points at each floor level for the horizontal distribution of power with insulated cables. Small busbar distribution systems can be used in retail premises and within raised floors in office buildings. These provide flexibility in the siting of outlet boxes. Overhead distribution in factories allows connections at any point along the busbar with plug-in boxes, allowing machinery to be moved at a later date. Conduit acts as the protective conductor. Three-phase 415 V supplies are distributed at current ratings of 100–600 A, branches being at 30, 60 or 100 A. An armoured TPS cable can supply the incoming end of the busbar. Where the system passes through a fire-resistant partition in the building, a fire barrier is fitted across the inside of the conduit.

Conduit

Conduit and trunking systems are used to carry insulated cables and should last the service period of the building. The space occupied by the cable must not exceed 40% of the cross-sectional area of the inside of the conduit to allow for ventilation to remove the heat generated by cable resistance. Materials used are light- or heavy-gauge steel, plastics or flexible metal depending upon exposure to damp or explosive fumes. The external conduit will be galvanized. Lug grip connections are used for light-gauge pipe but screwed joints for heavy-gauge pipe. Pipe sizes are 16, 20, 25 and 32 mm. The conduits are used as the protective conductor. TPS conduit using solvent weld joints is lighter and easier to handle and does not corrode but requires the cable to incorporate the protective conductor. Its upper temperature limit is 60°C.

Rectangular galvanized sheet steel or TPS trunking is used where large cable-carrying capacity is needed. These must not be filled to greater than 45% of their cross-sectional area with cables. Open cable trays are used within service shafts and tunnels to allow heat release and easy access. Surface-mounted conduit can be incorporated into the interior decoration and up to three separate cable compartments are used for different services, including telecommunications, computer, power and lighting cables. Conduit may be installed under raised timber flooring, within the concrete floor slab or screed, in a grid, branch duct or perimeter distribution arrangement. Outlets that are raised or flush with the floor are provided to suit either fixed or movable office layouts.

Testing

Inspection and testing of an electrical installation are carried out before it is put into service and at regular intervals during use. The main reasons for this are to ensure that its operation will be entirely safe, in accordance with the demands put upon it, and energy efficient. The work entails the following tests.

Power measurement

Power consumed by a single- or three-phase electrical system or item of plant, such as a refrigeration compressor, fan or pump motor, a lighting or power sub-circuit, is measured with a portable instrument, three openable jaw cable clamps for a three-phase measurement and a voltage measurement clamp. A single-phase cable has its line conductor temporarily separated from the neutral and earth cables for measurement with one clamp. Newly installed plant is checked for equal phase currents, voltage and power consumed during commissioning. The energy auditor measures these some years after installation, as commissioning data may no longer be relevant or plant may have been changed. The logger should also measure power factor so an assessment can be made of whether to install power factor capacitors to reduce incoming current. Each phase line conductor has one of the openable clamps fitted. Voltage difference across the circuit is measured with additional cables and attachment clips. Both useful power in kW and apparent power in kVA are measured. All work on live electrical wiring is only done by a registered electrician.

Verification of correct polarity

A visual inspection is carried out of all fuses and switches to check that they are fitted into a line conductor. The centre contact of each screw lamp holder is connected to the line conductor. Plugs and sockets must be correctly connected and wire rigidly held.

Tests of effective earth

There are four separate tests:

1. *Test of the protective conductor*: 40 V 50 Hz supply of up to 25 A is injected into the earth conductor. Its resistance is not to exceed 1 Ω. An impedance test meter is used.
2. Earth loop impedance test: a line-earth loop impedance test meter is attached to a 13 A three-pin plug. This is plugged into each power socket and the meter injects a current into the earth protective conductor. The current flows along the supply authority's cable sheath to the local transformer and back to the power socket along the line conductor.
3. *Test of residual current devices*: a test transformer providing 45 V is connected to a socket outlet. A short-circuit current is passed from the neutral to the protective conductor, causing the residual current device to trip instantaneously.

4. Measurement of consumer earth electrode resistance: where this is used, a test electrode is put into the ground and a steady 50 Hz current is passed between the electrode and the consumer's earth electrode to determine its circuit resistance.

Further tests are:

1. *Insulation resistance tests*: An insulation test meter is connected between the line and protective conductors. A 500 V direct current is applied to this circuit by the meter and an electrical resistance of 0.5 M Ω or more must be shown.
2. *Test of ring circuit continuity*: Each ring circuit is tested for resistance at the distribution board with an ohmmeter. Probes are connected to each side of the line conductor ring and a zero resistance proves a continuous circuit. The test is repeated on the neutral and protective conductor circuits and spur branches.

Tests on installations must be carried out in accordance with the appropriate wiring regulations by a competent person, who is a professionally qualified electrical technician having installation experience, and a formal completion and inspection certificate is issued.

Telecommunications and networks

Cables between the switchboard and socket for each telephone are accommodated within vertical and horizontal service ducts spaced 50 mm from alternating current cables to avoid speech interference. Alternatively, a partitioned chamber can be reserved throughout the cable conduit. Wireless networks are increasingly taking over the functions of cable systems in buildings with small devices and BMS controller sensors having their own TCP/IP internet address. A building may have an underground or overhead fibre optic cable connection to a service provider system in the street. Large entities can have a microwave communications dish on the roof for satellite or line-of-sight cross-country transmissions, with fenced enclosures around them to avoid radiation damage to personnel. Fibre optic cabling connects the service or dish to a server computer owned by the system user. A server can be anything from a small PC on a desk up to a large room full of computer racks. Each business leasing space in a large building has their own server within their rented floors. Heat output from large server computers can be considerable and requires 24-hour cooling from dedicated refrigerated packaged units. Server activity is continuous and not necessarily related to office hours of the users. Ethernet fibre optic cable networks may extend throughout the building from a router switch box providing a Local Area Network (LAN) for all wired communications. One or more routers provide Wireless Area Network (WAN) within the building probably using the *IEEE 802.11 WLAN* Standard operating radio waves at around 2.4 GHz frequency with transfer rates ranging from 2–54 Mbits/s over a limited distance of up to 70 m. Data transfer speed deceases with distance from the router. Wireless Personal Area Networks (WPAN) connect devices within an area usually within physical reach of the user such as infrared controllers and Bluetooth radio devices. We are also covered by the wireless Wide Area Networks (WAN), for mobile telecommunication devices from local region base transmitter/receivers wherever we are. We are now the full communication people who could hardly have been imagined by our grandparent's generation.

Lightning conductors

A protection system against lightning is required depending on building construction, degree of isolation, height of the structure, topography, consequential effects and lightning prevalence, also for temporary structures and buildings under construction. A copper and aluminium 10 mm rod, 25 mm × 3 mm strip, *TPS*-covered strip, copper strand and copper braid are used for conductors. The air terminal is above the highest point of the structure. A down conductor is bolted to the outside of the building so that side flashing between the lightning conductor and other metalwork will not occur.

Ground termination is with a series of earth rods driven to depths of up to 5 m, cast iron or copper plates 1 m square horizontally or vertically oriented 600 mm below ground, or a copper lattice of flat strips 3 m × 3 m at a depth of 600 mm. Where large floor areas containing earth rods are to be concreted, a precast concrete inspection pit is built over the rod location. Electrical resistance to earth of the whole system is not to exceed 10 Ω Calculation of the ground earth resistance requires a knowledge of the earth type and resistivity. Typical values of earth resistivity are 10 Ω m for clay, 50 Ω m for chalk, 100 Ω m for clay shale and 1000 Ω for slate shale. The resistance of a rod electrode in earth is:

$$R = \frac{0.37\rho}{l} \log\left(\frac{4000l}{d}\right) \ \Omega$$

where l is the earth rod length (m), d is the earth rod diameter (mm) and ρ is the resistivity of the soil (Ω m). A number of rods are connected in parallel and spaced 3.5 m apart to provide the required resistance.

EXAMPLE 9.8

Design a lightning conductor system for a building 30 m high in an area where thunderstorms are expected. The ground has high chalk content and rod electrodes 4 m long are to be used. The conductors are to be 25 mm × 3 mm copper strip. The specific resistance ρ of copper is 0.0172 $\mu\Omega$ m.

Length of conductor is air terminal, down conductor and ground lead, 40 m.

$$\text{resistance of conductor } R = \rho\frac{l}{A} \ \Omega$$

Where A is the conductor cross-sectional area (m^2).

$$R = 0.0172 \times 10^{-6} \ \Omega m \times \frac{40\,\text{m}}{(0.025 \times 0.003)\,\text{m}^2}$$

$$= 0.092 \ \Omega$$

$$\text{resistance } R \text{ of earth electrode} = \frac{0.37\rho}{l} \log\left(\frac{4000l}{d}\right) \ \Omega$$

where the earth resistivity ρ is 50 Ω m, the electrode length l is 4 m and the electrode diameter d is 10 mm; hence:

$$R = \frac{0.37 \times 50\Omega\,m}{4\,m} \times \log\left(\frac{4000 \times 4}{10}\right) \ \Omega$$

$$= 4.625 \times \log 1600$$

$$= 14.819\Omega$$

The resistance of one electrode in the ground plus the down conductor is greater than the 10 Ω allowed, and so we find the combined resistance of two electrodes connected in parallel.

$$\frac{1}{R} = \frac{1}{R_1} + \frac{1}{R_2} = \frac{1}{14.8} + \frac{1}{14.8} = 0.135$$

$$R = \frac{1}{0.135} = 7.4\,\Omega$$

Two electrodes and the down conductor connected in series have a total resistance of:

$$R = (7.4 + 0.0092)\Omega = 7.4\,\Omega$$

This is less than the 10 Ω allowed and is satisfactory. The resistance of the lightning conductor is negligible in relation to that of the earth electrodes. The calculations have been made on the assumption that the lightning discharges in a direct current. Lightning energy can produce a current to earth of 20000 A for a few milliseconds. It causes physical damage to building structures, starts fires in combustible materials and injury to people, sometimes fatal.

Questions

1. Explain how electricity is generated and transmitted to the final user.
2. List the sources of energy used for the generation of electricity and state their immediate and long-term benefits.
3. Explain, with the aid of sketches, the meaning of the terms 'single-phase' and 'three-phase' electricity supplies and show how they are used within buildings.
4. What does 'balancing the phases' mean?
5. Calculate the electrical resistance per metre length at 20°C of a copper conductor of a 10 mm^2 cross-sectional area.
6. Find the electrical resistance of a copper conductor 1.5 mm^2 in cross-sectional area if its total length is 25 m and its temperature is 20°C.
7. 28 m of copper conductor 4 mm^2 in cross-sectional area is covered with thermal insulation, which causes the cable temperature to rise to 45°C. Calculate the percentage increase in electrical resistance compared with its value at a cable temperature of 20°C.
8. Sketch the methods of connection used for measurements of current, voltage drop and power consumption in an electrical resistance heater on an alternating current circuit.
9. State the function of power factor correction in alternating current circuits.
10. Calculate the apparent power, in kilovolt-amperes, of an electric motor which is connected to a 415 V AC three-phase supply and has a current flow of 17.5 A.
11. Calculate the resistance of a 3 kW immersion heater on a 240 V AC circuit.
12. What current, in milliamperes, would flow to earth during an insulation resistance test when a 500 V DC e.m.f. is applied between the line and protective conductors and the resistance is found to be 1.75 MΩ?
13. Show by sample calculation why smaller cables can be used for long-distance power transmission when very high voltages are used.
14. A 33 kV supply to a factory carries 250 A per phase or line. Calculate the usable electrical power in the factory if the average power factor is 0.68.
15. Find the maximum length of 6 mm^2 cable that can be used if the maximum current-carrying capacity is to be utilized on a 240 V circuit.

16. A building site is to have the following electrical equipment in use each day:

    1. concrete mixer, 5 kW, 4 h, 415 V;
    2. sump pump, 1.5 kW, 6 h, 240 V;
    3. 20 lamps, 150 W each, 4 h, 240 V;
    4. 5 flood lamps, 300 W each, 4 h, 240 V;
    5. 6 hand tools, 750 W each, 5 h, 110 V.

    The power factor of the machinery is 0.8. Site work takes place 5 days per week for 28 weeks. Electricity costs 6p per unit. Find:

    1. total kilovolt-amperes and the line current of the required temporary incoming supply system;
    2. cost of the electricity used on site during the contract.

17. Sketch and describe the safety precautions taken to avoid contact with both overhead and underground electricity cables during site construction work.
18. Sketch a suitable arrangement of temporary wiring, control and safety equipment on a site where the following items are employed: tower crane, sump pump, five flood lamps, security lighting and circuits on each of three floors for hand lamps and tools.
19. List the site programme for the main contractor in the installation and operation of temporary site electrical services.
20. Sketch and describe the characteristics of rewirable and cartridge fuses and residual current devices.
21. Show how an underground electrical service cable enters a building. Sketch the arrangement of electricity distribution within a typical residence.
22. List the cable and conduit systems used for electricity distribution and state their applications.
23. Briefly describe the methods of testing electrical installations.
24. State the requirements of telecommunications installations.
25. Sketch and describe a lightning conductor installation for a city centre office block.
26. Design a lightning conductor system for a 60 m-high building on clay shale using 4.5 m earth rods. The down conductor is to be a copper rod 10 mm in diameter.
27. Why is power factor in electrical systems an issue for concern? More than one correct answer is possible.

    1. It is not of any concern.
    2. Low power factor means electrical energy is used inefficiently.
    3. High power factor means electrical energy is wasted.
    4. 100% power factor is not usually attainable at a justifiable cost.
    5. Low power factor means electrical supply system becomes oversized.

28. What is a low power factor?

    1. Zero.
    2. 0.95.
    3. 0.85.
    4. 1.0.
    5. 0.8 and below.

29. Which is the correct formula for power factor?

    1. $PF = \dfrac{kVA}{kVAh}$

    2. $PF = \dfrac{kVAh}{kWh}$

3. $PF = \dfrac{kW}{kVA}$

4. $PF = \dfrac{input\ energy}{kilovoltampere}$

5. $kW = \dfrac{PF}{kVA}$

30. How does a 30 mA residual current circuit breaker work? More than one correct answer is possible.

    1. Line and neutral conductors wrap around an electromagnetic core and no current flows through the core as magnetic flux from each wire cancels the other.
    2. Line and neutral conductors wrap around an electromagnetic core, current flows through the core due to magnetic flux from each wire.
    3. Line and neutral conductors wrap around an electromagnetic core, current flows through the core due to magnetic flux induced from line conductor.
    4. Line and neutral conductors wrap around an electromagnetic core, magnetic flux circulates harmlessly through the core due to the alternating current line and neutral conductors.
    5. Current leakage to earth from the protected appliance loses neutral current, causing imbalance between line and neutral magnetic fluxes in core; imbalance flux generates 30 mA in trip solenoid circuit breaker.

31. How can cartridge fuses cope with high starting currents at electric motors?

    1. They cannot.
    2. Only micro-circuit breakers are used to protect motors driving compressors.
    3. High rupturing capacity cartridge fuses regularly pass 500% of normal running current.
    4. High rupturing capacity cartridge fuses have silver elements and packed with silica to allow high starting currents for a known duration.
    5. High rupturing capacity cartridge fuses have bimetallic elements and are packed with carbon granules to allow high starting currents.

32. Which are correct about mineral-insulated copper-sheathed electrical cables? More than one correct answer is possible.

    1. Fragile and easily damaged.
    2. Withstand most fires and remain operational.
    3. Supplied in hard copper fixed lengths, like plumbing pipes.
    4. Screwed gland joints exclude water from the hygroscopic insulant.
    5. Cannot be installed outdoors.

33. What is meant by busbar? More than one correct answer is possible.

    1. Communications bus, C-bus, in a computer.
    2. Ethernet communications cabling system around a large network.
    3. TPS-insulated circular bar conductors at high level through an industrial manufacturing plant.
    4. Bare copper or aluminium bar conductors carried on insulators within sheet metal conduit.
    5. Up to 600 ampere three-phase vertical or horizontal distribution allowing off-takes along length.

34. What is the meaning of inverter drive?

    1. An electric motor installed in an inverted position.
    2. Three-phase electric motor.

3. Three-phase motor running in single phase.
4. Digitally driven motor.
5. Motor driven at variable alternating current frequencies.

35. Which is correct about electrical systems?

1. Watts = Volts × Amps.
2. Three-phase current = 3 × single phase current.
3. Watts and Volt-Amperes are always the same value.
4. Current = Voltage × Resistance.
5. Current measured in Mega-Ohms.

36. How is power consumed by an electrical item measured?

1. Wattmeter cut into the line conductor.
2. Voltmeter and current meter both cut into line conductor to the load.
3. Ammeter connected into the line conductor to the load and a voltmeter connected in parallel with the load.
4. Voltmeter connected into line conductor to the load and an ammeter connected in parallel with the load.
5. Wattmeter connected in parallel with the load.

37. Which system of public electricity is provided to small to medium-sized non-domestic buildings in the UK and Australia?

1. 50 Hz, 240 volt, AC.
2. 210 volt, 1 phase, 55 Hz, alternating current.
3. 2 phase, 60 Hz, 200 volt.
4. 440 volt, 3 phase, 60 Hz.
5. AC, 50 Hz, 3 phase, 415 volt.

38. How is three-phase electricity created at a power station?

1. Three single-phase alternators have interconnected output power circuits.
2. Single-phase generators with rectifiers creating three phases.
3. Multiple transformers create three phases.
4. Each alternator has three stator coils so one revolution of the rotor generates three separate sine wave outputs.
5. Three single-phase alternators each power one line voltage; each building takes all three lines to have three-phase power.

39. Which of these does single-phase electricity current look like on an oscilloscope screen? (Hint, draw them.)

1. Zero phase angle zero current, 90° phase angle maximum positive current, 180° phase angle zero current.
2. Zero phase angle maximum positive current, 90° phase angle zero current, 180° phase angle maximum negative current.
3. Zero phase angle 50% maximum positive current, 90° phase angle 100% positive current, 180° phase angle zero current.
4. 135° phase angle zero current, 225° phase angle maximum negative current, 315° phase angle zero current.
5. Zero phase angle zero maximum negative current, 90° phase angle zero current, 180° phase angle maximum positive current.

40. Which correctly describes the relationship between voltage and current in an alternating current system?

1. Voltage and amperes are synchronized.
2. Voltage peak occurs behind peak value of current.
3. Current always follows voltage producing it by exactly one phase.
4. Inefficient generators produce a current flow lagging voltage.
5. Current always follows fractionally behind the voltage that produces flow of electricity.

41. What is the advantage of a three-phase system?

1. Widely variable phase current meets variable demands.
2. Continuously stable power supply.
3. More easily generated.
4. Can generate at any desired frequency.
5. Quieter than single phase.

42. How should the services in a building take power from a three-phase supply?

1. Each phase serves a different part of the building.
2. The mechanical services distribution board always takes all its power from the yellow phase.
3. Single-phase circuits take current from each phase.
4. Equal current taken from each phase.
5. 240 volt circuits for lighting and small power equipment each connect to all three phases.

43. Where are cartridge fuses or MCBs always installed?

1. Live phase cable.
2. Neutral wire.
3. Earth conductor.
4. External to the building.
5. Within a fire-resistant switchboard.

44. Which is true about three-phase motors?

1. Run hot.
2. Need built-in cooling fan.
3. Generate more noise than an equivalent single-phase motor.
4. Quieter than single-phase motors.
5. Provides more power output for same line current as a single-phase motor.

45. What opposes flow of electrical current into an electric motor?

1. Resistance of the motor coils.
2. Inductance.
3. Temperature coefficient of resistance increases circuit electrical resistance.
4. Mechanical feedback from forces on motor output shaft.
5. Capacitance of motor control system.

46. How is electrical energy consumed by an item of plant, equipment or a whole building, measured?

1. Magnetic field-sensing data logger is strapped to the single- or three-phase cable.
2. Kilowatt meter measures instantaneous current flow and applied voltage.
3. Integrating data logger multiplies output signal from a magnetic current transducer with voltage applied at the same time.
4. Ammeter reading multiplied by voltmeter reading divided by the time in seconds of their duration is calculated and added to a running total of energy consumed.
5. Integrating meter multiplies instantaneous current and voltage with the time duration and records kWh consumed.

47. What does self-induced electromotive force do to an electrical circuit?

    1. Nothing.
    2. Speeds up current flow.
    3. Opposes incoming current and causes it to lag behind voltage in real time.
    4. Opposes incoming current and causes it to appear to be leading the cyclic pattern of the driving voltage frequency.
    5. Supports the incoming current frequency and increases the current.

48. What does inductance do to an electrical system?

    1. Speeds up current.
    2. Multiplies available power by a percentage.
    3. Reduces current.
    4. Reduces available voltage.
    5. Causes current to lag behind applied voltage.

49. What is the time difference between voltage and current in an AC system called?

    1. Lead angle.
    2. Microsecond gap.
    3. Phase.
    4. Peak difference.
    5. Lag.

50. How is electrical power factor raised?

    1. Cannot be improved after equipment installation.
    2. Replace with more efficient specification motor.
    3. Install digitally controlled AC/DC rectifiers at mechanical switchboard.
    4. Install capacitor banks in parallel with each plant item.
    5. Renegotiate electrical supply contract.

51. Which is a typical time interval for a residual current device to open a 60 amp circuit breaker double pole switch when a fault current occurs?

    1. 6 minutes.
    2. 1 minute.
    3. 1 second.
    4. 20 milliseconds.
    5. Less than 0.001 seconds.

52. What does RCD stand for?

    1. Residual circuit device.
    2. Residual current device.
    3. Resistance circuit design.
    4. Radio carbon dating.
    5. Ratio circuit device.

53. Which is the most common form of mortality from electric shock?

    1. Burns.
    2. Ventricular fibrillation.
    3. Muscle spasm.
    4. Pain.
    5. Bleeding.

54. Which is not a normal application for use of MICC wiring?

    1. Fire alarms systems.
    2. Public buildings such as theatres.
    3. Tunnels.
    4. Temporary buildings.
    5. Power stations and heavy industrial buildings.

55. Why are cables installed within fixed conduit?

    1. Hides ugly cables.
    2. Conduit is a permanent fixture of the building while cables require replacement when aged.
    3. Conduit becomes earth continuity conductor.
    4. Reduce heat emission from cables.
    5. Protects PCV cables from heat gain from environment.

56. State what is meant by the term Wi-Fi, who owns it and how it is used.
57. Explain what is meant by the term internet.
58. State the levels of wireless communication technology ranging from personal items such as those only used by the wearer, household, computer-to-computer, commercial building, campus, city and global. Which systems use each level? How do they affect buildings and their services? How do you envisage future advances in these technologies?
59. What are the known risks from using TCP/IP cable and wireless communication systems? Will wireless communication be inherently safer or more secure than hard-wired systems? How is data security maintained? Give examples to illustrate your answers.
60. Explain what is meant by on-site generation. Discuss its merits with practical examples. Is it financially worthwhile for homes, large commercial or industrial sites; stating your reasons? Is it only really needed for emergency use? Give examples of how it is used in emergencies.

10 Lighting

Learning objectives

Study of this chapter will enable the reader to:

1. explain the use of day and artificial lighting;
2. use lux and lumen;
3. state normal lighting levels;
4. explain general and task illumination;
5. understand permanent supplementary artificial lighting of interiors (PSALI);
6. understand artificial lighting terminology;
7. assess the importance of the maintenance of lighting installations;
8. calculate the room index;
9. discuss the problem of glare;
10. calculate the number of lamps needed to achieve a design illumination level;
11. calculate lamp spacing for overall design;
12. calculate the electrical loading produced by the lighting system;
13. understand the use of air-handling luminaires;
14. understand lamp colour-rendering and colour temperature;
15. know the range of available lamp types;
16. understand the working principles of lamp types and their starting procedures.

Key terms and concepts

artificial and natural illumination 230; colour-rendering index 238; daylight factor 231; efficacy 239; glare 230; heat generation 237; illuminance 231; lamp types 239; lumen 234; luminaire 237; luminance factor 234; lux 231; maintenance 234; observed illumination pattern 233; power consumption 240; room index 235; spacing to height ratio 236; starting arrangements 239; task illumination 231; utilization factor 235; working plane 236.

Introduction

Artificial illumination for architectural, functional and decorative purposes is a major consumer of primary energy. Developed civilizations have become used to very high illumination standards with high electricity consumption. It might be claimed that artificial illumination allowed the modern world to develop technology; try studying, writing and building intricate components at night without it and compare with what went before electric lighting. It may also be offered that artificial illumination is a major direct and indirect cause of global warming. Use of natural daylight is encouraged to reduce fuel consumption for lighting but this occurs at the expense of heating and cooling energy consumption at the building outer envelope which is in contact with the external environment. A compromise solution is inevitable, and the building services engineer is at the centre of the calculations needed to minimize total energy consumption for all usages. The factors involved in determining illumination requirements are discussed in relation to lighting levels for various tasks and the possible use of daylight. Lighting terms are introduced as are glare considerations. The lumen design method is demonstrated for office accommodation. Lamp colour rendering is discussed, and the use of luminaires with air-conditioning systems. Lamp types, their uses and control arrangements are explained.

Natural and artificial illumination

Natural illumination by penetration of direct solar and diffuse sky visible radiation requires correctly designed passive architecture. Large glazed areas may provide sufficient day lighting at some distance into the building but can also cause glare, overheating and high heating and cooling energy costs. The building style revolution of the 1960s brought large wide areas of plane glazing into modern homes and office developments. It was a fashion change and not one brought about by a desire to improve energy efficiency. The other extreme of vertical narrow slot windows limits energy flows while causing very unequal lighting levels near the room's perimeter. Reflected illumination from other buildings, particularly from those having reflective glazing or metallic architectural features, may cause annoyance. A careful consideration of all the largely conflicting variable elements is necessary if a comfortable internal environment is to be produced.

Artificial lighting is provided to supplement daylight on a temporary or permanent basis. Local control of lights by manual and/or automatic dimming and switching aids economy in electricity consumption. The colours rendered by objects on the working plane should match the colours under daylight. The working plane may be a desk, drawing board or display area.

Illumination intensity, illuminance, measured in lux on the working plane, is determined by the size of detail to be discerned, the contrast of the detail with its background, the accuracy and speed with which the task must be performed, the age of the worker, the type of space within which the task is to be performed and the length of time continuously spent on the task. The working plane is the surface being illuminated. Other areas are lit by overspill from it and by reflections from other room surfaces. Table 10.1 gives some typical values for illuminance commonly encountered and used for design.

Higher levels of illuminance may be provided for particularly fine detail tasks at the area of use by local, or task, illumination: for example, up to 3000 lx for inspection of small electronic components and 50000 lux on a hospital operating table. Bright sunlight provides up to 100000 lux. Local spot lighting for display purposes and exterior illumination are used to accentuate particular features of the working plane.

Permanent supplementary artificial lighting of interiors became common in office accommodation, shops and public buildings during the 1960s. Figure 10.1 shows the constituents of the

Table 10.1 Typical values of illuminance.

Application	Illuminance (lux)
Emergency lighting	0.2
Suburban street lighting	5
Dwelling	50–150
Corridors	100
Rough tasks with large-detail, storerooms	200
General offices, retail shops	400
Drawing office	600
Prolonged task with small detail	900

Note: Data is approximate and only to be used with examples in this book.

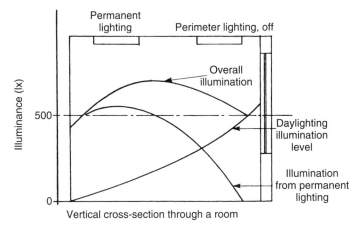

10.1 Permanent supplementary artificial lighting.

overall design illumination. Localized task illumination is provided in addition to a background lighting scheme but may not necessarily produce a reduction in total power consumption. Combinations of overall office lighting can be provided from ceiling level white fluorescent lamps having light diffusers, plus indirect background lighting.

Heat generated by permanent lighting can be extracted from the light fitting, luminaires, by passing ventilation extract air through it, raising the air temperature to 30–35°C, then supplying this heated air to perimeter rooms in winter. Further air heating with finned-tube banks and automatic temperature control would be part of normal ventilation or air-conditioning system. This can be termed a heat reclaim system, incorporating regenerative heat transfer between the outgoing warm exhaust air and incoming cold fresh air.

Penetration of daylight into a building can be enhanced with north-facing roof-lights, skylights with motorized louvres which are adjusted to suit the sun's position and weather conditions, or mirrored reflectors which direct light rays horizontally into the building. An example is shown in Figure 10.2.

Desk-mounted lamps produce a range of up to 9:1 in illuminance values across the working surface and form strong contrasts between surrounding and working surfaces which may result in discomfort glare. Direct dazzle from unshaded lamps, reflected glare, shadows around objects and hands, and heat radiation can cause discomfort.

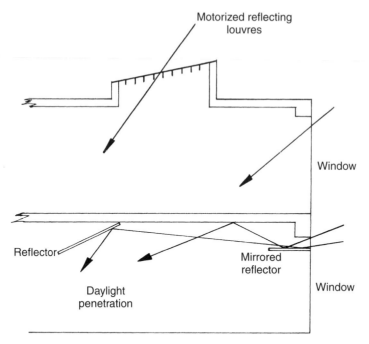

10.2 Use of daylight reflectors.

Lighting is a visual subject and it may be treated as such for design purposes. Light distribution follows the laws of geometry and is modelled in software applications, but it is not necessarily helpful to study lighting only as applied mathematics. Designs are an artistic and engineering combination of architecture, interior design, decoration, illumination functionality, and economy of use of electrical energy, maintainability, safety, environmental health, controllability, prestige, and the overall and specific requirements of the user. Architects and engineers coordinate their skills to create an acceptable visual and technical solution. To consider the design of lighting from a visual standpoint, either use computer simulation software, or, more simply, find two or more cardboard boxes. These boxes are to be used to represent rooms. Dimensions are unimportant as the principles of lighting apply to volumes of any size. They can be used to develop answers to questions 15–21 at the end of this chapter. This may be seen as a crude method of higher education; however, the use of scale models is a well-established artistic and scientific discipline. Computer-generated illumination plots are available for a similar design purpose but at much higher cost. The interior of the shoebox, or similar, can be covered or painted with dark or light colours. The source of light for the interior of the box is the indoor artificial illumination or daylight through a window of the user's location. Cut windows or roof-lights into the box in proportion to the room design that is to be modelled. Each window should be hinged so that different combinations of window opening, and night-time, can be reproduced. Battery lamps could be utilized for the illumination arrangements. Rectangular slots in the flat ceiling of the box can represent linear fluorescent luminaires. Small circular holes in the ceiling would model spot lighting. Objects can be placed within the scaled room to establish the effects upon desks, furniture, partitions and artefacts. Cut a small viewing window in the end of the box. Use this aperture to observe and sketch the areas of light and shade within the room. Try various combinations of surface colour, day, artificial and permanent supplementary artificial lighting. Figure 10.3 is an example of an observed lighting pattern in a model room that has

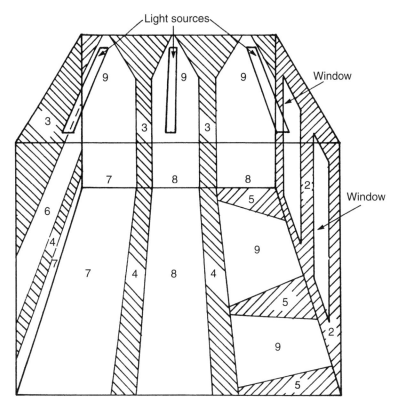

Shadow and brightness on a scale of dark 1 to bright 10

10.3 Observed illumination pattern.

both side windows and a representation of recessed tubular fluorescent luminaires in the ceiling. Alternatively, sketch the interior of the room where you are. Shade the surfaces where shadows occur and describe the brightness level with a numbered scale. Draw the areas where bright light concentrates near to windows or light fittings. Describe how the variation of light and dim areas creates the appearance of the room. Notice what happens during the day and evening as daylight varies and more artificial lights are used. Could you improve the appearance of the room? Are more window shading devices needed? How does the room make you feel on a scale of homely, warm, safe, clinical, insecure, industrial or glaring? How do colours interact with the artificial and day lighting system? Are you getting the point that lighting is more than a simple engineering task (CIBSE, 2011)? Look at the very good illustrations of room lighting in this guide and compare with rooms familiar to you. How about the external environment of buildings, car parks and streets? Apply the same approach to analyse outdoor lighting. Ask a grandparent about the lighting they had in the early 1920s; compare with today's standards and realize it is a different world now. Go to the tenth floor in any city of the world and look out of a window at night. What do you see? An ocean of megawatts of artificial lighting; commercial buildings almost unoccupied but lit; ribbons of moving lights along every illuminated road and around every statue; we burn primary energy at a phenomenal rate. This is the world we create and it has little to do with reducing emissions (Figure 10.3).

Definition of terms

Some of the terms that are used in lighting system design are:

- *Contrast*: difference in the light and dark appearance of two parts of the visual field seen simultaneously.
- *Colour rendering*: effect of an illuminant on the colour appearance of an object relative to day light.
- *Cylindrical illuminance*: total luminous flux falling on a vertical cylindrical photocell sensor representing a human head.
- *Daylight factor*: ratio of the natural illumination on a horizontal plane within the building to that present simultaneously from an unobstructed sky, discounting direct sunlight.
- *Efficacy*: luminous efficacy is the lamp light output in lumens per watt of electrical power consumption.
- *Glare*: discomfort or impairment of vision due to excessive brightness.
- *Illuminance*: luminous flux density at a surface in lumens per square metre, l/m^2, lux, normally on the working plane.
- *Indirect lighting*: fraction of luminous flux reflected from room surfaces.
- *Installed loading*: electrical power used for lighting the working plane, W/m^2.
- *Light meter*: current-generating photocell calibrated in lumens per square meter, lux.
- *Lumen, lm*: unit of luminous flux. It is the quantity of light emitted from a source or received by a surface. A 100 W tungsten filament lamp emits around 1200 lm.
- *Lux, lx*: unit of illuminance; $1\ lx = 1\ lm/m^2$.
- *Light loss factor (LLF)*: overall loss of light from the dirtiness of the lamp (0.8), luminaire (0.95) and the room surfaces (0.95). Clean conditions LLF may be 0.7 but 0.5 when equipment and room become soiled.
- *Luminous intensity, I (candela)*: power of a source or illuminated surface to emit light in a given direction.
- *Luminaire*: complete apparatus that contains the lamp, light emitter and electrical controls.
- *Maintenance factor (MF)*: allowance for reduced light emission due to build-up of dust on a lamp or within a luminaire. Normally 0.8 but 0.9 if the lamps are cleaned regularly or if a ventilated luminaire is used. Light loss factor is preferred.
- *Utilization factor (UF)*: ratio of the luminous flux received at the working plane to the installed flux.

Maintenance

A planned maintenance schedule will include regular cleaning of light fittings and the lamp to ensure the most efficient use of electricity. Ventilated luminaires in air-conditioned buildings remain clean for quite long periods as the air flow through the building is mechanically controlled and filtered. The lamp also operates at a lower temperature, which prolongs its service and maximizes light output. Gradual deterioration of light output from all types of lamps after their design service period, lamp efficacy could fall to half its original figure. Phased replacement of lamps after two or three years maintains design performance and avoids breakdowns. Figure 10.4 shows a typical illuminance profile for a tubular fluorescent light fitting with 6-monthly cleaning and a 2-year lamp-replacement cycle.

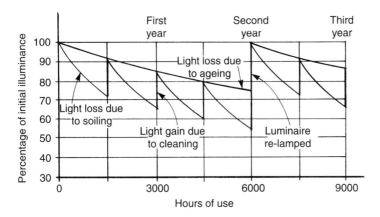

10.4 Overall fluorescent light fitting performance with maintenance.

Utilization factor

The utilization factor is provided by the manufacturer and takes into account the pattern of light-distribution from the whole fitting, its light-distributing efficiency, the shape and size of the room for which it is being designed and the reflectivity of the ceiling and walls. Values vary from 10% where indirect distribution is employed, the room has poorly reflecting surfaces and all the light is upwards onto the ceiling or walls, to 90% for the most energy-efficient designs. Spot lighting can have a utilization factor of nearly unity on the target surface. Configuration of the room is found from the room index:

$$\text{room index} = \frac{IW}{H(I+W)}$$

where I is the room length (m), W is the room width (m) and H is the height of the light fitting above the working plane (m). The ability of a surface to reflect incident light is given by its luminance factor varying from 10% for a floor, 50% for light coloured walls and 80% for a white ceiling.

Utilization factors for a light fitting comprising a white metal support batten and two 58 W white fluorescent lamps 1500 mm long, bare fluorescent tubes suspended under the ceiling as used in commercial buildings vary from 10% to 90% depending on values of the room index and surface luminance factors. Enclosing tubes with a plastic diffuser to improve its appearance usually lowers the utilization factor. This luminaire operates at 240 V consumes 140 W and has a power factor of 0.85 and a running current of 0.68 A.

Disability glare is when a bright light source prevents the subject from seeing the necessary detail of the task. Veiling reflections can be formed on windows and visual display unit (VDU) screens from nearby lamps. A limiting glare index is recommended for each application. To maximize contrast on the working plane, luminaires should be placed in rows parallel to the direction of view. The rows should be widely spaced to form work areas between them. The zone of the ceiling that would cause glare or veiling reflections can be viewed with a mirror on the working plane from the normal angle of work. A luminaire or direct sunlight should not appear in the mirror. The number of light fittings is found from the total lumens needed at the working plane and the illumination provided by each fitting using the formula:

$$\text{number of fittings} = \frac{\text{lux} \times \text{working plane area m}^2}{LDL \times UF \times MF}$$

where *LDL* is the lighting design lumens produced by each lamp, *UF* is the utilization factor and *MF* is the maintenance factor. A typical high output luminaire with two fluorescent lamps, 1500 mm long, emits 5100 lumens measured at 2000 h of use. This is known as its lighting design lumens (*LDL*).

EXAMPLE 10.1

A drawing office 16 m × 11 m and 3 m high has a white ceiling and light-coloured walls. The working plane is 0.85 m above the floor. Double-lamp luminaires emitting 5100 lumen are to be used and their normal spacing-to-height ratio (SHR) is 1.75. Calculate the number of luminaires needed and draw their layout arrangement. Find the electrical power consumption of the lighting system. Luminance factors are 70% for the ceiling and 50% for the walls. A high standard of maintenance will be assumed, giving a maintenance factor of 0.9. The lighting design lumens are taken as 5100 lm for the whole light fitting. Illuminance required is 600 lm/m^2. Electrical power consumption of each luminaire is 140 W.

Height *H* of fittings above the working plane is:

$$H = (3 - 0.85)\,\text{m}$$

$$= 2.15\,\text{m}$$

$$\text{room index} = \frac{IW}{H(I+W)} = \frac{16 \times 11}{2.15 \times (16+11)} = 3.03$$

For a room index of 3:

$$\text{utilization factor} = 79\% = 0.79$$

$$\text{number of fittings} = 600\frac{\text{lm}}{\text{m}^2} \times \frac{16\,\text{m} \times 11\,\text{m}}{0.79 \times 0.9} \times \frac{\text{luminaire}}{500\,\text{lm}} = 29.12$$

Ratio of the spacing *S* between rows to the height *H* above the working plane is:

$$SHR = \frac{S}{H} = 1.75$$

$$S = 1.75\,H$$

$$= 1.75 \times 2.15\,\text{m}$$

$$= 3.76\,\text{m}$$

If it is assumed that windows are along one long side of the office and that rows of luminaires will be parallel to the windows, this will produce areas between rows where drawing boards,

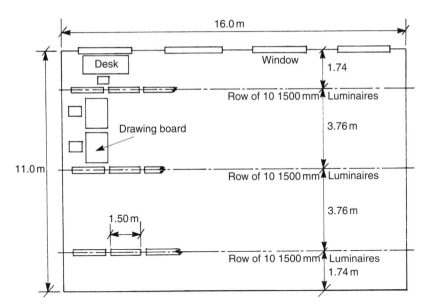

10.5 Arrangement of the luminaires in a drawing office in Example 10.1.

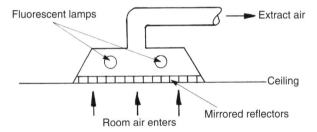

10.6 Ventilated luminaire.

desks and VDU terminals can be sited to gain maximum benefit from side day lighting without glare and reflection. Perimeter rows of luminaires are spaced at about half of S, 1.74 m, from the side walls. Three rows of 10 luminaires are required, as shown in Figure 10.5, giving 30 luminaires and a slightly increased illuminance.

The power consumption will be $30 \times 140\,W$, i.e. $4200\,W$, which is:

$$\frac{4200\,W}{16\,m \times 11\,m} = 23.86\,W/m^2 \text{ floor area}$$

Air-handling luminaires

Luminaires that are recessed into suspended ceilings are ideally placed to be extract air grilles for the ventilation system. Heat generated is removed at its source and the lamp can be maintained at its optimum operating temperature to maximize light output and colour-rendering properties. Dust build-up should also be less in an air-conditioned building where all the circulating air is filtered in the plant room. Figure 10.6 shows the air flow through a luminaire that has ventilation openings and mirrored reflectors.

Up to 80% of the electrical energy used by the light fitting can be absorbed by the ventilation air as it passes through. Air flow rates are around 20 l/s through a 1500 mm twin tube fluorescent luminaire and a temperature increase of about 8°C is achieved. Extract air at about 30°C can be produced and the heat it contains can be reclaimed for use in other parts of the building.

Colour temperature

Colour temperature is a term used in the description of the colour-rendering property of a lamp. Colours of surfaces under artificial illumination are compared with the colours produced by a black body heated to a certain temperature and radiating in the visible part of the spectrum between the ultraviolet and infrared bands. Any colour that matches that shown by the heated black body is said to have a colour temperature equal to the temperature of the black body. A candle has a colour temperature of 2000 K and blue sky has a colour temperature of 10000 K. Correlated colour temperature is that temperature of the heated black body at which its colour most closely resembles that of the artificial source. A colour-rendering index (CRI) is used to compare the colour-rendering characteristics of various types of lamp compared to those illuminated by a reference source; a black body radiator of 5000 K correlated colour temperature or 'reconstituted' daylight if more than 5000 K is needed. Colours are then illuminated by the test lamp. Colour differences produced between the source and test lamps provide a measure of the colour-rendering properties of the test lamp. A CRI of 50 corresponds to a warm white fluorescent lamp. A CRI of 100 is produced by an incandescent-halogen lamp that radiated identically with the reference source.

Lamp types

Incandescent, electromagnetic ballast fluorescent and mercury vapour lamps mainly used in residential and commercial applications are being phased out up to 2017. Retrofit and new designs will use higher energy efficiency lamps. Changing from incandescent lamps and halogen down lights to light emitting diode (LED) equivalents reduces electricity consumption by 80% as well as heat output. LED tubes can replace fluorescent tubes in industrial applications and reduce energy use by 50%. Electroluminescence of a charged semiconductor produces light in a LED. Fluorescent tubes controlled with electronic ballasts and run at 20000 Hz are flicker-free, use less energy and are controllable from computer networks and sensors for occupancy and light level. LED and high frequency fluorescent lamps are dimmable. A summary of lamp types, their performances and applications is given in Table 10.2. Refer to manufacturers' information resources such as Phillips, GE and Osram.

General lighting service (GLS) tungsten filament lamps are inexpensive, but are being phased out, give good colour matching with daylight and last up to 2000 h in service. They can be controlled by variable-resistance dimmers and are used in a supplementary role to higher-efficacy illumination equipment. Tungsten halogen spot or linear lamps have a wide variety of display and floodlighting applications. Miniature fluorescent lamps can be used as energy-saving replacements for GLS lamps. Folded-tube and single-ended types are available. A typical folded lamp, SL18 18 W, produces 900 lumens at 100 h, has a correlated colour temperature of 2700 K, *CRI* of 80 and a service period of 5000 h. Its lumen output is equivalent to a 75 W GLS filament lamp.

Low-pressure mercury vapour-filled fluorescent tubular lamps are the most common. The tube diameter is 16–38 mm. Electrical excitation of the mercury vapour produces radiation, which causes the tube's internal coating to fluoresce. The colour produced depends on the chemical

Table 10.2 Lamp data.

Lamp	Lamp designation	Efficacy lm/W	Colour temperature K	CRI	Application
Semiconductor	LED	60+	2700	80	Interiors
Incandescent	GLS, tungsten	18	2900	100	Interiors
Incandescent	GLS tungsten-halogen	22	3200	100	Interior displays, outdoors
Fluorescent	White	80	3500	50	Industrial
Fluorescent	Natural	85	4000	85	Commercial
Tri-phosphor fluorescent	Cool white	85	4080	89	Interiors
Low-pressure sodium	SOX	180	2000	–	Roads, car parks, floodlighting
High-pressure sodium	White SON	112	2700	80	Floodlighting, large interiors
Metal halide	HID	85	5400	96	Industrial, floodlighting
Mercury fluorescent	MBF	50	4300	47	Roads, floodlights, industrial
Mercury-blended tungsten fluorescent	MBTF	25	3700	–	Industrial, floodlighting, roads

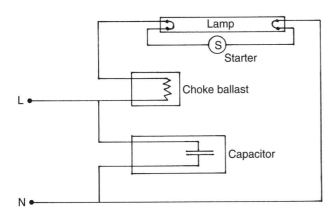

10.7 Switch-start circuit diagram for fluorescent and discharge lamps.

composition of the internal coating. High-efficacy 26 mm diameter lamps are filled with argon or krypton vapour and have a phosphor internal coating. A circuit diagram for a switch-start fluorescent lamp is shown in Figure 10.7.

The glow-switch starter S has bimetallic electrodes which pass the lamp electrode, preheating the current when starting cold. Upon becoming warmed, the bimetallic electrodes move into contact and establish a circuit through the lamp electrodes. Making this contact breaks the starter circuit, whose electrodes cool and spring apart after about a second, subjecting the lamp to mains voltage. Twin lamp and starter-less controls are used in new installations to minimize running costs. This old technology is being replaced with electronic ballasts at refurbishment or in new installations. Electronic controllers start the lamp, raise mains frequency to run the lamp at 20000 Hz, raise the power factor, control the starting and running current, allow control from occupancy and light level sensors and provide dimming, all at improved energy efficiency and lower operating cost. Luminaires are designed to run in air of 5–25°C d.b., and their service period will be reduced at higher temperatures. Circuit breakers or high rupturing capacity fuses, rather than re-wirable fuses, should be used for circuit protection. All fluorescent luminaires make an operating noise, which may be noticeable against very low background noise levels. Some radio interference is inevitable but this diminishes with distance from the radio set.

Low-pressure sodium discharge lamps emit light from electrically excited sodium vapour in a glass discharge tube. High-pressure sodium discharge lamps comprise a discharge tube of sintered aluminium oxide containing a mixture of mercury and sodium vapour at high pressure.

Mercury fluorescent lamps consist of electrically excited mercury vapour in a quartz discharge tube, which emits ultraviolet radiation, an infrared component from the mercury arc, and visible light. A phosphor internal coating fluoresces to produce the desired colour.

Metal halide high intensity discharge lamps contain metal halides in a quartz discharge tube, which gives a crisp white light. MBTF lamps are the mercury fluorescent type with additional tungsten filaments, which need no control gear and give light immediately. Other discharge lamps incorporate current-limiting ballasts and power-factor-correcting capacitors in a similar arrangement to that in Figure 10.7. Electronic controls are replacing such old technology.

EXAMPLE 10.2

A windowless office is to be illuminated for 15 h per day, for 6 days per week for 50 weeks per year. The floor is 20 m long and 12 m wide. An overall illumination of 450 lx is to be maintained over the whole floor. The total light loss factor for the installation is 70%. The designers have the choice of using 100 W tungsten filament lamps, which have an efficacy of 12 lm/W and need replacing every 3000 h, or 65 W tubular fluorescent warm white lamps, which have an initial output of 5400 lm and are expected to provide 12000 h of service. The room layout requires an even number of lamps. Electricity costs 8 p/kWh. The tungsten lamps cost £1 each while the fluorescent tubes cost £10 each. Compare the total costs of each lighting system and make a recommendation as to which is preferable, stating your reasons.

For the lighting system:

$$\text{lighting hours} = 15 \times 6 \times 50 = 4500 \text{ h/yr}$$

$$\text{floor area} = 20 \text{ m} \times 12 \text{ m} = 240 \text{ m}^2$$

$$\text{installed lumens} = \frac{450 \times 240 \times 100}{70}$$

$$= 154\,286 \text{ lm}$$

For the tungsten lamps:

$$\text{input power} = \frac{154\,286}{12 \times 1000}$$

$$= 12.857 \text{ kW}$$

$$\text{number of lamps} = \frac{12.857 \times 1000}{100}$$

$$= 129 \text{ lamps}$$

Next even number is 130 lamps.

$$\text{installed power} = \frac{130 \times 100}{1000}$$

$$= 13 \text{ kW}$$

$$\text{electricity cost} = \frac{13 \times 4500 \times 8}{100}$$

$$= £4680/\text{yr}$$

The average annual cost for replacing the lamps can be found by multiplying the number of installed lamps by the anticipated hours of use, dividing by the lamp manufacturer's rated average life hours and then multiplying by the replacement cost for each lamp. In this case:

$$\text{lamp cost} = \frac{130 \times £1 \times 4500}{3000}$$

$$= £195/\text{yr}$$

These tungsten lamps expire within a year, so there will be annual expenditure. A new lighting system that has lamps providing reliable service for more than a year will not produce replacement expenditure in the first year or two. The owner needs to budget for the eventual replacement costs by assessing the average annual cost. A planned maintenance programme will have lamps replaced, and luminaires cleaned, prior to expiry. This work may be performed out of normal working hours for the building to avoid disturbance to its normal functions.

$$\text{total annual cost} = £(4680 + 195) = £4875/\text{yr}$$

For the fluorescent lamps:

$$\text{number of lamps} = \frac{154286}{5400}$$

$$= 29 \text{ lamps}$$

Next even number is 30 lamps.

$$\text{input power} = \frac{30 \times 65}{1000}$$

$$= 1.95 \text{ kW}$$

$$\text{electricity cost} = \frac{1.95 \times 4500 \times 8}{100}$$

$$= £702$$

$$\text{lamp cost} = \frac{30 \times £10 \times 4500}{12\,000}$$

$$= £112.5/yr$$

$$\text{total annual cost} = £(702 + 112.5) = £814.5/yr$$

Both methods of lighting require at least the annual cleaning of lamps and luminaires. Reasons for using fluorescent lamps are:

1. They produce an annual cash saving of $f(4875 - 814.5) = £4060.5/yr$.
2. They only need changing (12000/4500) after 2 years 8 months of use.
3. Tungsten lamps need changing ($12 \times 3000/4500$) every 8 months.
4. There is less heat emission from fluorescent lighting, particularly radiant heat, so the air-conditioning cooling load is lower.
5. They give better colour rendering.
6. They give better diffused light distribution.

Control of lighting services

The energy that is consumed by artificial lighting systems is both an expensive use of resources and a high monetary cost. A minimum level of illumination may be desirable for the security of personnel or monitoring of the building and its contents for the detection of intruders. The changes from day lighting to full artificial lighting and then to security illumination can be achieved with manual and automatically timed operation of switches. This usually leaves unoccupied areas illuminated. A light-sensitive photocell can be used to detect illumination level and an automatic controller may be programmed to reduce the use of the electrical lighting system. The presence of the occupants, or an intruder, can be detected by passive infrared, acoustic, ultrasonic or microwave-radar-based systems.

The detector and control system needs to be sufficiently fast-acting and sensitive so that the occupant is not stranded within a darkened room and suffers injury or fear. It is equally important that only those lights that are actually needed are switched on, and not for the entire room or space to be illuminated when only one person enters to use a small area. Local control of the light switching may be preferable to a system that is operated from a remote energy-management system computer. The local system should be faster in operation and will be less subject to distribution system or computer breakdown.

The design of a control scheme for an occupied space may include a minimum number of luminaires, which are switched on from a time switch or by the occupant to provide safe access. Groups of luminaires that are near windows may be controlled from local photocell detectors to ensure that the perimeter lighting remains off as long as possible. The internal parts of the space may be operated from automatic presence detectors. Data on the length of operation of each lighting unit can be transmitted to the computer-based building management system so that real-time usage and electrical power consumption can be recorded. Timed-off controllers avoid lights left on excessively after working hours and when nobody is present.

Questions

1. Explain, with the aid of sketches, how interiors can be illuminated by daylight. State how natural illumination is quantified.
2. State the relationship between the visual task and the illuminance required, giving examples.

3. Sketch and describe how supplementary artificial lighting is used to achieve the desired illuminance.

4. Discuss the use of localized task-illumination systems in relation to the illumination level provided, reflection, energy conservation, shadows and user satisfaction.

5. Define the terms used in lighting system design.

6. Draw a graph of illumination provided versus service period for an artificial illumination installation to show the effect of correct maintenance procedure.

7. Calculate the room index for an office 20 m × 12 m in plan, 3 m high, where the working plane is 0.85 m above floor level.

8. State the luminance factors for a room having a cream ceiling and dark grey walls.

9. Find the utilization factor for a bare fluorescent tube light fitting having two 58 W, 1500 mm lamps in a room 5 m × 3.5 m in plan and 2.5 m high. The working plane is 0.85 m above floor level. Walls and ceiling are light stone and white respectively.

10. Sketch satisfactory arrangements for natural and artificial illumination in modern general and drawing offices, a library and a lounge. Comment particularly on how glare and reflections are controlled.

11. A supermarket of dimensions 20 m × 15 m and 4 m high has a white ceiling and mainly dark walls. The working plane is 1 m above floor level. Bare fluorescent tube light fittings with two 58 W, 1500 mm lamps are to be used, of 5100 lighting design lumens, to provide 400 lx. Their normal spacing-to-height ratio is 1.75 and total power consumption is 140 W. Calculate the number of luminaires needed, the electrical loading per square metre of floor area and the circuit current. Draw the layout of the luminaires.

12. Discuss how the use of air-handling luminaires improves the performance of the lighting installation and makes better use of energy.

13. Explain how the rendering of colours by illumination systems is measured.

14. Compare the energy efficiency and colour-rendering of different lamp types, stating suitable applications for each.

15. Use the cardboard box small-scale models of rooms to investigate the visual design of lighting systems.

    1. Cut different shapes and locations of windows and roof lights such that they all have the same open area.
    2. Colour the internal surfaces differently by means of dark, light, removable and reflective sheets of materials.
    3. Cut slots and holes into the ceiling to model different designs of strip fluorescent and filament lighting layouts; replaceable ceilings with different designs are helpful.
    4. View the interior of the room under various day lighting and artificial lighting arrangements.
    5. Make three-dimensional sketches of what you see of the lighting layouts produced, showing the shading. Calculate the lighting level found on each area on a scale of 1 (dark) to 10 (bright).

16. Using the models created for Question 15, answer the following.

    1. What is the effect of *quantity* of daylight on the *quality* of the day lighting system created?
    2. What effect do the colours of the room interior surfaces have on the quality of the lighting produced?
    3. What colours should the room surfaces be?

    Justify your views in relation to the use of the room, its maintenance costs and design of the decoration.

17. Using the models created for Question 15, answer the following.
    1. What patterns of illumination are produced on the end walls by differently spaced rows of strip lighting?
    2. What are the best spacing arrangements between rows of strip or circular lamps? These depend upon what is being illuminated, so state the objectives of the lighting design first.

18. Create an approximate scale model of the interior of a room that is known to you. Experiment with three combinations of day lighting and artificial lighting to find the best overall lighting scheme for the tasks to be performed in the room.

19. Write a technical report to explain how the reflectance of room surfaces, the location, dimensions and shape of glazing, the spacing of rows of luminaires and their height above the working plane are related to the efficient use of electrical energy in the overall lighting design.

20. Put small boxes and partitions into a scale model of a room to represent furniture, desks, horizontal and vertical working planes. Carry out an experimental investigation of the problems that arise for the lighting designer.

21. State how the lighting design can be made to feature particular parts of the interior of the building and the parts that should be featured for safety and appearance reasons.

22. Analyse the costs of these competitive lighting systems and recommend which is preferable, stating your reasons. A heavy engineering factory is to be illuminated for 15 h per day, for 5 days per week for 50 weeks per year. The floor size is 120 m long and 80 m wide. An overall illumination of 250 lx is to be maintained over the whole floor. The overall light loss factor for the installation is 63%. The designers have the choice of using 150 W tungsten-halogen lamps, which produce 2100 lm and need replacing every 2000 h, 80 W tubular fluorescent lamps, which produce 6700 lm and are expected to provide 12000 h of service, and 250 W high-pressure sodium lamps, which produce 27500 lm and are expected to last for 24000 hours. The lighting layout needs an even number of lamps. Electricity costs 7.2 p/kWh. The tungsten lamps cost 90p each, the fluorescent tubes cost £10.50 each and the sodium lamps cost £61 each. Replacing any lamp takes two people 2 min. and their combined labour rate is £17 per hour. The hire cost of scaffolding is £120 per 8 hour day.

23. A lecture theatre is to be illuminated for 8 h per day, for 5 days per week for 30 weeks per year. The floor is 32 m long and 16 m wide. An overall illumination of 350 lx is to be maintained over the whole floor. An even number of lamps is to be used. Utilization factor for the installation is 0.73 and the maintenance factor is 0.7. Designers have the choice of using 100 W tungsten filament lamps, which have luminous efficacy of 10 lm/W and need replacing every 2000 h, 100 W quartz halogen low-voltage lamps in reflectors, which have an efficacy of 95 lm/W and provide 23000 hours' use, and 65 W tubular fluorescent lamps, which have an efficacy of 57 lm/W and are expected to provide 7500 h of service. Electricity costs 8 p/kWh. The tungsten lamps cost 85 p each, the halogen cost 29 p each and the fluorescent tubes cost £11.25 each. Compare the total costs of each lighting system and make a recommendation as to which is preferable, stating your reasons.

24. When 900 lumen falls onto a 2.0 m² surface from a fluorescent lamp, and a light meter finds that 360 lumen are reflected, the luminance of the surface is:
    1. 0.71.
    2. 0.4.
    3. 0.2.
    4. Insufficient information.
    5. 1.25.

25. What is illumination intensity?

    1.   Luminosity.
    2.   Light level.
    3.   Lumens.
    4.   Lux.
    5.   Lumen per watt.

26. Illumination intensity provided is related to what?

    1.   Contrast between detail and background on the working plane.
    2.   Cleanliness of luminaires.
    3.   Task lighting needed.
    4.   Lighting colour temperature.
    5.   Provision of shadow-free lighting.

27. Which illuminance levels are correct? More than one correct answer is possible.

    1.   Small electronic component inspection 3000 lux.
    2.   Corridor 250 lumen/m^2.
    3.   Bright sunlight 100000 lux.
    4.   Rough tasks 500 lumen/m^2.
    5.   Hospital operating theatre table 10000 lux.

28. What is lighting design?

    1.   A primarily scientific application.
    2.   Engineering design.
    3.   A unique combination of technology and visual effects.
    4.   Entirely a mathematical application.
    5.   Best achieved by an iterative design technique.

29. Efficacy of a lamp means?

    1.   Efficiency at converting colour temperature into illuminance on the working plane.
    2.   Percentage of lamp globe or tube surface area that emits light.
    3.   Proportion of lamp light output directed toward the working plane.
    4.   Lumens light output per watt of electrical input power to lamp.
    5.   Luminous efficacy is total illuminance produced divided by lamp electrical input power including all control equipment consumption.

30. Light system glare means?

    1.   Whether discomfort is created in a particular viewing direction.
    2.   Impairment of vision due to excessive brightness.
    3.   Sedentary position includes direct view of sunshine.
    4.   Florescent indoor lighting systems do not cause glare.
    5.   All lighting causes glare when viewed directly.

31. What is an air-handling luminaire?

    1.   A sealed light fitting inside an air handling unit to allow servicing work.
    2.   A light fitting open to the room air allowing cooling.
    3.   Luminaire passing air returning to the ductwork system from the conditioned room.
    4.   Lamp designed for low temperature operation.
    5.   Sealed luminaire to keep out moisture.

32. What is meant by lamp colour temperature?

     1.   Description of colour-rendering property.
     2.   Colour shift in surface caused by the type of lamp.
     3.   Colour of the lamp when viewed directly.
     4.   Infra-red fraction of the sun produced by the lamp.
     5.   Calculated average temperature from each of eight wavelength bands in the light
          produced by a lamp.

33. Which are correct descriptions of a lighting system sensor operation? More than one correct
    answer is possible.

     1.   Sensor detects light level in room and switches on rows of luminaires to maintain set
          lux level.
     2.   All sensor types are used by the building management computer system to switch rows
          of luminaires on and off.
     3.   Groups of luminaires switched on from a sensor detecting occupancy within controlled
          space.
     4.   Microwave sensor detects use of electrical equipment within room and switches
          lights on.
     5.   Microwave sensor detects any small movement within controlled space and switches
          luminaires on and then off when no movement is detected for a set time interval.

34. Which statement is correct about lighting?

     1.   Light and heat energy are mutually convertible.
     2.   Light radiation is the same as heat radiation.
     3.   In the absence of humans or animals, there would be no light.
     4.   Light energy only exists within a visible spectrum.
     5.   Low power laser beams are for heat energy.

35. Which statement is correct about units for lighting?

     1. W/m^2.
     2.   Light intensity is 5.67×10^{-8} W/m^2 K^4.
     3.   Lumens.
     4.   Flux in Webber/m^2.
     5.   Light intensity increases with distance from source.

36. Which statement is correct about lighting?

     1.   Light intensity varies inversely with the square of the distance from the source.
     2.   Light intensity varies inversely with the distance from the source.
     3.   Light intensity reduces with the distance from the source.
     4.   Light intensity varies inversely with the square root of the distance from the source.
     5.   Light intensity varies inversely with the distance cubed from the source.

37. Which is the recommended design illumination for office work?

     1.   40 lux.
     2.   400 lux.
     3.   800 lumen/m^2.
     4.   2600 lux.
     5.   200 lumen/m^2.

38. Which is a valid technical reason for burning primary energy resources to provide artificial
    lighting?

     1.   Overcomes affect of dull sky.
     2.   Makes outside of building appear more impressive.

3.   Human task illumination.
4.   Good colour rendering of exteriors.
5.   Increases lighting available during daytime.

39. What has been the most significant benefit to humans since the time of ancient Rome?

1.   Air conditioning in hot climates.
2.   Refrigeration.
3.   Central heating and cooling providing comfort.
4.   Computers, mobile telephones and television.
5.   Indoor electric lighting.

40. What is supplementary daylight?

1.   Manually operated blinds or curtains.
2.   Mechanically operated shading devices.
3.   Perimeter fluorescent lighting.
4.   Artificial lighting temporarily or permanently used to stabilize variations in day lighting.
5.   Artificial lighting having the same colour rendering temperature as daylight.

41. What is illuminance?

1.   Proportion of available light flux received on working plane.
2.   Lux received on working plane.
3.   Colour brightness index.
4.   Luminous flux intensity reflected from a surface.
5.   Brightness of perceived detail relative to background.

42. What is a light meter?

1.   Easily portable data logger.
2.   1 lumen/m^2 projected for 1 metre.
3.   Current-generating photocell calibrated in lux.
4.   Current-generating photocell that activates an alarm in a BMS security control room.
5.   Photocell that receives a laser beam from an emitter as part of a security, smoke or heat detection system.

43. What is a lumen?

1.   Unit of lighting not presently in use.
2.   1000 candela/m^2
3.   SI unit of light output or received.
4.   A directional measurement of light.
5.   Lighting power of a source.

44. What are lux?

1.   Total light output from a source.
2.   A measurement of glare.
3.   31.4 candela.
4.   1 lumen/m^2.
5.   A measurement of reflected light.

45. What is a luminaire?

1.   A part of the lamp control system.
2.   The complete light fitting.
3.   A unit of light output flux.

4.   A unit of reflected light.
5.   Those parts of the lighting fitting other than the lamp.

46. Which is not true about fluorescent lamps?

1.   Low cost and reliable.
2.   Poor colour rendering compared to daylight.
3.   Visible flicker may be observed from 50 Hz single phase.
4.   Energy efficiency around 85 lumen per Watt.
5.   Colour temperature 4000 K and RA8 of 85.

47. Where are sodium discharge lamps used?

1.   Low pressure SOX sports facilities.
2.   Not used for street lighting.
3.   High pressure SON produces industrial indoor white lighting.
4.   Where good colour rendering preferred.
5.   Regular replacement installations.

48. How does a fluorescent lamp work?

1.   Fluorescing powder-filled tube is electrically heated.
2.   Fluorescent lining of tube becomes heated.
3.   Vapour within tube fluoresces.
4.   Vapour within tube is electrically charged.
5.   Tube interior coating fluoresces when irradiated.

49. Which is a feature of an energy-efficient lighting control system?

1.   BMS reports maintenance of all lighting system usage.
2.   All rows of luminaires remain switched on during working day as frequent starting uses more energy.
3.   Lamp deterioration increases with frequent switching so should remain on continuously.
4.   Occupancy sensors are programmed to keep lights on for half an hour after occupants leave.
5.   Timed switch-off controller minimizes lighting use after occupation ceases.

50. Explain how combinations of natural and artificial lighting with automatic sensing and computer-based control systems can help to meet the *HM Government Carbon Plan, 2011* (DECC, 2011a). In your opinion, are modern lighting systems that are already installed and are being installed today ever going to contribute to carbon emission reduction? How? Give examples of good and average practices known to you.

11 Condensation in buildings

Learning objectives

Study of this chapter will enable the reader to:

1. identify the moisture content of humid air by its vapour pressure;
2. understand dew-point temperature;
3. identify the sources of moisture within a building;
4. understand the flow and storage characteristics of moisture flows found in habitable building;
5. explain the causes of condensation;
6. discuss the damage which can be caused by condensation;
7. calculate vapour diffusion resistance;
8. calculate vapour flow;
9. calculate air vapour pressure;
10. calculate air dew-point temperature;
11. understand atmospheric pressure terms;
12. use the e^x calculator function;
13. use the $\log_e$ calculator function;
14. convert from e^x to $\log_e$ forms;
15. draw thermal temperature gradients;
16. draw dew-point temperature gradients;
17. identify condensation zones;
18. discuss surface and interstitial condensation;
19. understand where to install thermal insulation and vapour barriers in relation to condensation risk and thermal and structural integrity requirements.

Key terms and concepts

condensation 251; dew-point temperature gradient 260; diffusion 257; exponential and logarithmic functions 252; mould growth 251; partial pressure 252; surface and interstitial

Introduction

Condensation risk is analysed during design of a building, when retrofit measures such as additional thermal insulation, double glazing or ventilation control are being considered or where damage from condensation has been discovered. Anti-condensation measures are linked to temperature control systems and ventilation provision in that they determine the size of plant and resulting operating costs.

The fundamentals of air and water vapour mixtures are introduced and then the moisture diffusion properties of building materials are analysed. A convenient form of equation to enable the air dew-point temperature to be found using a student's scientific calculator is derived and this saves the need to refer to charts or tables.

Thermal and dew-point temperature gradients are calculated, allowing moisture flow rate to be found. The rate of moisture deposition within the structure can then be assessed for its damage potential.

Sources of moisture

Air is a mixture of dry gases and water vapour. The water vapour exists in the form of finely divided particles of superheated steam at the air dry-bulb temperature. Total atmospheric pressure consists of the sum of the partial pressures of the two main constituents.

Typically, one standard atmosphere exerts a pressure of 1013.25 mb at sea level. When the air conditions are 25°C d.b. and 20°C w.b., the standard atmosphere is made up of 993.08 mb dry gas and 20.17 mb water vapour pressure. If this air is allowed to come into contact with a surface at a temperature of 17.6°C, the air becomes saturated with moisture and can no longer support all the water in its vapour state. This temperature is known as the air dew-point t_{dp}, and is shown in the psychrometric chart in Figure 5.4 on p. 90. Further data on the properties of humid air can be obtained from the CIBSE (2007) and the psychrometric chart that may be used to find data. Sources of water vapour in an occupied building are:

1. people, upwards of 0.7 kg per 24 h;
2. cooking;
3. washing, bathrooms, drying clothes;
4. humidifiers and open water surfaces;
5. animals (dogs exhale more moisture than people produce overall);
6. combustion of paraffin (complete combustion of 1 kg of C_9H_{20} produces 1.4 1 kg of water vapour).

Porous structural surfaces, furniture and fabrics within the building absorb moisture and then release it into the internal atmosphere when the temperature and humidity allow this. Some moisture travels through the structure and evaporates externally unless it is prevented from doing so by an impervious layer or vapour barrier. The majority of internally produced humidity is removed by ventilation.

The warm internal atmosphere is able to hold more moisture than the cooler external air; thus the partial pressure of the vapour, i.e. the vapour pressure p_s, is higher inside than outside. This vapour pressure difference causes mass transfer of moisture out of the building through the porous structure and by the ventilation air flow. When dense materials such as precast concrete or impervious barriers form a major part of the construction, moisture removal from the normal habitation is slow and condensation occurs on cold surfaces. Brickwork or masonry that is already saturated has the same effect.

Condensation and mould growth

Condensation and mould may readily form on window sills and in the corners of rooms, where the surface temperatures are lower compared with large areas. Gloss paint stops the absorption of moisture into an otherwise porous plaster or wooden component and water droplets form on the surface. The production of surface condensation in heated rooms is generally avoided in structures with a U value lower than 1.4 W/m^2 K.

Dampness in the timber of roofs, not caused by the ingress of rain, may be due to condensation on low-temperature surfaces. A well-insulated flat plaster ceiling may produce these low-temperature conditions in the roof construction. Natural ventilation through gaps between the tiles and roofing felt is normally sufficient to stop rot and damp patches on the plasterboard. Well-sealed roofs, with boarding under the felt, should have their ventilation increased by means of openings in the soffit of the eaves after extra thermal insulation. Humid air enters the roof space through gaps in the fiat plaster ceiling around access hatches and pipes. Sealing these substantially reduces condensation risk. Roof insulation should stop at the wall head and not be pushed into the eaves, or ventilation will be restricted.

Condensation forms within a structure or inside a solid material wherever the temperature falls below the dew-point of the moist air at that location. This is known as interstitial condensation. Vapour diffusion through building materials is calculated in a similar manner to the calculation of heat flow.

Vapour diffusion

Flow of water vapour through a porous building material or composite slab is analogous to the flow of heat through the structure. Convection currents transfer heat and moisture at the fluid-solid boundary. Conduction heat transfer is similar to vapour diffusion through a porous material, and its resistance to moisture flow varies with density, as does thermal resistance but in the opposite sense. The mass flow rate of moisture through a composite structure consisting of a number of plane slabs and surfaces in series is:

$$G = \frac{P_{s1} - P_{s2}}{R_v}$$

where G is the mass flow rate of vapour (kg/m^2 s), R_v is the vapour resistance of the structure (N s/kg) and P_{s1} and P_{s2} are the vapour pressures on surfaces 1 and 2 on each side of the slab (N/m^2 or Pa). The total vapour resistance R_v is given by:

$$R_v = r_v \times l$$

where R_v is the total vapour resistance of a slab of homogeneous material (GN s/kg), r_v is the vapour resistivity of a material (GN s/kg m), and l is the thickness of material (m). Surface films and air cavities have only slight resistance to the flow of vapour and they are not normally included. Some typical values of vapour resistivity are given in Table 11.1. The complete resistance to the flow of vapour through some typical vapour barrier films is given in Table 11.2.

Table 11.1 Vapour resistivity.

Material	Vapour resistivity (GN s/kg m)
Brickwork	40
Dense concrete	200
Aerated concrete	30
Glass fibre wool	10
Foamed urea formaldehyde	30
Foamed polyurethane, open cell	1000
Foamed polystyrene	500
Hardboard	520
Insulating fibreboard	20
Mineral fibre wool	6
Plaster	50
Plywood	520
Wood wool/cement slab	15
Wood	50

Note: Data is approximate and is only to be used for examples in this book.

Table 11.2 Vapour resistances of films.

Material	Vapour resistance (GN s/kg)
Aluminium foil	over 4000
Double-layer Kraft paper	0.35
Gloss paint	8
Interior paint	3
Polythene film, 0.1 mm	200
Roofing felt	4 (and up to 100)

Note: Data is approximate and is only to be used for examples in this book.

Further data are available in the CIBSE (2007) and in the BRE Digest (1992). Note that the values quoted in the BRE Digest are in MN s/g m (mega Newton seconds per gram metre) for vapour resistivity and MN s/g for the vapour resistance of films. These units of measurement are the same as GN s/kg m (giga Newton seconds per kilogram metre) and GN s/kg as both the numerator and denominator have been reduced by 1000 times.

Values of vapour pressure are available in CIBSE (2007) but can be calculated with sufficient accuracy from the following curve fit to the saturation conditions data:

$$p = 600.245 \exp (0.0684 \, t_{dp}) \text{Pa}$$

where t_{dp} is the air dew-point temperature from the psychrometric chart (°C) and $\exp(x) = e^x$ where e = 2.71828 is the exponential operator. Tabulated vapour pressures are in millibars (mb), and since 1 bar = 100000 N/m^2 = 100000 Pa, 1 mb = 100 N/m^2 = 100 Pa.

EXAMPLE 11.1

State the vapour pressure for air at 25°C d.b., 20°C w.b. in Pascals.

The vapour pressure for air under these conditions is 20.17 mb.

$$p_S = 20.17 \text{ mb} \times \frac{100 \text{ Pa}}{1 \text{ mb}} = 2017 \text{ Pa}$$

EXAMPLE 11.2

Calculate the vapour pressure for saturated air at 25°C d.b., 20°C w.b. from the dew-point.

The dew-point is $t_{dp} = 17.6°C$. Therefore,

$$p_S = 600.245 \times \exp(0.0684 \times 17.6) \text{ Pa}$$

Scientific calculators have an e^x function, and in this calculation:

$$x = 0.0684 \times 17.6 = 1.20384$$

Consequently,

$$p_S = 600.245 \times e^{1.203\,84}$$

Having left 1.20384 in the displayed x register, execute the e^x function, producing 3.33289, and multiply by 600.245 to obtain:

$$p_S = 2000.6 \text{ Pa}$$

This is less than 1% different from the tabulated vapour pressure and is of sufficient accuracy bearing in mind the other figures involved in the problem. Dew-point temperature can be found from the saturation vapour pressure by rearranging the equation:

$$p_S = 600.245 \times \exp(0.0684 \, t_{dp}) \text{ Pa}$$

$$\frac{p_S}{600.245} = \exp(0.0684 \, t_{dp})$$

This is a logarithmic equation of the form:

$$y = e^x$$

Where y is the number whose logarithm to base e is x. Logarithms to base e are called natural logarithms and are expressed as follows:

$$\log_e y = x \quad \text{or} \quad \ln y = x$$

EXAMPLE 11.3

Compute the natural logarithm of 2 and then raise e to the power of this logarithm.

Enter 2 into the calculator x display and press the *In* key; the answer is 0.6931. Therefore:

$$\ln 2 = 0.6931 \quad \text{or} \quad \log_e 2 = 0.6931$$

Now, as $x = 0.6931$ is displayed in the calculator, execute e^x. This results in:

$$e^{0.6931} = 2$$

Thus it is seen that e^x is the antilogarithm of $\ln x$, and the two expressions $y = e^x$ and $\ln y = x$ are interchangeable to suit the problem. Thus, for:

$$\frac{p_s}{600.245} = \exp(0.0684\, t_{dp})$$

we can write:

$$\ln\left(\frac{p_s}{600.245}\right) = 0.0684\, t_{dp}$$

and,

$$t_{dp} = \frac{1}{0.0684} \times \ln\left(\frac{p_s}{600.245}\right) \,^{\circ}C$$

EXAMPLE 11.4

Calculate the dew-point for saturated air with a vapour pressure of 2000.6 Pa.

$$t_{dp} = \frac{1}{0.0684} \times \ln\left(\frac{2000.6}{600.245}\right) \,^{\circ}C$$

$$= \frac{1}{0.0684} \times \ln 3.33$$

$$= 17.6\,^{\circ}C$$

An alternative general equation is stated in BRE (1992) for the calculation of vapour pressure:

$$p_s = 0.6105 \times \exp\left(\frac{17.269 \times t_{dp}}{237.3 + t_{dp}}\right) \text{kPa}$$

This is less convenient to use when a dew-point temperature is to be calculated from a known vapour pressure. The curve-fit equation that has been demonstrated here is of sufficient accuracy for most manual estimations of condensation risk. Typical values of vapour pressure are given in Table 11.3. These will accommodate some applications without reference to tables or equations.

Table 11.3 Vapour pressures.

Air condition t_a (°C d.b.)	% Saturation	Dew-point t_{dp} (°C)	Vapour pressure (Pa)
−5	100	−5	402
−3	80	−5.6	381
0	80	−2.7	489
1	80	−1.8	526
2	80	−0.9	565
5	80	1.9	699
10	80	6.7	984
14	60	6.5	965
15	60	7.4	1030
20	50	9.4	1182
22	50	11.3	1339

Note: Data is approximate and is only to be used for examples in this book.

EXAMPLE 11.5

Calculate the total vapour resistance of a cavity wall constructed from 13 mm plaster, 100 mm aerated concrete, 40 mm mineral wool, 10 mm air space and 105 mm brickwork.

For each layer:

$$R_V = r_V \times l$$

For the whole structure:

$$\Sigma(R_V) = \Sigma(r_V \times l)$$

From Table 11.1 the vapour resistivities are plaster 50, aerated concrete 30, mineral wool 6 and brickwork 40. The surface films and air space have no resistance to the flow of vapour. Material thicknesses are used in metres.

$$\Sigma(R_V) = 50 \times 0.013 + 30 \times 0.1 + 6 \times 0.04 + 40 \times 0.105$$
$$= 8.09 \text{ GNs/kg}$$

EXAMPLE 11.6

The cavity wall in Example 11.5 is to be used for a dwelling exposed to an external environment of −1°C d.b., 80% saturation, where the heating system is designed to maintain the internal air at 22°C and 50% saturation. If the wall has a surface area of 110 m², find the moisture mass flow rate taking place through the wall.

From the CIBSE psychrometric chart, the internal and external air dew-point temperatures are found to be:

internal $t_{dp} = 11.5°C$

external $t_{dp} = -3.5°C$

The internal air vapour pressure is:

$p_{s1} = 600.245 \times \exp(0.0684 \times 11.5)$ Pa

$= 1318.09$ Pa

External air pressure is:

$p_{s2} = 600.245 \times \exp[0.0684 \times (-3.5)]$ Pa

$= 472.45$ Pa

Using $R_v = 8.09$ GN s/kg from Example 11.5, we obtain the moisture mass flow rate through the wall:

$$G = \frac{p_{s1} - p_{s2}}{R_v} \text{ kg/m}^2 \text{ s}$$

$$= (1318.09 - 472.45)\frac{N}{m^2} \times \frac{kg}{8.09 \text{ GN s}} \times \frac{1 \text{ GN}}{10^9 \text{ N}}$$

$$= 1.045 \times 10^{-7} \text{ kg/m}^2 \text{ s}$$

Temperature gradient

Heat flows through a structure from an area of high temperature to one of lower temperature. Homogeneous materials have a linear temperature gradient through their thickness, as shown in Figure 11.1. Temperature drops 1–2 and 3–4 are caused by the internal and external surface film resistances. To determine the surface and intermediate temperatures, the overall rate of heat, flow through the whole structure is equated with the individual heat flows in each slab:

$$Q_f W = U \frac{W}{m^2 \text{ K}} \times A m^2 \times (t_1 - t_5) \text{ K}$$

This same rate of heat flow Q_f also passes through the internal surface film, the concrete and the external surface film. Therefore:

$$Q_f = \frac{1}{R_{si}} \frac{W}{m^2 \text{ K}} \times A m^2 \times (t_1 - t_2) \text{ K}$$

$$Q_f = \frac{\lambda}{l} \frac{W}{m^2 \text{ K}} \times A m^2 \times (t_2 - t_4) \text{ K}$$

$$Q_f = \frac{1}{R_{so}} \frac{W}{m^2 \text{ K}} \times A m^2 \times (t_4 - t_5) \text{ K}$$

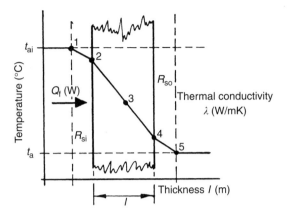

11.1 Temperature gradient through a solid construction.

The heat flow rate Q_f can easily be evaluated from U, t_{ai} and t_{ao}. The only unknowns in the other equations are the second temperatures t_2 and t_4. An intermediate temperature t_3 can be calculated at half the concrete thickness by using $l/2$:

$$Q_f = \frac{2\lambda}{l} \frac{W}{m^2\, K} \times A\, m^2 \times (t_2 - t_3)\, K$$

If the wall area is taken as $1\, m^2$:

$$t_2 = t_1 - Q_f\, R_{si}$$

$$Q_f = \frac{\lambda}{l}(t_2 - t_4)$$

$$t_4 = t_2 - \frac{l}{\lambda} Q_f$$

$$Q_f = \frac{1}{R_{so}}(t_4 - t_5)$$

$$t_5 = t_4 - Q_f\, R_{so}$$

Calculating the outdoor air temperature t_5 is a check on the accuracy of the calculations and method. It should agree with the original value used in finding Q_f to within $\pm 1\%$.
To find t_3, use:

$$Q_f = \frac{2\lambda}{l}(t_2 - t_3)$$

$$t_3 = t_2 - \left(\frac{l}{2\lambda} Q_f\right)$$

EXAMPLE 11.7

Calculate the temperature gradient through a medium-weight concrete block wall 100 mm thick λ 0.51 W/m K, $R_{si} = 0.12\,\text{m}^2$ K/W, $R_{so} = 0.06\,\text{m}^2$ K/W, internal and external air temperatures are 20°C d.b. and −1°C d.b.

Thermal transmittance is:

$$U = \frac{1}{R_{si} + l/\lambda + R_{so}}$$

$$= \frac{1}{0.12 + 0.1/0.51 + 0.06}\,\text{W/m}^2\,\text{K}$$

$$= 2.66\,\text{W/m}^2\,\text{K}$$

For a wall area of 1 m^2:

$$Q_2 = 2.66 \times [22 - (-1)]\,\text{W}$$

$$= 61.16\,\text{W}$$

Using the numbered locations in Figure 11.1:

$$t_1 = 22°C\,, t_5 = -1°C$$

$$t_2 = 22 - 61.16 \times 0.12 = 14.66°C$$

$$t_4 = 14.66 - \frac{0.1}{0.51} \times 61.16 = 2.67°C$$

$$t_5 = 2.67 - 61.16 \times 0.06 = -1°C$$

Which agrees with the input data:

$$t_3 = 14.66 - \left(\frac{0.1}{2 \times 0.51} \times 61.16\right) = 8.66°C$$

This should be the temperature midway between t_2 and t_4 i.e. $(14.66 \pm 2.67)/2 = 8.67°C$, which it is.

EXAMPLE 11.8

Calculate the temperature gradient through a cavity wall consisting of 13 mm lightweight plaster, 100 mm lightweight concrete block, 40 mm mineral fibre slab, 10 mm air space and 105 mm brickwork. Internal and external air temperatures are 22°C and 0°C. Thermal conductivities are as follows: plaster, λ_1 0.16 W/m K; concrete, λ_2 0.19 W/m K; mineral fibre, λ_3 0.035 W/m K; brickwork, λ_4 0.84 W/m K; $R_{si} = 0.12\,\text{m}^2$ K/W; $R_{so} = 0.06\,\text{m}^2$ K/W; R_a 0.18 m^2 K/W.

$$U = \left(R_{si} + \frac{l_1}{\lambda_1} + \frac{l_2}{\lambda_2} + \frac{l_3}{\lambda_3} + R_a + \frac{l_4}{\lambda_4} + R_{so} \right)^{-1}$$

$$= \left(0.12 + \frac{0.013}{0.16} + \frac{0.10}{0.19} + \frac{0.04}{0.035} + 0.18 + \frac{0.105}{0.84} + 0.06 \right)^{-1} \text{W/m}^2 \text{ K}$$

$$= 0.45 \text{ W/m}^2 \text{ K}$$

For a wall area of 1 m²:

$$Q_f = 0.45 \times (0.22 - 0) = 9.9 \text{ W}$$

Using the notation from Figure 11.2, temperatures are:

$$t_2 = 22 - (0.12 \times 9.9) = 20.81°C$$

$$t_3 = 20.81 - \left(\frac{0.013}{0.16} \times 9.9 \right) = 20.01°C$$

$$t_4 = 20.01 - \left(\frac{0.01}{0.19} \times 9.9 \right) = 14.8°C$$

$$t_5 = 14.8 - \left(\frac{0.04}{0.035} \times 9.9 \right) = 3.49°C$$

$$t_6 = 3.49 - (0.18 \times 9.9) = 1.71°C$$

$$t_7 = 1.71 - \left(\frac{0.105}{0.84} \times 9.9 \right) = 0.47°C$$

$$t_8 = 0.47 - (0.06 \times 9.9) = 0.12°C$$

A slight error of 0.55% has occurred as a result of rounding all the results to two decimal places. Notice the larger temperature drops across the two main insulating materials.

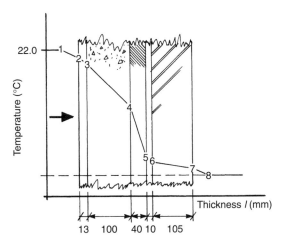

11.2 Temperature notation for Example 11.8.

Dew-point temperature gradient

Moist air passing through the structure from the high internal air vapour pressure to the lower external air vapour pressure will form a gradient of vapour pressures. The vapour pressure at any location is calculated from the mass flow rate of vapour G and vapour resistances in the same manner as for the thermal gradient.

The moist air dew-point temperature is calculated for each of these vapour pressures and another temperature gradient is drawn on the structural cross-section. If the thermally produced structure temperature equals or falls below the local air dew-point, then condensation will commence at that location. This information is used to decide whether a wall or roof will remain dry and whether a vapour barrier should be installed. The vapour barrier is fitted on the warm side of any zone of interstitial condensation.

Once the internal and external air vapour pressures, total vapour resistance and mass flow rate of vapour are known, the equation is:

$$G = \frac{p_{si} - p_{s2}}{R_v}$$

$$p_{s2} = p_{s1} - R_v G$$

Note that this is of the same form as:

$$t_2 = t_1 - RQ_f$$

Surface air films and cavities offer negligible resistance to the flow of moisture.

EXAMPLE 11.9

Calculate the dew-point gradient for the cavity wall in Example 11.8. Determine whether surface or interstitial condensation will take place. Internal and external percentage saturations are 50 and 100 respectively.

Referring to Figure 11.2:

$$p_{s1} = p_{s2}$$

$$p_{s5} = p_{s6}$$

$$p_{s7} = p_{s8}$$

$$p_{s3} = p_{s1} - (\text{vapour resistance of plater} \times G)$$

p_{s3} and p_{s1} are the vapour pressures on each side of the plaster, and

$$\text{vapour resistance of plaster} = r_v \times l$$

Similar equations are written for the other materials, with appropriate resistivity's r_v and thicknesses l. The resistivities are plaster 50, concrete 30, mineral wool 6 and brickwork 40.

From the CIBSE psychrometric chart (Figure 5.4), for internal air at 22°C d.b., 50% saturation, $t_{dp} = 11.5°C$, and for external air at 0°C d.b., 100% saturation, $t_{dp} = 0°C$. Then,

$$p_{s1} = 600.245 \times \exp(0.0684 \times 11.5) \, Pa$$

$$= 1318.09 \, Pa$$

$$p_{s8} = 600.245 \times \exp(0.0684 \times 0) \, Pa$$

$$= 600.245 \, Pa, \text{ because } (e^0 = 1).$$

Vapour resistance R_v for this wall was calculated in Example 11.5:

$$R_v = 8.09 \, GN \, s/kg$$

Mass flow of vapour per m^2 of wall area:

$$G = (1318.09 - 600.245)\frac{N}{m^2} \times \frac{kg}{8.09 \, GNs} \times \frac{1 \, GN}{10^9 \, N}$$

$$= 88.7 \times 10^{-9} \, kg/m^2 \, s$$

Vapour pressure and corresponding dew-point temperature can now be calculated for each numbered point through the wall:

$$p_{s3} = 1318.09\frac{N}{m^2} - 50 \times 0.013\frac{GN \, s}{kg} \times \frac{10^9 \, N}{1 \, GN} \times \frac{88.7}{10^9} \frac{kg}{m^2 \, s}$$

$$= (1318.09 - 57.7) \, N/m^2$$

$$= 1260.4 \, Pa$$

$$t_{dp3} = \frac{1}{0.0684} \times \ln\left(\frac{p_{s3}}{600.245}\right) °C$$

$$= \frac{1}{0.0684} \times \ln\left(\frac{1260.4}{600.245}\right) °C$$

$$= 10.85°C$$

$$p_{s4} = 1260.4 - 30 \times 0.1 \times 88.7 \, Pa$$

$$= 994.3 \, Pa$$

$$t_{dp4} = \frac{1}{0.0684} \times \ln\left(\frac{994.3}{600.245}\right) °C$$

$$= 7.38°C$$

$$p_{s5} = 994.3 - 6 \times 0.04 \times 88.7 \, Pa$$

$$= 973 \, Pa$$

$$t_{dp5} = \frac{1}{0.0684} \times \ln\left(\frac{973}{600.245}\right) °C$$

$$= 7.06°C$$

$$p_{s7} = 973 - 40 \times 0.105 \times 88.7 \text{ Pa}$$

$$= 600.5 \text{ Pa}$$

$$t_{dp7} = \frac{1}{0.0684} \times \ln\left(\frac{600.5}{600.245}\right) {}^{\circ}\text{C}$$

$$= 0.006{}^{\circ}\text{C} \text{ shows small calculation inacuracy}$$

$$= 0{}^{\circ}\text{C} \text{ the significant value}$$

Dew-point temperature gradient ends with the input data and is superimposed upon a scale drawing of the thermally induced gradient, shown in Figure 11.3.

Owing to the high thermal resistance and very low vapour resistance of the mineral fibre slabs fixed in the cavity, under the design conditions as stated, the material temperature drops below the moist air dew-point temperature midway through its thickness. Interstitial condensation will occur within the mineral fibre, air space and external brickwork. Ventilation of the remaining wall cavity allows evaporative removal of the droplets. Variation of internal and external air conditions will limit the duration of such temperature gradients. Periods of condensation will be very intermittent. Solar heat gains to the external brickwork will raise the temperature of the structure and help to reduce condensation periods. The walls most at risk are those always shaded from direct sunlight and having reduced wind exposure due to the proximity of nearby buildings.

When condensation takes place, a change of phase occurs as the water vapour turns into liquid. The calculations of water vapour transfer end at this discontinuity. The reason for undertaking the calculations was to establish if and where condensation is formed. The whole thickness of

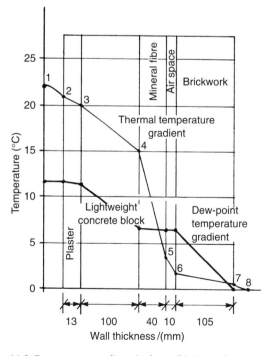

11.3 Temperature gradients in the wall in Example 11.9.

the layer where liquid forms is likely to be dampened owing to capillary attraction within porous solid material. The quantity of condensation that may be formed during a 60-day winter period can be assessed from the expected average air conditions. This aids the prediction of the physical damage that may be caused.

Installation note

Added thermal insulation can cause conditions where condensation will take place owing to the lowered structural temperatures. The installation of a vapour barrier raises the overall vapour resistance and may be able to keep the dew-point gradient below the thermally produced temperatures.

1. Materials with a low thermal resistance but a high vapour resistance, such as aluminium foil, sheet plastic, roofing felt and gloss paint, are placed on the warm side of the structure.
2. Materials of high thermal and vapour resistance can be placed anywhere.
3. Materials of moderate vapour resistance but high thermal resistance should be placed on the cold side of the structure. Such materials will require the addition of a weather-resistant coating when applied to the outside of a building.

Thermal or acoustic insulation applied to industrial roofs can increase the possibility of the occurrence of condensation. Alternative schemes are shown in Figures 11.4 and 11.5.

In the cold deck design, the roofing sheets remain at little more than the external air temperature. Ventilation through gaps at the eaves, junctions where the sheets overlap and holes around steel supports provide passageways for the ingress of moist air. Condensation on the underside of the cold deck will cause water to run down the steelwork and wet the ceiling. It is unlikely that a tight seal can be made between the vapour barrier and the steel supports to

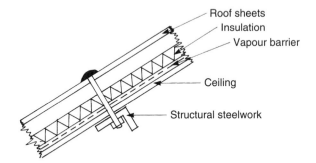

11.4 Cold deck roof.

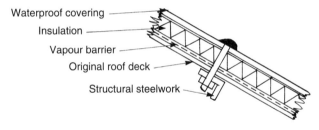

11.5 Warm deck roof.

avoid this happening. Before insulation of an existing roof, its underside is maintained at above the local dew-point except during severe weather or when the heating plant is off.

Warm deck design improves the weather resistance of the roof and raises the original underside surface temperature even further. Interstitial condensation is unlikely because of the vapour barriers.

EXAMPLE 11.10

Calculate the external air temperature that will cause condensation to form on the underside of a factory roof constructed from 10 mm corrugated asbestos cement sheet. Internal air conditions near the roof are 23°C d.b. and 30% saturation. Thermal conductivity of asbestos cement sheet is 0.4 W/m K. The roof has a severe exposure. Internal air dew-point is 5°C, R_{si} 0.1 m² K/W, R_{so} 0.02 m² K/W.

Let the unknown external air temperature be t_o. Then for a roof area of 1 m²:

$$Q_f = U(23 - t_o)\,\text{W/m}^2$$

For the external surface film:

$$Q_f = \frac{1}{R_{si}}(23 - 5)\,\text{W/m}^2$$

$$= \frac{1}{0.1} \times 18\,\text{W/m}^2$$

$$= 180\,\text{W/m}^2$$

For the roof:

$$U = \frac{1}{R_{si} + l/\lambda + R_{so}}$$

$$= \frac{1}{0.1 + 0.01/0.4 + 0.02}$$

$$= 6.9\,\text{W/m}^2$$

Rate of heat flow through roof sheets = rate of heat flow through internal surface film.

$$6.9\frac{\text{W}}{\text{m}^2\,\text{K}} \times (23 - t_o)\,\text{K} = 180\frac{\text{W}}{\text{m}^2}$$

$$26.0 = 23 - t_o$$

$$t_o = -3.09°\text{C}$$

Questions

1. Describe how the following forms of condensation occur: temporary, permanent and interstitial.

2. List the sources of moisture in buildings.
3. What is the purpose of installing a vapour barrier and what effect does it have on the dew-point temperatures within a structure?
4. Discuss the use of thermal insulation in reducing the likelihood of condensation in walls and roofs.
5. State examples of thermal insulation increasing the risk that condensation occurs.
6. List the actions that could be taken to reduce the water vapour input to a dwelling.
7. Discuss the use of heating and ventilation in combating condensation problems.
8. Why might prefabricated concrete buildings suffer more from condensation than other constructions?
9. What sources of moisture would you look for when consulted about mould growth on a building?
10. Describe the constituent parts of the atmospheric pressure.
11. What drives water vapour from one area to another?
12. Describe the way in which moisture is alternatively stored and released by porous building materials.
13. What is the flow of vapour through a composite structure analogous to?
14. State the conditions under which water vapour will condense on or within a construction.
15. Calculate the temperature gradients through the following structures. Internal and external air temperatures are to be taken as 21°C d.b. and −1°C d.b. Assume that $t_a = t_e$. The answers should be expressed as the surface or interface temperatures in descending order from the warm side. Outside surfaces are taken as having a high emissivity and normal exposure. All air spaces are ventilated. The thermal conductivity of glass is 1.05 W/m K.

    1. 6 mm single-glazed window.
    2. 6 mm double-glazed window.
    3. Cavity wall of 15 mm dense plaster, 100 mm lightweight concrete block, air space and 105 mm brick.
    4. An industrial roof of 10 mm asbestos cement corrugated sheet which has been given an external coating of 50 mm phenolic foam. The thermal conductivity of asbestos sheet is 0.4 W/m K.

16. A shop window consists of 6 mm plate glass in an aluminium frame. The display area air temperature is expected to be 15°C d.b. and to have a dew-point of 7°C. Find the external air temperature that will start to produce condensation on the inside of the window. The window has normal exposure.
17. If a double-glazed window is to be fitted in the shop in Question 16, what could the external air temperature drop to before condensation starts?
18. A hospital ward is to be maintained at 24°C d.b. and 80% saturation. The air dew-point is 20.5°C. Thermal insulation is to be added to the inside of the existing wall to avoid surface condensation when the external air temperature falls to −5°C d.b. The U value of the original wall is 1.9 W/m² K. Calculate the thickness of insulation material required if its thermal conductivity is 0.06 W/m K.
19. Calculate the thermal transmittance, temperature and dew-point gradients through a flat roof consisting of 25 mm stone chippings, 10 mm roofing felt, 150 mm aerated concrete, 75 mm wood wool/cement slabs, a ventilated 50 mm air space and 12 mm plasterboard. The roof has a sheltered exposure. Internal air conditions are 22°C d.b., 50% saturation. External conditions are 2°C d.b., 80% saturation. The stone chippings have a high emissivity when weathered and offer no resistance to the flow of water vapour. Plot a graph of the two temperature gradients and find if condensation is likely to occur.

20. Calculate the thermal transmittance, temperature and dew-point gradients through a wall consisting of 15 mm dense plaster, 100 mm medium-weight concrete block work, 40 mm glass fibre quilt, a ventilated 10 mm air space and 105 mm brickwork. The wall has a severe exposure. The average internal air conditions are 14°C d.b., 60% saturation when the external conditions are 1°C d.b., 80% saturation. Plot a graph of the two temperature gradients and find if condensation is likely to occur.

21. Where are indoor condensation problems are most likely found?

    1. Hot dry ambient air locations.
    2. Hot humid ambient air locations.
    3. Temperate maritime climates, such as the UK where outdoor air humidity remains high.
    4. Below zero ambient air locations.
    5. In any building anywhere.

22. Why does condensation occur within a building?

    1. Users leave taps running, baths full, water evaporates and deposits onto room surfaces, making air moist.
    2. Kettles, cooking, fish tanks and open bowls of water evaporate more water vapour into room than ventilation can remove.
    3. Porous building materials provide pathways for cold moisture to ingress a warm building and make indoor surfaces damp.
    4. Evaporated water within the building meets surfaces at below dew-point temperature.
    5. Cyclic variation of indoor surface temperatures always produces below dew-point locations.

23. What always combats condensation problems within occupied buildings?

    1. Thermal insulation.
    2. Impervious building materials.
    3. Removing open water surfaces.
    4. Air conditioning.
    5. Heating and ventilation.

24. From where does water vapour originate within a building?

    1. Atmospheric rain.
    2. Wind-driven atmospheric humidity.
    3. Occupants and their activities.
    4. Lack of sufficient natural and mechanical ventilation.
    5. Refrigeration systems and food storage.

25. What is the relationship of building materials to moisture mass transfer?

    1. There is none, building materials do not leak water.
    2. Good design and construction remove all moisture issues.
    3. Modern building materials have zero permeability.
    4. Porous structural materials absorb and pass moisture.
    5. Vapour barriers isolate brick, concrete and thermal insulation materials from moisture generated within a building.

26. What drives moisture flow through a structure?

    1. Vapour pressure difference in Pascals.
    2. Air temperature difference in °C d.b.
    3. Percentage saturation difference.

4.   Air moisture content difference in kg H_2O/kg dry air.
5.   Wet bulb air temperature difference in °C w.b.

27. Which is correct about condensation and mould growth?

1.   Always occurs in buildings over 20 years old.
2.   Cannot happen with current design standards.
3.   Readily forms in the UK in room corners, on window sills, in cupboards on external walls and within structures having an overall thermal transmittance of over 1.4 W/m^2 K.
4.   Must be removed, the surface gloss painted and outdoor air ventilation minimized.
5.   Impervious external surface materials need to be matched with an outdoor vapour barrier to stop moisture flowing into the wall, floor or roof structure.

28. Vapour diffusion into a structure that then condenses is called what type?

1.   Adiabatic.
2.   Complex.
3.   Leakage.
4.   Interstitial.
5.   Intermediate.

29. What is happening in a building during vapour diffusion?

1.   Odours and gases produced indoors slowly percolate through the building structure to outdoors.
2.   Steam from water boiling, cooking and hot water washing become absorbed into furniture, furnishings, carpets and surface plaster unless removed by ventilation.
3.   Internally sourced water vapour migrates through porous building structures.
4.   Low indoor vapour pressure drives moisture towards higher outdoor moist air vapour pressure.
5.   Liquid water passes through the building structure.

30. In the context of condensation in building, what does r_v symbolize?

1.   Vapour resistivity of a material.
2.   Vapour resistivity of a structure.
3.   Resistance to vapour flow of a structure.
4.   Volumetric resistivity of a structure.
5.   Volumetric resistance of a material to moisture transfer.

31. Which is true?

1. 10^3 kN s/g m $=$ GN s/kg m.
2. 10^3 N s/g $=$ MN s/g.
3.   MN s/g m $=$ GN s/kg m.
4. 10^6 N s/kg $=$ MN s/g.
5.   kN s/kg $=$ MN s/g.

32. What is the vapour resistivity of a structural material measured in?

1.   kN/m^2 s.
2.   MN s.
3.   kN s/kg.
4.   GN s/kg m.
5.   kN kg/m^2 s.

12 Gas

Learning objectives

Study of this chapter will enable the reader to:

1. calculate the flow of gas required by an appliance;
2. know how to measure and calculate gas pressure;
3. choose suitable gas pipe sizes;
4. describe gas service entry;
5. understand the working of a gas meter;
6. identify the space requirement for gas meters;
7. describe gas flue systems;
8. understand how gas combustion is controlled and regulated;
9. know how gas systems are operated safely.

Key terms and concepts

flue systems 273; gas burner controls 275; gas flow rate 269; gas meter 273; gas service entry 272; gross calorific value 269; ignition and safety controls 275; pipe size 272; pressure 269; U-tube manometer 269; ventilation 273.

Introduction

Gas services are provided to most buildings, and safety is of paramount importance. Economy is also important, and the versatility and controllability of gas are appreciated. It is used for heating, hot-water production, refrigeration for small and large cooling loads, electrical power generation, cooking and decorative heating.

Gas is converted from its primary fuel state into useful energy at the point of use. Its distribution energy loss is accounted for in the standing charge to the final consumer and, although it is charged for, does not appear to be related to the load, as for water pipes or electricity cables. The use of natural gas and, ultimately, an artificially produced substitute from coal and oil is a highly efficient use of primary resources, and all efforts are directed at continuing this trend. This chapter introduces the calculation of gas flow rate into a load and the gas pressure measurement that is used to monitor the flow rate and condition. The sizing of pipe work depends on the gas pressure of the incoming service and that required by the final gas-burning appliance in order to maintain the correct combustion rate. Incoming gas service provisions, metering and particular flue systems are explained. Gas is more suitable than any other fuel for low-level flue gas discharge, provided that sufficient dilution with fresh air is available. Typical methods of gas burner control and pressure reduction are explained.

Gas pipe sizing

The gas flow rate required for an appliance can be found from the manufacturer's literature or calculated from its heat output and efficiency:

$$\text{appliance efficiency } \eta = \frac{\text{heat output into water or air}}{\text{heat input from combustion of gas}} \times 100\%$$

Most gas appliances have an efficiency of 75%. The gross calorific value (GCV) of natural gas (methane) is $39 \, MJ/m^3$. If the appliance heat output is SH kW, then the gas flow rate Q required is,

$$Q = \frac{SH \, kW}{\eta} \times \frac{1 \, kJ}{1 \, kWs} \times \frac{1 \, MJ}{10^3 \, kJ} \times \frac{1}{GCV} \frac{m^3}{MJ} \times \frac{10^3 \, l}{1 \, m^3}$$

$$= \frac{SH}{\eta \times GCV} l/s$$

The maximum allowable gas pressure drop due to pipe friction between the gas meter outlet and the appliance will normally be 75 Pa. Gas pressure in a pipe is measured with a U-tube water manometer as shown in Figure 12.1. The water in the U-tube manometer is displaced by the gas pressure until the gas and water pressures in the two limbs are balanced. The atmosphere acts equally on both limbs under normal circumstances and all pressures are measured above atmospheric. These are called gauge pressures.

Methane has a specific gravity (SG) of 0.55 and so its density is:

$$\rho_{gas} = SG \times \rho_{air}$$

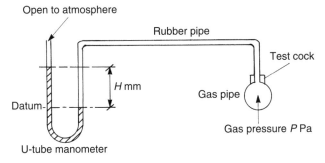

12.1 Measurement of gas pressure.

The standard density of air at 20°C d.b. and 1013.25 mb barometric pressure is 1.2 kg/m^3. Thus,

$$\rho_{gas} = 0.55 \times 1.2 \text{ kg/m}^3$$

$$= 0.66 \text{ kg/m}^3$$

The water column H in the manometer shows the net effect of the gas pressure P and the different density of the column H of gas in the right-hand limb. At the datum, the pressure exerted by the left-hand column equals the pressure exerted by the right-hand column.

$$P_{water} = P - \text{pressure of column } H \text{ of gas}$$

The pressure at the base of a column of fluid is ρgH where $g = 9.807$ m/s^2. The density of water ρ_w is 10^3 kg/m^3. Therefore,

$$\rho_w gH = P - \rho_{gas} gH$$

$$P = \rho_w gH - \rho_{gas} gH$$

$$= gH(\rho_w - \rho_{gas})$$

Thus:

$$P = 9.807 \frac{m}{s^2} \times H\text{ mm} \times \frac{1 \text{ m}}{10^3 \text{ mm}} \times (1000 - 0.66)\frac{kg}{m^3}$$

$$= \frac{9.807 \times 999.34}{1000} \times H \frac{kg}{m s^2} \times \frac{1 \text{ N} s^2}{1 \text{ kg m}} \times \frac{1 \text{ Pa m}^2}{1 \text{ N}}$$

$$= 9.801 \, H \text{ Pa}$$

A sufficiently good approximation is $P = gH$ Pa, where H is in millimetres of water gauge.

EXAMPLE 12.1

A gas-fired boiler has a heat output of 280 kW and an efficiency of 82%. Calculate the flow rate of gas required.

$$Q = \frac{SH}{\eta \times GCV} = \frac{280}{0.82 \times 39} = 8.76 \text{ l/s}$$

EXAMPLE 12.2

Calculate the gas pressure if a water manometer shows a level difference of 148 mm.

$$P = 9.807 \times 148 \text{ Pa}$$

$$= 1450 \text{ Pa} \times \frac{1 \text{ kPa}}{10^3 \text{ Pa}}$$

$$= 1.45 \text{ kPa}$$

EXAMPLE 12.3

A gas-fired warm-air unit requires a gas pressure of 475 Pa at the burner test cock. What reading will be found on a water manometer?

$P = gH$ Pa

$$H = \frac{P}{g} = \frac{475}{9.807} = 48.5 \text{ mm}$$

Gas pressure may be expressed in millibars (mb) as:

1 bar = 100 000 Pa

1 bar = 100 kPa

$$1 \text{ mb} = \frac{100 \text{ kPa}}{1000} = 0.1 \text{ kPa}$$

1 mb = 100 Pa

The gas pressure loss rate $\Delta p / EL$ along a pipeline is needed to find the pipe diameter. EL is the equivalent length of the measured straight pipe, bends and fittings. An initial estimate can be made for the flow resistance of pipe fittings by adding 25% to the straight lengths L m of pipe.

EXAMPLE 12.4

Calculate the pressure loss rate in a gas pipeline from the meter to a water heater. The pipe has a measured length of 34 m.

$EL = (1.25 \times 34) \text{m} = 42.5 \text{ m}$

The allowable pressure loss Δp is 75 Pa. Thus

$$\frac{\Delta p}{EL} = \frac{75 \text{ Pa}}{42.5 \text{ m}} = 1.76 \text{ Pa/m}$$

EXAMPLE 12.5

A pressure loss rate of 4.2 Pa/m is to be used for sizing a gas pipeline in a house. Calculate the maximum length of pipe that can be used if the resistance of the pipe fittings amounts to 20% of the installed pipe run.

$$\frac{\Delta P}{EL} = \frac{75\,Pa}{EL\,m} = 4.2\,Pa/m$$

$$EL = \frac{75}{4.2} = 17.9\,m$$

$$EL = 1.2\,l$$

$$l = \frac{17.9}{1.2} = 14.9\,m$$

This is the maximum length of run.

The gas pressure of the incoming service will be up to 5 kPa and this is reduced by a governor at the meter inlet to give 2 kPa in the installation within the building. For large gas-burning equipment, the gas pressure may have to be increased with a booster. A boosting system comprises an electrically driven reciprocating compressor, a high-pressure storage tank and automatic pressure and safety controls. Pipe diameters can be found from Table 12.1.

EXAMPLE 12.6

Find the pipe size required for a gas service carrying 1.75 l/s and having an allowable pressure loss rate of 5.1 Pa/m.

From Table 12.1, using the nearest pressure drop below 5.1 Pa/m, in this case, 5 Pa/m, a 28 mm copper pipe can carry 1.8 1/s and would be suitable.

Gas service entry into a building

The gas service pipe from the road main should slope up to the entry point to the building, at right angles to the road main and entering the building at the nearest convenient place. Ground cover of 375 mm is maintained and new pipe work is made of plastic. When old steel services are renewed, the plastic pipe can be run inside the steel. A meter compartment can be built into the external wall in housing installations and the service clipped to the wall

Table 12.1 Flow of methane (natural gas) in copper pipes

$\Delta p/EL$ (Pa/m)	Gas flow rate Q l/s for pipe diameters			
	15 mm	22 mm	28 mm	32 mm
1	0.08	0.31	0.69	1.22
2	0.16	0.47	1.05	1.84
3	0.21	0.59	1.33	2.34
5	0.29	0.81	1.8	3.15
7	0.35	0.98	2.2	3.83
10	0.44	1.21	2.7	4.71

Note: Data in this table is approximate and is only for use within the examples of this book.

under a cover. This facilitates meter reading without entry to the property. Computer monitoring of energy meters using a telephone link to the supply authority will eventually replace manual reading. Where the meter compartment is inside the building, the service should pass through the foundations in a pipe sleeve, plugged to stop the ingress of moisture and insects but allowing for some movement. A 300 mm square pit is provided in a concrete floor to allow the service to rise vertically to the meter. The pit can subsequently be filled with concrete. The meter compartment must not be under the only means of escape in the event of a fire in a building where there are two or more storeys above the ground floor, unless it is located in an enclosure having a minimum fire resistance of half an hour.

Gas service pipes, meters and appliances should always be in naturally ventilated spaces, as dilution with outside air is the best safety precaution against the accumulation of an explosive mixture with air. Early detection of leaks is essential, but ventilation assists the dilution of leaks. Gas detectors can be provided as an additional precaution.

Domestic credit meters pass up to 10 l/min, 0.17 l/s and are 212 mm wide, 270 mm high and 155 mm deep. Their overall space requirement is approximately double the width and height measurements for pipe work, valve and filter. Industrial meters have flanged steel pipe work up to 100 mm in diameter and a bypass to allow uninterrupted gas flow in the event of meter breakdown. A 500 l/s meter is 2 m wide, 2.25 m high and 1.6 m deep. Due allowance must be made for doorways and access for replacing the meter during the building's use. A separate meter room is recommended, which should be secure, accessible, illuminated and weatherproof with no hot pipes or surfaces.

Manufactured town gas came from the conversion of coal or oil. It had a high hydrogen content and flame speed but its cross-calorific value was half that of methane. In future, substitute natural gas (SNG) may be manufactured from hydrocarbons as indigenous reserves become exhausted. SNG will come from the chemical conversion of coal, tar sand or crude oil and will have characteristics similar to those of methane.

Gas pipes or meters should usually be spaced 50 mm from electrical cables, conduits, telecommunications cables or other conductors. Electric and gas meters may be accommodated in a single compartment if a fire-resistant partition separates them.

Positive displacement mechanically operated meters are used as the billing meter. These meters have three compartments with a horizontal valve plate near the top of the casing and a vertical division plate in the lower section. Bellows formed by a metal disc surrounded by a leather diaphragm are located on each side of the division plate to measure the gas flow. Rotary meters may be used for downstream gas flow metering for energy management purposes and these may be logged by a BMS.

Flue systems for gas appliances

Gas appliances can be flued by a wide variety of methods, as the products of combustion are mainly water vapour, carbon dioxide, nitrogen and oxygen, at a temperature of about 95°C after the draught diverter. The function of the draught diverter is to discharge flue products into the boiler room during a down-draught through the chimney. Such reverse flows occur infrequently for a few seconds during adverse wind conditions. Diversion ensures that the correct combustion process is not interrupted. It stops the pilot flame from being blown out, with consequential shut-down of the appliance until manual ignition is arranged. The draught diverter also dilutes the products of combustion by entrainment of room air into the flue. A carbon dioxide concentration of 4% by volume is found in the secondary flue after the diverter. The primary flue pipes are those before the diverter. Flue systems are described below.

Brick chimney

New masonry chimneys must be lined with vitrified clay or stainless steel pipe. Existing chimneys may incorporate a stainless steel flexible flue liner, which can be pulled through an existing chimney with a rope and rounded plug. The liner has the same diameter as the appliance flue outlet, often 125 mm for domestic appliances, and is built into the top of the chimney with a plate to form a sealed air space between the liner and the brickwork. This acts as thermal insulation to maintain flue gas temperature. If the flue gases were allowed to cool to below about 25°C, condensation of the water vapour would occur and deterioration of the metal and brickwork would reduce serviceability. Asbestos cement or glazed earthenware pipes can be built into new chimneys for protection of the brickwork. A cowl is fitted to the flue to reduce the ingress of rain and the possibility of down-draughts.

Free-standing pipe

A free-standing pipe flue for a gas appliance is double-walled stainless steel with thermal insulation between the inner and outer pipes. A flue pipe taken through a roof is fitted with a malleable slate to weatherproof the junction. The terminal should be 600 mm from the roof surface and clear of windows or roof-lights. An internal flue from a small domestic appliance can be connected to a ridge terminal. An externally run asbestos cement flue pipe has a branch tee junction at its emergence through the wall. A 25 mm copper drain pipe takes condensation to a drain gulley.

Balanced flue

Figure 12.2 shows the balanced-flue system used for boilers, warm air units, convectors and water heaters. It is used for appliance ratings up to around 30 kW. External wind pressure is applied equally to the combustion air inlet and the flue gas outlet parts of the combined terminal. The only pressure difference causing air flow through the appliance is that caused by combustion. The flue terminal should not be underneath a window or within 0.5 m of a doorway or openable window. It should not be located in a confined corner where external air flow might be restricted. Fan-assisted balanced flues have been used and these allow more flexibility

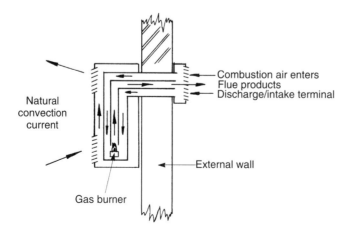

12.2 Balanced-flue gas appliance.

in siting the appliance further away from the terminal. Balanced-flue appliances are also called room-sealed appliances.

Built ducts

Room-sealed appliances in multi-storey flats are connected to a vertical precast concrete shaft extending from the fresh air inlet grille at ground level to a terminal on the roof. Combustion air is taken from the duct by each heater and its flue products are passed into the shaft. The duct is sized so that sufficient ventilation is provided for the whole installation. With a U-duct a separate combustion air inlet duct takes air from the roof downwards to the lowest appliance, and then the upward duct acts in the same manner as the Se-duct. A shunt duct has precast concrete wall blocks, 100 mm wide, with a rectangular flue passage built into partition walls or the inner leaf of a cavity wall. A continuous flue way is formed for each heater, often a gas fire, to ceiling level. A pipe then connects to a ridge terminal. Several flues built into a wall side by side are called a shunt duct system.

Fan-diluted flue

Fan-diluted flues are mainly used in commercial buildings where a conventional flue pipe and terminal could not be used or would be unsightly: for example, in a shopping precinct. Fresh air enters a galvanized sheet metal duct, which passes through the boiler plant room and discharges back into the atmosphere. A centrifugal or axial flow fan in the duct is started before the boilers are ignited and an air flow switch cuts the burners off in the event of fan failure. Secondary flue pipes from the boilers are connected into the duct on the suction side of the fan. Dilution of the combustion products takes place and the discharge air from the system may contain as little as 0.5% carbon dioxide and be down to 30°C. Any condensing moisture is carried by the high-velocity air stream and is dispersed as steam into the atmosphere. Air inlet and discharge louvres should be positioned on the same external wall to balance wind pressures on each. The discharge can be made into a shopping arcade or covered walkway to make use of the available heat. Careful fan selection is essential to avoid creating a noise nuisance. Figure 12.3 shows a fan-diluted flue installation for one boiler.

Boosted flue

A domestic boiler may have a booster centrifugal fan fitted into its flue pipe to allow a long horizontal run or even a downwards run. The pipe diameter can be smaller than the boiler flue outlet diameter and the fan pressure rise is used to overcome the frictional resistance. A typical installation is shown in Figure 12.4.

Ignition and safety controls

Natural gas is burnt in an aerated burner in which half the air needed for combustion is entrained into the gas pipeline by a nozzle and venturi throat. This premixed gas and air goes to the burner, which is often a perforated plate through which the mixture passes. Further mixing occurs above the plate and the flame is ignited by a permanently lit pilot jet. A sheet of clear blue flame is established over the top of the burner plate or matrix. Large gas boilers, over 45 kW, use forced-draught burners in which gas and air are blown under pressure into a swirl chamber where the flame is established, with a fair amount of noise.

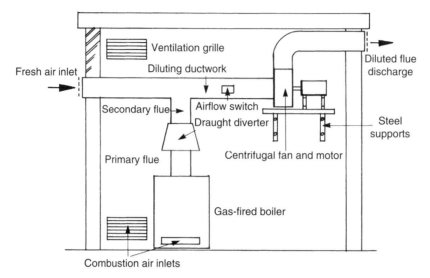

12.3 Fan-diluted flue.

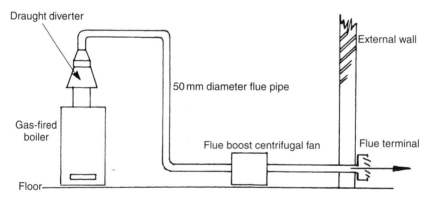

12.4 Boosted flue.

   Gas burner control of appliances less than 45 kW is achieved as follows. The pilot flame, which is ignited manually or with a piezoelectric spark, heats a thermocouple circuit whose electrical voltage and current holds open the flame-failure solenoid valve. In the event that the pilot flame is extinguished, the flame-failure solenoid becomes de-energized because the thermocouple is no longer heated, and the main gas supply is stopped. As long as the pilot is alight, the control thermostat in the boiler waterway or warm-air unit outlet duct is able to ignite the burner by opening its own solenoid valve. Figure 12.5 shows the diagrammatic arrangement of the control system for a gas burner of less than 45 kW. A combination gas valve is used to incorporate some of these functions into one unit.

   The governor maintains a constant gas pressure to the burner by means of a synthetic nitrile rubber diaphragm which rises when the inlet pressure is increased. The diaphragm is connected to a valve, which closes when the diaphragm rises. This action increases the resistance to gas flow and maintains the outlet gas pressure at the set value. An adjustable spring is used to set the downstream pressure appropriate to the gas flow rate required by the burner.

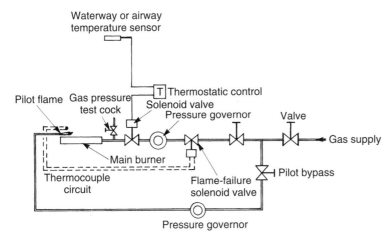

Waterway or airway
temperature sensor

Pilot flame Gas pressure┌─T Thermostatic control
          test cock   Solenoid valve
                       Pressure governor   Valve

Gas supply

Main burner

Thermocouple   Flame-failure   Pilot bypass
circuit   solenoid valve

Pressure governor

12.5 Gas burner controls.

Questions

1. A gas-fired water heater has a heat output of 30 kW at an efficiency of 75% and a gas pressure of 1225 Pa. Calculate the gas flow rate required at the burner and the reading on a U-tube manometer in millimeters water gauge at the outlet from the pressure governor.

2. Express gas pressures of 55 mm H_2O, 350 N/m^2, 75 Pa, 1.5 kPa and 1.05 bars in millibars.

3. The pipe from a gas meter to a boiler is 18 m long and has elbows that cause a resistance equivalent to 25% of the measured length. Calculate the maximum allowable pressure loss rate for the pipeline.

4. If the maximum allowable pressure loss rate in a pipeline can be 2.3 Pa/m and the resistance of the pipe fittings amounts to 20% of straight pipe, what is the maximum length of pipe that can be used?

5. A gas boiler of 43 kW heat output and 75% efficiency is supplied from a meter by a pipe 23 m long. The resistance of the fittings amounts to 25% of the pipe length. Find the gas supply pipe size needed.

6. Calculate the actual gas pressure drop through a 22 mm pipe carrying 0.81 l/s when the pipe length is 12 m and the fittings resistance amounts to 20% of its length.

7. Sketch and describe the gas service entry and meter compartment arrangements for housing.

8. Explain how a gas meter measures gas flow rate and total quantity passed during a year.

9. List the methods of flueing gas appliances and compare them in relation to their application, complexity and expected cost.

10. Explain, with the aid of sketches, the sequence of operation of safety and efficiency controls on gas fired appliances.

11. Around what pressures do natural gas and liquefied petroleum gas run at in pipe distribution systems in buildings? More than one correct answer is possible.

    1. Up to 40 atmospheres.
    2. Usually above 200 kPa.
    3. Up to 3,000 mm water gauge.
    4. 20 to 200 mm water gauge.
    5. Less than 100 kPa.

12. How is natural gas metering for billing carried out?

    1.   Rotary gas meter monitored by the BMS.
    2.   Flow rate measured from the pressure drop through an orifice plate in the pipeline.
    3.   Positive displacement gas consumption meter.
    4.   Pitot-static tube in pipeline and a data logger.
    5.   Rotary vane anemometer in pipeline and an integrating revolution-counter meter.

13. What happens to water vapour in flue gas?

    1.   Condenses when flue gas cools to water vapour dew point temperature.
    2.   Remains as vapour at all times.
    3.   Cools and appears as steam discharging into atmosphere.
    4.   Becomes absorbed into flue system materials and drains.
    5.   Combines with other flue gases.

13 Plant and service areas

Learning objectives

Study of this chapter will enable the reader to:

1. identify the actions necessary prior to commencement of a construction, in order to facilitate the correct provision of utility services;
2. coordinate utility services under public footpaths;
3. design suitable routes for utility services;
4. calculate the areas buildings need for services plant;
5. find the sizes of plant room needed for all services from preliminary building information at the design stage;
6. allocate routes for services through the building structure;
7. understand the need for fire barriers in service ducts and correctly locate them;
8. choose suitable sizes for pipe service ducts;
9. understand the need to allow space for pipes crossing over each other;
10. estimate sizes for service shafts carrying air ducts;
11. know the requirements for walkway and crawl way ducts;
12. understand the need for expansion provision in pipes;
13. identify the ways in which pipes can be supported with allowance for thermal movement;
14. understand the ways in which thermal movement is accommodated;
15. know how thermal, fire, support and vibration measures are applied to pipes air ducts passing through fire barriers;
16. apply flexible connections to plant;
17. appreciate the need for coordinated drawings;
18. understand the use of services zones;
19. be able to design boiler house ventilation.

Key terms and concepts

air-duct support 295; air-handling plant 284; boiler room 283; cold- and hot-water storage 283; combustion air ventilation 295; computer servers 284; cooling plant 284; coordinated drawings 295; coordinated service trench 282; crawl ways and walkways 288; data centres 285; electrical substation 286; fire compartment 287; flexible connection 296; foam seal 293; footpath 282; fuel handling 283; highway 281; lifts 285; noise and vibration 296; pipe anchor, guide, roller, loop expansion and insulation 292; pipe bellows, articulated and sliding joint 293; public utilities 280; service ducts 286; service plant area 288; services identification 281; services zones 296; standby generator 286; temporary works 280; underfloor pipe ducts 286.

Introduction

The building design team needs information on the size and location of services plant spaces and their interconnecting ducts before the engineer has sufficient data on which to base calculations. There may be only general definitions of the spaces to be heated or air conditioned and preliminary design drawings. Past experience of similar constructions reveals the likely plant and service duct requirements. The use of such data is explained and worked examples are used. The planning of external utility supplies is shown, together with typical arrangements for internal multi-service ducts. Support and expansion provision for distribution services is demonstrated, as is the design of combustion air ventilation for boilers.

Mains and services

The planning and liaison with public utilities must be included in the initial application made by the developer for planning permission. Each utility requires detailed information at the estate design stage in order to facilitate:

1. siting of plant or governor houses, substations, service reservoirs, water towers and other large items of apparatus, and also early completion of associated easements and acquisition of and early access to land in order to ensure service to the development by the programmed date;
2. design of mains and service layouts;
3. location of and requirements for road crossings;
4. provision and displacement of highway drainage;
5. programming of cut-offs from existing premises that are to be demolished;
6. arrangements for protecting and/or diverting existing plant and services;
7. provision of supplies to individual phases of the development, including temporary works services;
8. acquisition of materials and manpower resources;
9. siting of service termination and/or meter positions in premises and service entry details;
10. provision of meter-reading facilities;
11. provision of public lighting.

Developers must provide information on:

1. the intended position of public carriageways, verges, footpaths, amenity areas and open spaces;

2. existing and proposed ground levels;
3. position and level of proposed foul and surface-water sewers and any underground structures.

The utility will inform the developer of the need to close or restrict roadways, and these matters will then be discussed with the local authorities. All main services to more than one dwelling should be located on land adopted by the Highway Authority. The location on private property of a main designed to serve a number of dwellings can lead to friction between residents if excavation for repairs or maintenance is needed, and also makes it difficult for the utilities to gain ready access in an emergency. With the exception of road crossings, mains and services other than sewers should not be placed in the carriageways. The routes chosen should be straight and on the side of the carriageway serving most properties. Any changes of slope should be gradual. The prior approval of the utilities must be sought if landscaping will alter the levels of underground services. Public sewers must be laid to appropriate levels and gradients in straight lines between manholes, usually under the carriageway. An underground clear width of 1.8 m is needed between a private boundary and the kerb foundations but extra allowance should be agreed for:

1. fire hydrants;
2. inspection covers and manholes;
3. large-radius bends for pipes and cables;
4. fuel oil distribution pipes;
5. district heating pipes and manholes;
6. through-services not connected to the development;
7. cross-connections between services to form ring mains rather than dead ends;
8. imposed loads from adjacent buildings – medium-pressure gas mains must be 2 m from the building line.

Protective measures are taken where there is a risk to pipes or cables from vehicles that may park on soft ground. Where special paving is used, early consultation with the utilities will help to avoid subsequent defacement due to maintenance work. Footpaths should be used for the utilities. Sewers need to be laid in conjunction with the early stages of road building. Utilities operate under statutory powers and will not carry out work as a subcontractor. On completion of site construction, a copy of the plans showing the installed routes and details of the mains and services is sent to each utility to enable permanent records to be established. The minimum dimensions for the locations of mains and services under a pavement are shown in Figure 13.1. Brick tiles, concrete covers or yellow marker tapes are put over 11 kV and 415 V electricity cables. The 11 kV cables have a red PVC over-sheath and 415 V cables have a black PVC over-sheath.

Mains and services are surrounded with selected back-fill that is free of sharp or hard objects, and the trench is filled and compacted with earth that is free of rubble or site debris.

Plant room space requirements

General provision

Coordination of the services with architectural and structural design is required at the earliest possible moment during conception of the project. Plant rooms are places of physical work as well as equipment locations. Access to them is restricted to approved people under the control of the building manager. Safety, noise control, natural and mechanical ventilation, heating and

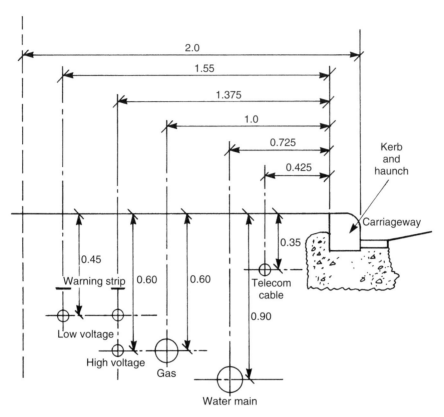

13.1 Positions of main services in straight routes under footpaths on residential unit.

cooling, ease of access from stairways and lifts, equipment and exit signage, adequate lighting and low voltage power outlets all make for a successful building. Many plant rooms remain almost unchanged for 25 years or more, so a well thought-out facility that can be maintained is more than an afterthought. Keeping them clean, decorated and well lit is an important part of the workplace. Telephone communication with the building manager from each plant room is a life-safety issue as well as a vital means of communication when work is being conducted or faults occur; mobile phones may be out of contact due to the heavy presence of concrete, steel and noise. Plant rooms must not be used to store flammable materials, cleaning fluids, cleaning machines, furniture; they are not junk or store rooms, they are specialist and dangerous work places, often hot and noisy. Inappropriate access is not permitted.

New plant rooms must be designed to provide full access for maintenance and repair. Existing plant rooms often appear overcrowded to the maintenance engineer and energy auditor. Adequate head room for safe access and movement is necessary. Noise from refrigeration compressors, hospital medical air compressors and hospital medical suction vacuum pumps is always uncomfortable; hearing protection is often advisable. Large fans generally do not create excessive noise within plant rooms; if they did, duct-borne noise would be troublesome downstream and possibly elsewhere. Fuel-fired water heaters and steam boilers usually do not make excessive noise that leaves the plant room. Large main plant rooms house the maintenance engineer's office, record drawings and commissioning manuals of everything installed, equipment storage, duplicated water pumps, isolating valves, extensive pipe systems, water treatment tanks, anti-corrosion dosing tanks, mechanical switch boards and building management system field panels. This is where the building management system front end computer is often located as continuous access use is made of it by the maintenance team and contractors.

The design stage of a new building is too early for heating and cooling loads, plant sizes and system types to be known with any certainty, but reliable information is required to form the basis for decisions. Building Services Research and Information Association (BSRIA) surveys (Bowyer, 1979) of existing buildings have shown that their plant room requirements can be expressed as a percentage of floor area. A simple factory or warehouse may need 4%, most other types of building around 9% and a well-serviced hospital around 15% of building floor area for plant, excluding the plant areas; meaning that a 5000 m^2 small hospital may require 750 m^2 for all mechanical and electrical plant, including hospital plant such as medical vacuum, medical gas tanks and incinerator, making a gross floor area of 5750 m^2. Some of the outline requirements for services plant rooms quoted by Bowyer (1979) are given here.

Cold-water storage

Volume of cold water to be stored to cover a 24 h period is calculated from the building's occupancy and type. An incoming break tank may be required at ground or basement level for pneumatically boosted systems. Fire-fighting services may need water storage at ground level. Tanks can be 1, 2 or 3 m high, with 1 m clearance allowed around them for insulation, pipes and access.

Hot-water storage

Space needed for vertical indirect hot-water storage cylinders, secondary pumps, pipes, valves, controls and heater coil withdrawal is:

$$\text{plant room floor area m}^2 = 1.7 \times \text{cylinder volume m}^3 + 10$$

Room heights of 3–4.8 m are needed depending on cylinder height.

Boilers

A rough assessment of the boiler power in watts can be made by multiplying the heated volume in metres cubed by 30. The required boiler plant room floor area is:

$$\text{plant room floor area m}^2 = 80.99 + 31.46 \times \ln (\text{boiler capacity MW})$$

Plant room heights are up to 5 m. Domestic and small commercial boilers are accommodated within normal ceiling heights and their floor areas are not predicted with this equation. The area calculated allows for two equally sized boilers, pipes and water treatment, pressurization and pumping equipment.

Fuel storage and metering

Electricity and gas meters are part of the incoming services accommodated within the plant room space calculated for other equipment. Partition walls and access for reading and removal are required. Two equally sized oil storage tanks, supports, tanked catch pit and access are accommodated in:

oil storage tank room area $m^2 = 22.52 + 0.64$ (oil storage volume m^3)

Plant room height is up to 4.5 m. Tanks are frequently located externally and stand on three brick or concrete block piers so that oil will flow by gravity to the burners. They are of mild steel and protected with bitumen paint.

Air handling

The air-handling plant room size is assessed by assuming that the mechanically ventilated parts of the building have between 6 and 10 air changes per hour. The expected supply air volume flow rate is:

$$Q = \frac{NV}{3600} m^3/s$$

V is the volume of ventilated space (m^3) and N is the number of air changes per hour (6–10). The plant room will be 2.5–5 m high depending on the sizes of fans, ducts, filters, heater and cooler coils, humidifiers and control equipment. The floor area is:

plant room floor area $m^2 = 6.27 + 7.8 \times Q \, m^3/s$

A fresh-air-only system, such as an induction system, is sized on one air change per hour.

Cooling plant

Refrigeration plant capacity may be as high as 30 W/m^3 of building volume. Plant room height is 3–4.3 m and the floor area is:

plant room floor area $m^2 = 80.49 + 35.46 \times \ln$ (cooling load MW)

The area is for two refrigeration machines, pumps, pipes and controls. Additional space on the roof is needed for the cooling tower.

Computer servers

Commercial buildings with workstation computers have a server room in continuous operation. An office of twenty or more workstations has a significant server requirement and external cable data modem link equipment to the internet. Intermittent work at the PC front end computer by the network operator is needed, as well as space for expansion of electrical equipment, more cabling, fault servicing, and storage of documents relating to the network. Continuous server computers with back-up data storage disk drives generate considerable heat gains, often within internal closed rooms having no natural ventilation or cooling. Dedicated 24-hour operation of mechanical ventilation, recovery of useful heat and mechanical cooling often do not coordinate easily with the general office air conditioning systems and require careful design and implementation. Electrical equipment is only rated for operation at up to 40°C, easily reached within an unventilated closed internal room full of computers and screens, making suitable temperature control vital for the functioning of the purpose of the building, reliability of computer systems and physical security of the server installation. There are no water or gas services within these rooms. Adequate lighting and low voltage 240 V power outlets are provided. Roof-mounted microwave communication transmitters, receivers and satellite dishes need clearly defined manual exclusion space due to high intensity radiation hazards. A secure fenced communication roof area separates regular daily access from that area to mechanical plant rooms and cooling towers.

Data centres are buildings serving one corporate structure, many businesses or a widely available internet service such as banking and online information systems to the public. A data centre can theoretically be anywhere in the world and serve everyone, such are our widespread wired and wireless communications systems today. Cool climate countries in northern Europe may be able to minimize refrigeration loads with mainly free cooling from low outdoor air temperature much of the year. Air temperature and humidity control are essential to remove the high heat outputs and avoid condensation occurring. These areas may also have access to nuclear, hydro and wind-generated power to claim they are low carbon dioxide emitters. Cloud computing services from such data centres provide on-demand services, reduce the need for users to have all the software and data they use, on their own PC. This makes small portable computers more usable, minimizes the need for every business to have their own server, IT staff, overhead costs and complexity. It may be claimed to be a green solution; watch the news items. Data centres age rapidly and expect to be current for 2–9 years. They are either lights-out, darkened and unattended, therefore can be located anywhere, or attended continuously. Uninterruptable power supply (UPS) from battery back-up and on-site diesel or gas engine generator plus security and fire suppression are all needed. The largest-scale data centres fill shipping containers with racks of computers (http://www.youtube.com/watch?v=zRwPSFpLX8I), also (http://en.wikipedia.org/wiki/Data_Center).

In conclusion, a computer and telecommunication server installation in the buildings we deal with range from one PC in a small office LAN or monitoring a BEMS, through a floor to ceiling rack of computer trays serving one or more floors of a large building, and up to an industrial-scale building filled with shipping containers of racks. Mechanical and electrical services for these ranges from no special provision in a small office, through a stand-alone packed air conditioner for an office floor server room, up to a vast dedicated industrial-scale server farm consuming MW. All server systems need UPS, power supply electronic conditioning to stabilize fluctuations in a public supply, software and hardware security, automatic monitoring, cooling, humidity control, frequent updating, repair, access, fire suppression and attendance.

Lifts

An early assessment of requirements is made in conjunction with the lift engineer. Each electrical lift has a lift motor room on top of the shaft that is several times the floor area of the shaft. Lifts are grouped close together and the room-level lift motor room covers all shafts plus control equipment. Older electrical lifts have mechanical switch control panels of considerable size. A lifting beam, lifting gear and lift motor concrete bases along with one or more spare lift motors are provided for repair work. Digital control with frequency inverters is normal for modern lift installations, requiring a large control panel. Lighting and access for regular inspections, maintenance and large item replacement add to the lift motor room size. A lift motor room on top of a tall air-conditioned building has significant internal and external heat gains. Natural ventilation is always provided but mechanical air flow and cooling are provided where there is a significant risk of the lift motor room exceeding a safe environmental working temperature for the staff and the electrical equipment. Smaller capacity electric lifts have the driving motor and control gear installed on the support frame of the top of each car, so only wire ropes and the counterweight are within the lift shaft and there is no lift motor room.

Hydraulic lifts may be used to service up to around six floors. The lift car is raised from hydraulic rams and cables within the lift shaft. They are silent and slow speed, some allowing glass-sided cars as there are no visible cables. Motive power comes from a hydraulic pump within the oil storage tank alongside the base of the lift shaft. Lowering the car pushes oil back into the tank while the car rising pumps it. The submerged pump is oil-cooled and uses minimum energy.

Electrical substation

The incoming high-voltage supply is located in a substation, which may be external or on an external wall of the building. The floor area needed is 35 m^2 for a 200 kVA load and up to 48 m^2 for a 2000 kVA load.

Standby diesel electric generator

A standby diesel electric generator supplies emergency electrical power of up to 100% of the connected load from the mains. It may only have the capacity to maintain essential lighting, a minimum lift service, stairway pressurization fans, hospital operating theatres and any other essential service for the site. The plant room will be adjacent to the substation and will be up to 4 m high. The floor area required is 18 m^2 for 50 kVA up to 37 m^2 for 600 kVA, plus a diesel oil storage tank for 7 days of continuous running.

Service ducts

Service ducts are passageways that traverse vertically and horizontally throughout a building, or between buildings, large enough to permit the satisfactory installation of pipes, cables and ducts, together with their supports, thermal insulation, control valves, expansion allowance and access for maintenance. Each service duct might be constructed as a fire compartment. An example of current practice is given in Figure 13.2.

Casings and chases of 100 mm diameter or less are fire-stopped to the full thickness of the wall or floor. The passage of a service must not reduce the resistance of the fire barrier. Plastic pipes can soften and collapse during a fire and allow the passage of flames and smoke. A galvanized steel sleeve with an intumescent liner can be used to surround the plastic pipe where a fire stop is needed. When its temperature rises to 150°C, the intumescent liner expands inwards to close the softened pipe and seal the wall aperture. Service duct sizes can be found from an estimate of ventilation air supply rate Q m^3/s, doubled to allow for the recirculation duct, with an assumed air velocity of 10 m/s for vertical ducts within brick or concrete enclosures or 5 m/s in false ceilings where quiet operation is important. At least 150 mm clearance is allowed between ducts and other surfaces for thermal insulation, jointing, supports and access.

Fan noise is contained within the air-handling plant room by acoustic attenuators, anti-vibration machine mountings and heavy concrete construction.

The total floor space taken by vertical pipe and cable routes will be up to 1% of the gross floor area. Horizontal service ducts and false ceilings 500 mm deep are used for air-conditioning ducts and other services. Recessed luminaires and structural beams encroach into the nominally available spaces.

Underfloor service ducts should be constructed to allow access for jointing and maintenance. The minimum standard for an underfloor duct is shown in Figure 13.3. The duct route is accurately

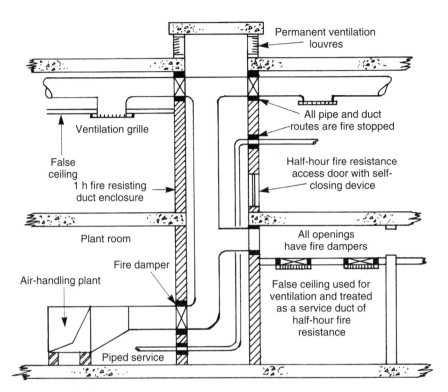

13.2 Service duct fire compartment.

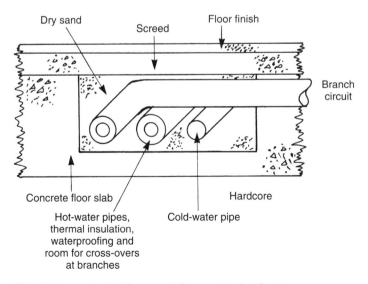

13.3 Minimum standard for an underfloor service duct for pipes.

Table 13.1 Minimum underfloor duct sizes for pipes

Hot-water flow and return nominal diameter (mm)	Underfloor duct dimensions width and depth (mm)
15	294
22	304
28	348
35	364
42	376
54	400

marked on installation drawings and access is gained by breaking the screed. Hot-water pipes are insulated with 50 mm thick rigid glass fibre and wrapped with polyethylene sheet, sealed with waterproof tape. The duct is filled with dry sand. Sufficient depth is allowed for branches to cross over the other pipes. Recommended sizes are given in Table 13.1.

Vertical and other service ducts can be sized in a similar manner by allowing a 50 mm gap between thermal insulation and other surfaces. Additional smaller pipes running in the same duct will require an increase in the width.

When builder's work holes for services are specified, the dimensions of the structural opening required should always be used, rather than the nominal pipe diameter. An underfloor, or underground, crawlway or walkway has the following features:

1. crawlway duct height 1.4 m;
2. walkway duct height 2 m;
3. 750 mm clear width between fixtures;
4. reinforced concrete construction;
5. floor laid to fall to a drainage channel along the length of the duct, with connections to the surface drainage system;
6. watertight access manhole covers at intervals with built-in galvanized steel stepladders;
7. watertight lighting fittings and power sockets;
8. services are painted to appropriate standard colours and clearly labelled;
9. control valves are numbered with an explanation list provided;
10. services branching into side ducts do not block through access.

EXAMPLE 13.1

A 6-storey office building of 30 m × 25 m is to be constructed on a suburban green field site. It is to be air conditioned with a low energy system using gas and electricity as energy sources. The main air handling unit, AHU, plant room is to be built on the roof. There will be 225 occupants. Calculate the plant room and service duct space requirements for the preliminary design stage. Only outside air is to be passed through the distribution ductwork system.

It is expected that the plant rooms will require 9% of the floor area of $(30 \times 25 \times 6)$ m^2, and so a first estimate is 405 m^2. This will be mainly on the roof as the heating system water heater, air conditioning cooling plant of chiller and cooling tower, domestic hot water heater and storage cylinder, pumps and main switchboards are normally located there also; thus, an oblong room of dimensions l and $2l$ could be used to establish the overall space requirement. This space can then be distributed around the building for the various services. Cold water storage tanks are often located on a mezzanine floor above the heating plant. Fire fighting water storage tanks are used in some buildings, where needed for the sprinkler system; these, and any oil storage tanks, are always located in basement or ground floor plant rooms; neither are needed in this case. The main electricity meters for the central building services and tenants' floors, the main electricity distribution board, the water and gas meters are usually located in a basement or ground floor plant room. The building management system computer, a server computer with screen, keyboard, printer and data back-up storage tape or disc, are sometimes located in a plant room, where only the BMS maintenance contractor ever uses it, but it should definitely be in the Building Manager's office where it can be used every day. The Building Manager should be using this computer for condition, fault and energy monitoring continuously.

$$\text{area} = l \times (2l) = 405 \text{ m}^2$$

$$2l^2 = 405 \text{ m}^2$$

$$l = \left(\frac{405}{2}\right)^{0.5} = 14.23 \text{ m}$$

A plant room of 14.23 m × 28.5 m could be accommodated on the roof. However, some of this space will be located lower down the building. A further estimate of the requirements can be made through consideration of each service.

1. Cold-water storage of 45 l per person:

$$\text{volume} = 225 \text{ people} \times 45 \frac{l}{\text{person}} \times \frac{1 \text{ m}^3}{10^3 \text{ litre}} = 10.1 \text{ m}^3$$

tank dimensions = 2.25 m × 4.5 m × 1 m

Add 1 m all round the tanks. Thus, the plant room floor space is 27.6 m^2. This could be reduced by stacking the tanks if there is sufficient headroom. The water main pressure will be sufficient to reach the roof; ground-level break tanks and pumps will not be needed.

2. Hot-water storage of 5 l per person:

$$\text{volume} = 225 \text{ people} \times 5 \frac{l}{\text{person}} \times \frac{1 \text{ m}^3}{10^3 \text{ litre}} = 1.125 \text{ m}^3$$

$$\text{volume of an indirect cylinder} = \frac{\pi d^2}{4} \times l \text{ m}^3$$

For a cylinder 1 m in diameter, its length l is:

$$l = 1.125 \times \frac{4}{\pi} = 1.43 \text{ m}$$

Floor area required is 1 m^2.

3. Boiler power is given by:

$$\text{boiler power} = (30 \times 25 \times 6 \times 3)\, m^3 \times 30\frac{W}{m^3} \times \frac{1\,MW}{10^6\,W}$$

$$= 0.405\,MW$$

$$\text{plant room floor area} = 80.99 + 31.46 \times \ln 0.405\, m^2$$

$$= 52.6\, m^2$$

4. Gas and electricity meters will be housed either in the roof plant room or in cubicles at the rear of the building on the ground floor.
5. The low energy system air-handling plant will pass only the fresh air supply. Volume flow rate of supply air will be a maximum of 10 litres per person.

$$Q = \frac{10\,l}{s} \times 225\,\text{people} \times \frac{1\,m^3}{10^3\,l}\, m^3/s$$

$$= 2.25\, m^3/s$$

$$\text{plant room area} = 6.27 + 7.8 \times 2.25\, m^2$$

$$= 23.8\, m^2$$

Air change rate from the provision of outdoor air is:

$$N = 3600 \times \frac{2.25\, m^3}{s} \times \frac{1\,\text{air change}}{30 \times 25 \times 6\, m^3}$$

$$N = 1.8\,\text{air changes/h}$$

This is acceptable and reasonable. There may be additional recirculated air within the floors from distributed terminal heating/cooling air control units such as variable air volume, induction or fan coil units.
6. Cooling plant capacity is $30\,W/m^3 = 0.405\,MW$, same as heating power calculated.

$$\text{plant room floor area} = 80.49 + 35.46 \times \ln 0.405\, m^2$$

$$= 48.4\, m^2$$

Floor space needed for chillers and cooling towers does not increase as fast as for heating plant with increasing capacity. Total plant room space requirements are estimated to be:

$$(27.6 + 1 + 52.6 + 23.8 + 48.4)\, m^2 = 153.4\, m^2$$

These two methods of estimation show that plant room space requirements are of the order of 153–$405\, m^2$. This is a wide spread of answers provided for the design concept stage. It will be refined by detailed design and better knowledge of the plant loads and locations to be used. A low energy building has the minimum of mechanical services plant for air handling and cooling. Heating systems have traditionally not required much plant room space and hot water radiators and convectors are very compact, mainly consuming wall space. The historical precedent for a 9% floor area requirement is likely to be too high for a modern system.

A vertical service duct is needed from the roof plant room to ground level carrying supply and exhaust air ducts, drainage and water pipes and cables. If the maximum air velocity in the air ducts is 6 m/s, their sizes will be:

$$\text{duct cross-sectional area} = \frac{Q\,m^3/s}{V\,m/s}$$
$$= \frac{2.25}{6}\,m^2$$
$$= 0.375\,m^2$$

If square ducts are used, they will be 612 mm × 612 mm or larger, such as with standard sizes of 700 mm × 600 mm. An estimated service duct arrangement is shown in Figure 13.4. This allows for thermal insulation and access to all the services.

False ceilings provide space for the horizontal distribution of services. The induction units will be located along the external perimeter under the windows or within the false ceiling. Holes 150 mm in diameter are needed in the floor slab, one for each unit, for the air-duct and, close by, two holes 50 mm in diameter for pipes.

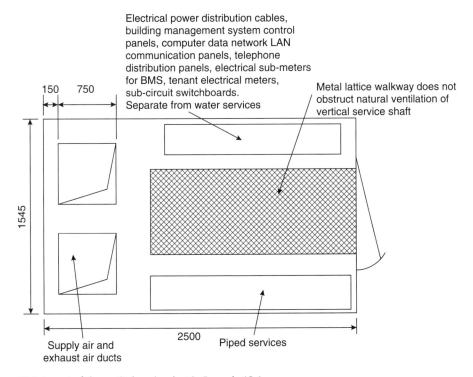

13.4 Layout of the vertical service duct in Example 13.1.

Pipe, duct and cable supports

Hot-water pipes can be supported with hardwood insulation rings clamped in mild steel brackets, as shown in Figure 13.5. Saddle pipe and cable clips, shown in Figure 13.6, are extensively used because of their low cost. They should be made of a material that is compatible with that of the service. A row of services may be bolted to a mild steel angle iron whose ends are built into the structure. The longitudinal spacing of supports depends on pipe size, material and whether the service is horizontal or vertical. Rollers, as shown in Figure 13.7, allow pipes to move freely during thermal movement.

   Expansion and contraction of short pipe runs are accommodated at frequent bends and branches, the pipes moving within their thermal insulation and non-rigid brackets. Spaces between pipes are sufficient to avoid contact. Long pipe runs need expansion devices, anchors and guides:

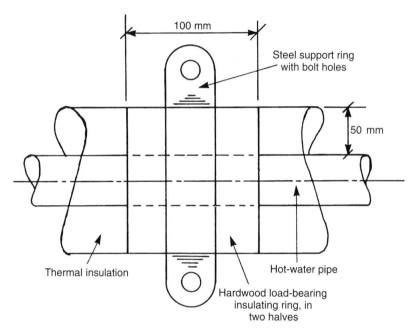

13.5 Insulated pipe support ring.

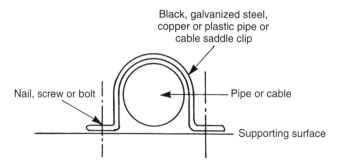

13.6 Pipe and cable saddle clip.

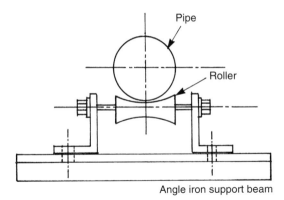

13.7 Roller pipe support.

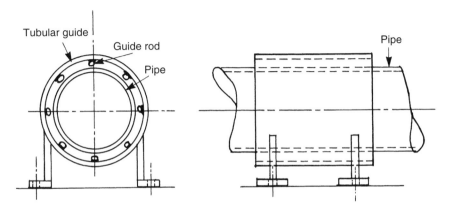

13.8 Tubular guide support.

1. With a pipe anchor the pipe is rigidly bolted or welded to a steel bracket which is firmly built into a brick or concrete structure.
2. Pipes can be supported by tubular guides, as shown in Figure 13.8, which allow longitudinal movement with minimum metal contact.
3. Several types of thermal expansion device are used, depending on application, space available and fluid pressure.

    (a) Pipe loops are least expensive in some cases and can be formed where external pipes pass over a roadway. They can be prefabricated with pipe fittings, welded, bent or factory formed, as shown in Figure 13.9.
    (b) Bellows are made of thin copper or stainless steel and have hydraulic pressure limitations. A complete installation is shown in Figure 13.10.
    (c) Articulated ball joints take up pipe movement at a change in direction, as shown in Figure 13.11.
    (d) Sliding joints are packed with grease and the pipe slides inside a larger diameter sleeve.

A fire stop unit is used where pipes pass through a fire compartment wall or floor and incorporates structural support, vibration insulation and fire resistance within a steel flanged sleeve which is in two halves, as shown in Figure 13.12. Silicone fire stop foam is used to seal the space around pipes. When it is exposed to heat, the foam chars to form a hard flame-resistant clinker.

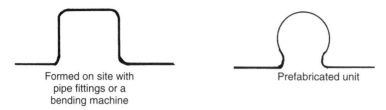

Formed on site with
pipe fittings or a
bending machine

Prefabricated unit

13.9 Pipe expansion loop.

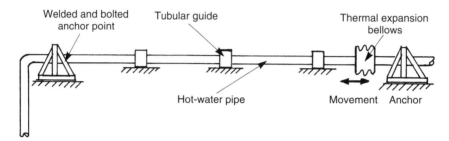

Welded and bolted
anchor point

Tubular guide

Thermal expansion
bellows

Hot-water pipe

Movement   Anchor

13.10 Complete pipe installation for thermal expansion provision.

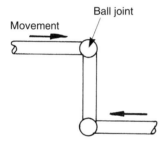

Ball joint

Movement

13.11 Articulated expansion joint.

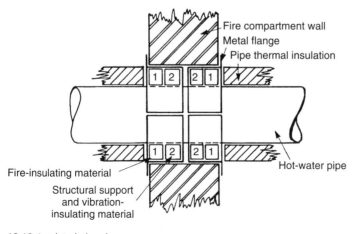

Fire compartment wall
Metal flange
Pipe thermal insulation

Fire-insulating material

Structural support
and vibration-
insulating material

Hot-water pipe

13.12 Insulated pipe sleeve.

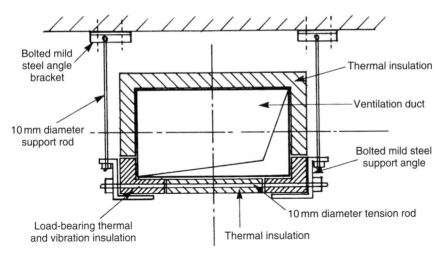

Bolted mild
steel angle
bracket

Thermal insulation

Ventilation duct

10 mm diameter
support rod

Bolted mild steel
support angle

10 mm diameter tension rod

Load-bearing thermal
and vibration insulation

Thermal insulation

13.13 Insulated duct support.

Ventilation ducts are fixed to the building with galvanized mild steel saddle clips for up to 300 mm diameter light-gauge metal; larger ducts have flanged joints, which are suspended with rods from angle brackets. Figure 13.13 shows a typical fixing.

Cables are supported along their entire length by the conduit or a perforated metal tray.

Plant connections

Connections to plant are made in flexible materials to reduce the transmission of vibration from fans, pumps and refrigeration compressors to the distribution services, or to allow greater flexibility in siting the equipment. A fan installation is shown in Figure 13.14. The discharge air duct has a sound attenuator to absorb excess fan noise. Polyurethane foam held in place by perforated metal sheet is used in the attenuator.

Coordinated service drawings

A master set of all service drawings is maintained under the overall charge of a coordinating engineer for the project. Drawings and a schedule of builder's work associated with the services are circulated round the construction team. The structural engineering implications of holes through floors and beams are checked at an early stage. Service space allocation is made on the basis of zones for particular equipment within structural shafts and false ceilings. Each engineering service is restricted to its own zone. Common areas are provided for branches, as shown in Figure 13.15.

Boiler room ventilation

Combustion appliances must have an adequate supply of outdoor air, otherwise the fuel will not burn properly and carbon monoxide will be produced. Under down-draught conditions, this will be a danger to occupants. Fatalities have occurred through improper appliance operation. Good installation practice is to introduce combustion air so that it does not cause a nuisance through draught, noise or poor appearance. Heat and fumes produced by the appliance are ventilated to outdoors through high-level openings. Any room containing a fuel-burning appliance may be

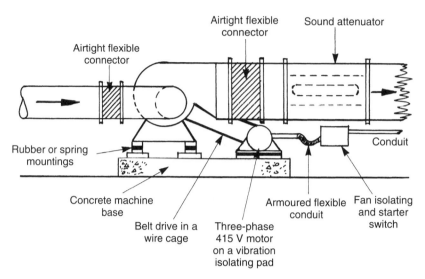

13.14 Flexible connections to an air-conditioning fan.

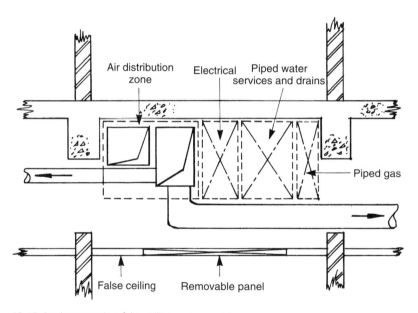

13.15 Service zones in a false ceiling over a corridor.

positively supplied with ventilation air by a fan if this is needed for some other purpose. An extract fan must not be used, as combustion products can be drawn into the occupied room. Natural inlet and outlet ventilation is predominant. The combustion air inlet for a domestic kitchen or living area can be:

1. through an external wall, just below ceiling level, to enable the incoming cold air to diffuse with the room air above head height; this avoids most draughts and occasional blockage by snow or debris;

2. through an external wall at low level behind a hot-water radiator or other heat emitter; a frost thermostat switches on the heating system at an internal air temperature of 5°C;
3. by direct connection of the combustion air from outside to the appliance casing, locality or enclosing cupboard with an underfloor duct. Two suitably sized air bricks are fitted into the external walls of a suspended timber floor on opposite sides of the building. Either a duct connection between the appliance casing and the floor space or a ventilation grille is put into the floor by the heater. A drain pipe or galvanized steel duct can be cast into a concrete floor slab for this purpose.

Questions

1. List the principal information and activities involved in the provision of main services throughout a housing estate.
2. Sketch a suitable arrangement for the services beneath the public highway and leading into a dwelling. Show the recommended dimensions and explain how the ground is to be reinstated.
3. Estimate the plant room and service duct space requirements of the following buildings, using the preliminary design information given.

    1. A naturally ventilated hotel with a hot-water radiator heating system. Roof and basement plant rooms are available. The hotel dimensions are 50 m × 30 m, with 10 storeys 3 m high. Total occupancy is 750. An oil-fired boiler plant is to be used.
    2. A single-storey engineering factory of dimensions 100 m × 40 m, using overhead gas-fired radiant heating. The roof height slopes from 3 m to 5 m at the central ridge. There are 300 occupants. Mechanical ventilators and smoke extractors will be fitted in the roof. A standby diesel electricity generator and an electrical substation are required.
    3. A 12-storey city centre educational building of 40 m × 20 m, 3 m ceiling height, with a single storey workshop block and laboratory area 40 m × 60 m × 4 m. The whole complex is to be mechanically ventilated with 4 air changes per hour. Hot-water radiators and fan con-vectors provide additional heating. Gas and electricity are to be used. The tower building has a basement with ramp access to ground level. A refectory is located at ground level. The total building occupancy is 2000.

4. Draw the installation of services in a vertical duct through a 3-storey office building. The duct is 2.5 m × 1.2 m. Boiler and ventilation plant is in the basement. There are false ceilings on all floors.
5. Sketch and describe how the spread of fire through a building is limited by the services installation.
6. A false ceiling over a supermarket contains recessed luminaires, sprinkler pipes and a single-duct air-conditioning system. The false ceiling is 400 mm and has structural steel beams 250 mm deep. Extract air from the shop passes through the luminaires. Draw the installation to scale.
7. A concrete floor with a wood block finish houses a service duct carrying two 35 mm heating services, two 28 mm hot-water services, a 42 mm cold-water service and 54 mm gas pipes. Side branches are required to carry a maximum of three 22 mm pipes. Continuous access covers are to be provided. The hot-water pipes are to have 50 mm thick thermal insulation, and at least 25 mm clearance is needed around the pipes. Draw a suitably detailed design showing dimensions, materials, pipe support, cover construction and pipe routes at the branch.

8. Describe, with the aid of sketches, how successful coordination between all the services can be achieved within builders' work ducts.

9. Explain how fuel-burning appliances fitted in kitchens, living rooms, cupboards and domestic garages can be adequately ventilated. Illustrate an example of each location and state the areas of ventilation openings required for appliances of 3 kW, 18 kW and 40 kW heat output.

10. Who are allowed into plant rooms?

    1. Everyone in the building.
    2. All contractors.
    3. Members of the public.
    4. Only employed maintenance engineers.
    5. Those approved by the building manager.

11. What are plant rooms designed to accommodate?

    1. Only mechanical plant.
    2. Only electrical plant.
    3. Plant and people.
    4. Wiring and pipes.
    5. Pumps and boilers.

12. What are plant room conditions supposed to be?

    1. Safe work places.
    2. Minimum size possible.
    3. Maximum size needed.
    4. Showcase for plant items.
    5. Always out of sight.

13. What do we know about computer server spaces?

    1. None needed, located beneath a desk.
    2. Located in roof plant space, out of sight.
    3. Any corner of a room will do.
    4. Secure accessible and safe room to work in.
    5. Always very hot places.

14. What do we know about computer server rooms?

    1. An inconvenient collection of electrical boxes.
    2. Vital hub of every business and office.
    3. Nobody ever works in there.
    4. Do not need ventilation.
    5. Provide useful heat into the building during winter.

15. What is essential for a computer server facility?

    1. Basement store room location.
    2. Empty internal office with lockable door.
    3. Interior secure work room with mechanical ventilation and temperature-controlled cooling 24 hours a day, 365 days a year.
    4. Partitioned space in basement car park as it is cool there.
    5. Any office or store room where enough space can be made available.

16. Where is the building maintenance manager's office likely to be located?

    1. Alongside the reception area.
    2. By the main plant room, often in the basement.

3. In the executive office suite on a high level floor.
4. In the basement car park.
5. In the entrance foyer.

17. Which are the most problematic noise sources in plant rooms for maintenance workers?

1. Gas-fired water heaters.
2. Toilet exhaust fans.
3. Air handling units.
4. Rotary and reciprocating compressors.
5. None of them are as plant is switched off when work is undertaken.

18. How often do technical workers enter large plant rooms?

1. Once a year.
2. Daily, several times.
3. Monthly service checks and fan belt changes.
4. Hourly logging of energy and operational data.
5. Once a week.

19. What is the space temperature control requirement for plant rooms?

1. None, they always remain cool.
2. Does not matter as air temperature never gets too hot for the mechanical plant.
3. Wall and roof ventilation openings.
4. Nobody works in there so it does not matter.
5. Natural and mechanical ventilation for workers and where necessary, cooling to limit room temperature for workers and electrical systems.

20. What is involved with lift motor rooms?

1. Driving motor and winding gear are located in basement plant areas.
2. Each lift has its driving motor and winding gear mounted above the shaft within a ventilated and cooled roof-level plant room.
3. All motors are located on top of each passenger car.
4. One electric motor drives all lift cars in a group from a roof-level plant room.
5. Sealed concrete plant room above each lift shaft.

14 Fire protection

Learning objectives

Study of this chapter will enable the reader to:

1. classify fire hazards;
2. identify the necessary ingredients for a fire;
3. describe the development of a fire;
4. recognize the hazards of smoke;
5. apply the correct fire-fighting system, or combination of systems, to different fire classifications;
6. understand the principles of portable fire extinguishers;
7. know the criteria for the use of hose reel, dry riser, wet riser, foam, sprinkler, carbon dioxide (CO_2), vaporizing liquid, dry powder and deluge fixed fire-fighting systems;
8. know the sources of water used in fire-fighting;
9. identify how fire development can be detected;
10. recognize the importance of smoke ventilation;
11. identify the locations for fire dampers in air ductwork and know how they operate.

Key terms and concepts

carbon dioxide 303; dry hydrant rise 303; dry powder 302; fire damper 309; fire-fighting system classification 302; fire risk classification 301; foam inlet 305; fuel, heat and oxygen 301; hazard, smoke and fire detectors 307; hose reels 303; hydrant valve 303; portable extinguishers 302; smoke 301; smoke ventilator 308; sprinkler systems 305; vaporizing liquid 303; water 302; water pumping 304; water sources 305; water storage 307; wet hydrant riser 304.

Introduction

The systems required to meet the needs of tackling small fires; evacuation and major fire-fighting both by the occupants and then the fire detection and suppression services are outlined.

Building management systems under computer monitoring and control will incorporate such systems, together with security functions. Integration of such equipment with the architecture, decor and other services is planned from the earliest design stage. In addition, others plan for safe evacuation and structural protection are discussed.

Fire classification

A building's fire risk is classified according to its fuel load, occupancy and use (CIBSE, 2010). Residential institutions pose risks from sleeping and not being made aware of smoke or fire and entrapment. Homes have a well-established and consistent fire load as well as accidental events occurring. Residential institutions have a high life risk from infirmity, lack of mobility and the number of occupants close together. Commercial offices have few deaths from fire as there are practised evacuations of able-bodied people, maintained fire-fighting systems and a known fire load. Shops and commercial premises have a historically low life risk but high risk potential, high fire load and some significant historical fires. Assembly and recreational buildings are capable of creating high life fire losses due to the number of occupants, unfamiliarity with escape routes and crowding. Industrial buildings have a significant history of fire losses, low occupancy, may have high fire loads, stored materials emitting hazardous fumes and smoke as well as potentially high commercial losses. Warehouses often have high fire loads from stored combustibles and chemicals, low occupancy but significant fires may have to extinguish themselves when the fuel load has exhausted; limiting the spread of the fire and smoke are the main concern. Farming and forestry may not seem to be a problem for the building services engineer, but field stubble burning, forest floor fuel reduction burning and waste timber burning after forest clearance can and do get out of control from wind changes or miscalculation, burning embers travel ahead of the flame front carried by the wind and start house fires that might destroy entire neighbourhoods. It is theoretically possible to protect homes or commercial buildings from exterior fires with water sprays and fire-resistant building materials. Fire sucks all the oxygen out of the air and it does not take long for those sheltering inside to be starved of air to breathe. Not being there is the first line of defence. Some fires are started deliberately, unfortunately, they can happen due to falling overhead live power lines, and even the hot exhaust from vehicles driven on dry grass. Each country has its own risk profile; while grass fires may not occur in a damp environment such as northern Europe, they are a real danger in such countries as Australia.

A fire is supported by three essential ingredients: fuel, heat and oxygen. The absence of any one of these causes an established fire to be extinguished. The fire-fighting system must be appropriate to the location of the fire and preferably limited to that area in order to minimize damage to materials, plant and the building structure. Radiation from a fire may provoke damage or combustion of materials at a distance. Structural fire protection can include water sprays onto steelwork to avoid collapse, as used in the Concorde aircraft production hangar.

The system of fire-fighting employed depends upon the total combustible content of the building (the fire load), the type of fire risk classification and the degree of involvement by the occupants. Fire escape design where children, the elderly or infirm are present needs particular care so that sufficient time is provided in the fire resistance of doors and partitions for the slower evacuation encountered.

Smoke contains hot and unpleasant fumes, which can be lethal when produced from certain chemicals and plastics. Visual obstruction makes escape hazardous and familiar routes become confusing. Packaging materials, timber, plastics, liquefied petroleum gas cylinders and liquid chemicals must not be stacked in passageways or near fire exits in completed or partially completed buildings. Each working site or building needs a safety officer responsible for general oversight.

Fires are classified according to what is burning and how it can be extinguished. Wood and textiles need water to cool the material below its combustion temperature and douse smoke. Petroleum fires are extinguished by excluding more oxygen from reaching the flame front. Gas, flammable metals and electrical fires need to have fresh oxygen supplies excluded. Electrical plant in a fire situation also requires that non-conducting fire fighting chemicals are used to avoid electrocution risk. The highest priority is that safety switches and fuses separate the circuit connections and eliminate further hazards to occupants, fire fighters and the building. Often water is used to fight fires and water conducts electricity.

Regular fire drills are conducted by the safety officer and employees are clearly notified of their responsibilities in an emergency. Staff duties will be to shepherd the public, patients or students out of the building to the rendezvous point, while maintenance personnel may be required to operate fire-fighting equipment while awaiting the fire brigade. While annual evacuation drills may seem tedious and unnecessary to many people in the middle of their working, lecturing or studying day, these are a vital part of employer responsibility, legal duty and are a confirmation to all users that they should be able to evacuate safely and save their own life. We do not want to be awarded a degree posthumously.

Portable extinguishers

Portable extinguishers are manually operated first-aid appliances to stop or limit the growth of small fires. Staffs are trained in their use and the appliances are regularly maintained by the suppliers. Fire blankets are provided in kitchens where burning pans of oil or fat need to be covered or personnel need to be wrapped to smother ignited clothing. Water extinguishers are red coloured canisters or fire buckets and used on wood, paper and textiles. Dry powder blue canisters can be used on all types of fire as powder interferes with flame chemistry. Foam extinguishers, cream, exclude oxygen for petroleum fires. Carbon dioxide, black, excludes oxygen for electrical fires and plant. Vaporizing liquid, green, sprays vapour that interferes with flames and is used for small fires, vehicles and some piped systems into electrical equipment rooms where there will not be any personnel.

Water

A 9 l water extinguisher is installed for each $210\,m^2$ floor area, with a minimum of two extinguishers per floor. A high-pressure CO_2 cartridge is punctured upon use and a 10 m jet of water is produced for 80 s. Water must not be used on petroleum, burning liquids or in kitchens as it could spread the fire.

Dry powder

Dry powder extinguishers contain from 1 to 11 kg of treated bicarbonate of soda powder pressurized with CO_2, nitrogen or dried air. A spray of 2–7 m is produced for 10–24 s depending on size. The powder interrupts the chemical reactions within the flame, producing rapid flame knockdown. The powder is non-conducting and does little damage to electric motors or appliances. A deposit of powder is left on the equipment.

Foam

Portable foam extinguishers may contain foaming chemicals that react upon mixing or a CO_2 pressure-driven foam. They cool the combustion, exclude oxygen, and can be applied to wood, paper, textile or liquid fires. Garage fires are a particular application. Sizes range from 4.5 to 45 l. A 7 m jet is produced for 70 s with a 9 l capacity model.

Vaporizing liquid

Halon vaporizing liquid bromochlorodifluoromethane (BCF) or bromotrifluoromethane (BTM) is a CFC. These are covered by the international agreement to cease their use and are no longer allowed in the UK.

Carbon dioxide

Pressurized CO_2 extinguishers leave no deposit and are used on small fires involving solids, liquids or electricity. They are recommended for use on delicate equipment such as electronic components and computers. The CO_2 vapour displaces air around the fire and combustion ceases. There is minimal cooling effect and the fire may restart if high temperatures have become established. Water-cooling back-up is used where appropriate once electrical connections are severed.

Fixed fire-fighting installations

Various fire-fighting systems are employed in a building so that an appropriate response will minimize damage from the fire and the fire-fighting system itself. Back-up support for portable extinguishers may be provided by a hose reel installation and this can be used by staff while the fire brigade is called. Some public buildings, shops and factories are protected by a sprinkler system, which only operates directly over the source of fire. This localizes the fire to allow evacuation. Where petroleum products are present, a mixture of foam and water is used.

Hose reels

Hose reels are a rapid and easy to use first-aid method, complementary to other systems and used by the building's occupants. They are located in clearly visible recesses in corridors so that no part of the floor is further than 6 m from a nozzle when the 25 mm bore flexible hose is fully extended. The protected floor area is an arc 18 m to 30 m from the reel, depending on the length of hose. A minimum water pressure of 200 kPa is available with the 6 mm diameter nozzle. This produces a jet 8 m horizontally or 5 m vertically. Minimum water flow rate at each nozzle is 0.4 l/s, and the installation should be designed to provide not less than three hose reels in simultaneous use: a flow rate of 1.2 l/s. Figure 14.1 shows a typical installation.

The local water supply authority might allow direct connection to the water main, and there may be sufficient main pressure to eliminate the need for pressure boosting. Pump flow capacity must be at least 2.5 l/s. The stand-by pump can be diesel-driven. Flow switches detect the operation of a hose and switch on the pump.

Dry hydrant riser

A dry hydrant riser is a hydrant installation for buildings 18–40 m high where prompt attendance by the fire brigade is guaranteed. A dry riser pipe 100 mm or 150 mm in diameter is sited within

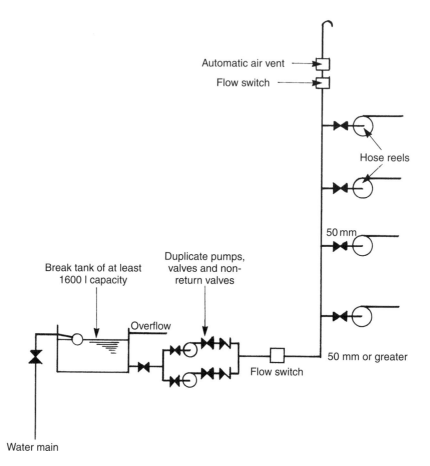

14.1 Hose reel installation.

a staircase enclosure with a 65 mm instantaneous valve outlet terminal at each landing. All parts of the building floor are to be within 60 m of the hydrant, measured along the line on which a hose would be laid. A test hydrant is fitted at roof level, and also a 25 mm automatic air vent. A double inlet breeching piece with two 65 mm instantaneous terminals is located in a red-wired glass box in an external wall, 760 mm above ground level and not more than 12 m from the riser. The inlet point is within 18 m of an access road suitable for the fire brigade pumping appliance. A brass blank cap and chain are fitted to each landing valve. The riser is electrically earthed. Landing valves are 1 m above floor level and are used by the fire brigade for their own hoses.

Wet hydrant riser

A permanently charged rising pipe 100 mm in diameter or greater supplies a 65 mm instantaneous valve outlet terminal at each floor. Water is maintained at a pressure of between 410 and 520 kPa. The upper pressure limit is to protect the fire brigade hoses from bursting and is achieved by fitting an orifice plate restriction before the landing valve on the lower floors of a tall building. The maximum static pressure in the system when all the landing valves are shut is limited to 690 kPa by recirculating water to the supply tanks through a 75 mm return pipe. Each hydrant valve is strapped and padlocked in the closed position. They are 1 m above floor level and are only used by the fire brigade for buildings over 60 m high which extend out of the reach of turntable ladders. The maximum normally permitted height is 60 m for a low-level break tank and booster set. Higher buildings have a separate supply tank and pump sets for each 60 m height. Pressure boosting of the water supply is provided by a duplicate pump installation capable of delivering at least 23 l/s. Pumps are started automatically on the fall of water pressure or water flow commencement. Audible and visual alarms are triggered to indicate booster plant operation.

A break tank capacity of 11.4–45.5 m³ is required and mains water make-up rate is 27 l/s or 8 l/s for the larger tank. Additionally, four 65 mm instantaneous fire brigade inlet valve terminals are provided at a 150 mm breeching fitting in a red wired-glass box in an external wall, as described for the dry hydrant riser. The box is clearly labelled. A nearby river, canal or lake may also be used as a water source with a permanently connected pipe from a jack well and duplicate pumps. Pneumatic pressure boosting is used to maintain system pressure in a similar manner to that for other water systems in tall buildings. The standby pump may be driven by a diesel engine fed from a 3–6 h capacity fuel storage tank providing a gravity feed to the engine.

Foam inlets

Oil-fired boiler plant rooms and storage tank chambers in basements or parts of buildings have fixed foam inlet pipe from a red wired-glass foam inlet box in an outside wall as for the dry hydrant riser. A 65 mm or 75 mm pipe runs for up to 18 m from the inlet box into the plant room. The fire brigade connect their foam-making branch pipe to the fixed inlet and pump high-expansion foam onto the fire. The foam inlet pipe terminates above the protected plant with a spreader plate. A short metal duct may be used as a foam inlet to a plant room close to the roadway. Vertical pipes cannot be used and the service is electrically bonded to earth. On-site foam-generation equipment is available and may be used for oil-filled electrical transformer stations. In the event of a fire, the electricity supply is automatically shut off, a CO_2 cylinder pressurizes a foam and water solution and foam spreaders cover the protected equipment.

Automatic sprinkler

High-fire-risk public and manufacturing buildings are protected by automatic sprinklers. These may be a statutory requirement if the building exceeds a volume of 7000 m³. Loss of life is very unlikely in a sprinkler-protected building. Sprinkler water outlets are located at about 3 m centres, usually at ceiling level, and spray water in a circular pattern. A deflector plate directs the water jet over the hazard or onto walls or the structure. Each sprinkler has a frame containing a friable heat-sensing quartz bulb, containing a coloured liquid for leak detection, which seals the water inlet. Upon local overheating, the quartz expands and fractures, releasing the spray. Water flow is detected and starts an alarm, pressure-boosting set and automatic link to the fire brigade monitoring station. Acceptable sources of water for a sprinkler system are:

1. a water main fed by a source of 1000 m³ capacity where the correct pressure and flow rate can be guaranteed;
2. an elevated private reservoir of 500 m³ or more depending on the fire risk category;
3. a gravity tank on site, which can be refilled in 6 h, with a capacity of 9–875 m³ depending on the fire risk category;
4. an automatic pump arranged to draw water from the main or a break tank of 9–875 m³ capacity;

5. a pressure tank: a pneumatic pressure tank source can be used for certain light fire risk categories or as a back-up facility to some other system.

Sprinkler installations are classified under four principal types:

1. water-filled pipes permanently charged with water;
2. dry pipe filled with compressed air and used where pipes are exposed to air temperatures below 5°C or above 70°C;
3. alternate system where pipes are filled with water during the summer and air in the winter;
4. a pre-action system is a dry pipe installation but has additional heat detectors which pre-empt the opening of sprinkler heads and admit water into the pipes, converting it to wet-pipe operation.

Different types of sprinkler head are used, depending on the hazard protected; their aim being to produce a uniform density of spray.

*   *Fusible link*: a soldered link in a system of levers holds the water outlet shut. At a predetermined temperature of 68°C or greater, the solder melts and water flow starts.
*   *Chemical*: similar to the fusible link but using a block of chemical, which melts at 71°C or greater, depending on the application.
*   *Glass bulb*: a quartz bulb containing a coloured fluid with a high coefficient of expansion, which fractures at 57°C or more.
*   *Open sprinkler heads*, called a deluge system, are used to combat high-intensity fires and protect storage tanks or structural steelwork. They are controlled by a quick-opening valve actuated from a heat detector or a conventional sprinkler arrangement. A drencher system provides a discharge of water over the external openings of a building to prevent the spread of fire to others. Homes and buildings in high fire-risk, hot dry rural climates can be protected with such systems from water storage tanks and open water storage pond, lake, stream or river. These are usually turned on manually prior to arrival of the flame front through forest or open grassland. Wind-driven fire fronts can travel at 30+ km/h and evacuation to a place of safety is essential.

Each sprinkler installation must be provided with:

1. a main stop valve, which is strapped and padlocked in the open position to enable the water flow to be stopped after the fire is extinguished;
2. an alarm valve: differential pressure caused by water flow through the valve opens a branch pipe to the alarm gong motor;
3. a water motor alarm and gong: water flow through a turbine motor drives a rotary ball clapper within a domed gong to give audible warning of sprinkler operation and commence evacuation of the building.

A satisfactory installation serving an automatic sprinkler distribution and hose reels is shown in Figure 14.2. Two 65 mm instantaneous fire brigade inlet pipes are provided at a clearly marked access box.

Carbon dioxide

Carbon dioxide is used in fixed installations protecting electrical equipment such as computer rooms, transformers and switchgear. Heat or smoke detectors sound alarms and CO_2 gas floods

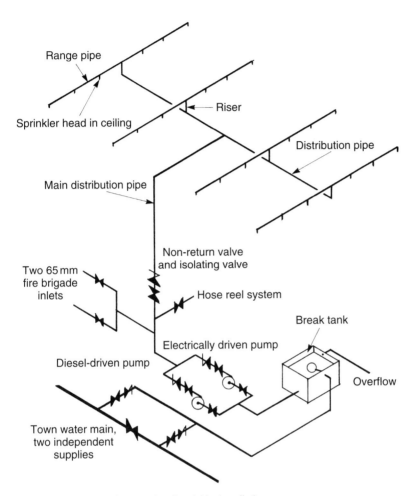

Range pipe

Sprinkler head in ceiling

Riser

Distribution pipe

Main distribution pipe

Non-return valve and isolating valve

Two 65 mm fire brigade inlets

Hose reel system

Break tank

Diesel-driven pump

Electrically driven pump

Overflow

Town water main, two independent supplies

14.2 Water supply to hose reel and sprinkler installation.

the room from high-pressure storage cylinders. Pipes transfer CO_2 to ceiling and underfloor distributors. System initiation can be manual or automatic but complete personnel evacuation is essential before CO_2 flooding is allowed.

Fire detectors and alarms

Detection of a potentially dangerous rise in air temperature or pressure or the presence of smoke is required at the earliest possible moment to start an alarm. Evacuation of the building and manual or automatic contact with the fire brigade monitoring switchboard should take place before people are at risk. Means of detection can be combined with security surveillance. Fire detection takes the following forms.

Hazard detectors

Hazard detectors give an early warning of the risk of a fire or explosion. A local rise in temperature leads to the melting of a fusible link in a wire holding open a valve on a fuel pipe to a burner, and thermal expansion of a fluid-filled bellows or capillary tube or movement of a bimetallic strip makes an alarm circuit. Flammable vapour detectors are used. For gas, oil, petrol or chemical vapour presence is detected by a catalytic chemical reaction. Diffusion detectors sense butane and propane vapour diffusing through a membrane. Explosion detectors respond to rise of local atmospheric pressure above a set value or at a fast rate.

Ionization smoke detectors contain a radioactive source of around 1 micro curie, typically americium-241, which bombards room air within the detector with alpha particles (ionization). Electrical current consumption is 50 μA. The presence of smoke reduces the flow of alpha ions; the electric current decreases and at a pre-set value an alarm is activated.

Visible smoke detectors use a source of light that is directed at a receiving photocell. Smoke obscures or scatters the light and an alarm is triggered.

A laser beam is refracted by heat or smoke away from its target photocell and an alarm is initiated. A continuous or pulsed infrared beam can be transmitted up to 100 m and can be computer-controlled to scan the protected area. It can also serve as an intruder alarm.

Closed-circuit television with manual security monitoring also acts as fire and smoke detection. Infrared imaging cameras reveal overheating of buried pipes and cables and can detect heat sources unseen by visual techniques.

Fire alarms are a statutory requirement. Audible bells, sirens, klaxons, hooters and buzzers are arranged so that they produce a distinctive warning. A visual alarm should also be provided throughout a building. Breakable glass call points are located 1.4 m above floor level within 30 m of any part of the premises.

The electrical system for fire detectors comprises alarms, a central control panel, an incoming supply and distribution board, emergency batteries, a battery charger and fire-resistant cable. A permanent cable or telephone line connection is made to the fire brigade and computer-controlled monitoring indicates any system faults. This system may be called an Emergency Warning and Intercommunication System (EWIS).

Smoke ventilation

Positively removed smoke through automatically opened roof ventilators can greatly aid escape and reduce smoke damage, often localizing a fire that would otherwise spread. The spread of smoke through ventilation ductwork is arrested by fire dampers where fire compartments within the building are crossed. Fire dampers may be motorized or spring-loaded multileaf, eccentrically pivoted flaps, sliding plate or intumescent paint-coated honeycombs which swell and block on heating. A typical arrangement of a pivoted flap damper is shown in Figure 14.3.

An air pressurization ductwork and fan system is switched on at the commencement of a fire to inject outdoor air into escape routes, corridors and staircases. The staircase static air pressure is maintained at 50 Pa above that of adjoining areas to overcome the adverse force caused by wind, mechanical ventilation and the fire-produced stack effect ventilation pressure. This ensures that clear air is provided in the escape route and smoke movement is controlled.

Questions

1. List the sources of fire within a building and describe how they may develop into a major conflagration. State how the spread of fire is expected to be limited by good building and services practice.
2. List the ways in which fire and smoke are detected and fire-fighting systems are brought into action.

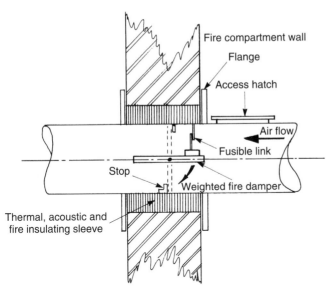

14.3 Hinged deadweight single blade fire damper in a ventilation duct.

3. Describe the methods and equipment used to fight fires within buildings in their likely order of use.
4. State the principal hazards faced by the occupants of a building during a fire. How are these hazards overcome? Give examples for housing, shops, cinemas, office blocks, single-storey factories and local government buildings.
5. Sketch and describe the fire-fighting provisions necessary in large industrial oil-fired boiler plant.
6. How are water and foam systems used to protect building structures from fire damage?
7. Compare a fixed sprinkler installation with other methods of fire-fighting. Give three applications for sprinklers.
8. Explain how sprinkler systems function, giving details of the alternative operating modes available. State the suitable sources of water for sprinklers.
9. Tabulate the combinations of fire classification and types of extinguisher to show the correct application for each. State the most appropriate fire-fighting system for each fire classification and show which combinations are not to be used.
10. Which are correct about fire-fighting services?

    1. Primarily are to minimize damage to the building and its services systems.
    2. Primarily to prevent and minimize danger to people.
    3. Secondary purpose is to save the building and its continued use.
    4. Are only used by professional fire fighters.
    5. Have both fixed and portable fire extinguishers.

11. Which are correct about hose reel fire-fighting systems?

    1. Are not suitable for office buildings.
    2. Never need to be tested.
    3. Should not be intrusively visible.
    4. Are for anyone to use to commence fire fighting until professionals arrive.
    5. Only used in public buildings.

12. Which are true about fire fighting systems in large buildings?

   1.   Never need to be tested after being commissioned.
   2.   Have a fire control panel identifying which parts of the system are activated.
   3.   Have an indicator board in the entrance to the building advising fire location.
   4.   Are all filled with water.
   5.   Are always connected to the computer-based building management system.

13. Which are true about sprinkler fire-fighting systems?

   1.   Have sprinklers located on a 3.0 m grid pattern covering the floors.
   2.   Can have sprinklers mounted on a sidewall 150 mm from the ceiling.
   3.   Have self-acting outlets heads that fracture on rise of air temperature.
   4.   Are part of a fire-fighting strategy.
   5.   Provide a high degree of security against fire damage.

14. Which is true about sprinkler fire-fighting systems?

   1.   Have sprinklers located on a 6.0 m grid pattern covering the floors.
   2.   Must be manually turned on.
   3.   Have self-acting outlets heads that fracture on rise of air temperature.
   4.   Are the only fire-fighting system a building needs to have.
   5.   Creates unnecessary water damage.

15. Which are correct about fire detection and alarm systems?

   1.   Thermal detectors in office ceilings.
   2.   Visible smoke laser light detectors in rooms.
   3.   Detailed drawings of detectors wired in series with each other.
   4.   Break-glass alarm call points in all areas.
   5.   Automatic connection to the fire brigade call centre.

16. Which are correct about smoke control during a building fire?

   1.   Is not a critical hazard for personnel.
   2.   Materials within buildings are sources of toxic chemicals.
   3.   Smoke control exhaust fans remove air from the building.
   4.   Escape routes are protected with water sprinkler systems.
   5.   Air pressure differentials across doorways control smoke movement to aid escape.

17. Identify the essential components of a fire.

   1.   Fuel and air.
   2.   Source of ignition and combustible material.
   3.   Paper, wood, solvents, air and warmth.
   4.   Fuel, oxygen and ignition temperature.
   5.   Fuel, air and high temperature radiated heat.

18. How does a fire commence?

   1.   A small flame radiates combustibility over a long distance.
   2.   Adjacent buildings cannot be set on fire.
   3.   Combustible material, liquid or gas becomes raised to its ignition temperature in the presence of oxygen.
   4.   Electrical services often start fire.
   5.   Any spark from a light switch or plant switch can start a fire.

19. Which is not a means of extinguishing fire?

    1. Deprive the fire of air.
    2. Cool the burning material.
    3. Cease the supply of more fuel.
    4. Calling the fire brigade.
    5. Closing doors and windows and evacuating the building.

20. Which is a means of extinguishing fire?

    1. Disconnect electrical supply from electrical service or item on fire.
    2. Switch lights off and leave that floor level.
    3. Cover burning photocopier with wool blanket and leave that floor.
    4. Spray water onto burning electrical heater and evacuate.
    5. Throw a fireproof blanket over a burning computer.

21. CO_2 is a greenhouse gas emission problem for the world (*HM Government Carbon Plan, 2011*). Why is it used in fire fighting?

    1. Global emissions will not diminish for hundreds of years.
    2. There is no inter-governmental agreement on ceasing its use.
    3. Amounts used are minimal.
    4. CO_2 cannot burn anything.
    5. Using it is a lesser problem.

15 Room acoustics

Learning objectives

Study of this chapter will enable the reader to:

1. know the potential sources of sound and vibration within buildings;
2. know what is meant by noise;
3. understand how sound travels through a building;
4. understand what is meant by sound pressure wave, sound power level and sound pressure level;
5. know how to calculate sound pressure levels for normal building services design examples;
6. use sound levels at the range of frequencies commonly used in building services engineering;
7. understand how sound and vibration are transmitted through buildings;
8. be able to identify the need for sound attenuation vibration isolation;
9. understand and use the decibel unit of measurement of sound energy;
10. know the meaning and use of direct and reverberant sound fields;
11. calculate the sound pressure level in a plant room, a space adjacent to the plant room, in the target occupied room and in the external environment outside the plant room;
12. use logarithms to base 10 in acoustic calculations;
13. understand the principle of sound absorption;
14. calculate the sound absorption constant for a room at different frequencies;
15. know the sound absorption coefficients for some common building materials and constructions;
16. understand and use reverberation time and attenuation;
17. calculate sound pressure levels at different frequencies within a plant room;
18. know what a reverberant room and an anechoic chamber are;
19. use directivity index sound absorption coefficients, mean absorption coefficient and room absorption constant;

20. understand the behaviour of equipment at resonant conditions and how to minimize or avoid its occurrence;
21. calculate and use the sound pressure level in a plant room;
22. calculate the sound pressure level experienced at an external location from a plant room;
23. calculate the sound pressure level generated in a room or space that is adjacent to a plant room;
24. calculate the sound pressure levels at different frequencies that are produced in the target occupied room;
25. understand, calculate and use noise rating data;
26. know how the acoustic design engineer relates the noise output from plant systems to the human response;
27. be able to calculate noise rating curves;
28. know the noise rating criteria used for building services design;
29. plot noise rating curves, plant and system sound pressure levels and find a suitable design solution;
30. know the formulae used in practical acoustic design work;
31. be able to carry out sound pressure level and noise rating deign calculations, try different solutions to attenuate plant noise and be able to produce a practical design to meet a design brief.

Key terms and concepts

absorption coefficient 316; acoustic barrier 334; acoustic energy 314; acoustic power 314; air ducts 315; air pressure 314; anechoic chamber 317; atmospheric pressure 314; attenuation 321; audible frequencies 315; Bel 315; compressors 321; decibel 315; directivity 324; direct sound field 315; ear response 314; engines 315; fan blades 314; fan noise spectrum 329; fans 314; flexible connections 315; frequency 315; hemispherical sound field 319; Hertz (Hz) 315; human ear 315; intermediate room 325; mechanical service equipment 321; molecular vibration 323; multiple reflection 317; natural vibration 317; noise 314; noise rating 327; outdoor environment 324; pipes 315; plant room 321; plant vibrations 321; porous material 315; pressure wave 314; pump blades 315; resonance 323; reverberant sound field 315; reverberation time 317; room absorption constant 316; rubber mountings 315; solid material 321; sound power level 315; sound pressure level 315; springs 315; structure 321; target room 326; total sound field 316; vibration 314.

Introduction

This chapter uses the worksheet file DBPLANT.WKS to find the noise rating that will be produced within an occupied room by direct transmission through the building from the noise-producing plant. The plant noise source creates sound pressure levels within the plant room. The plant room noise can pass through an intermediate space, such as a corridor, and then into the target occupied space. Sound can be transmitted from the plant room to a recipient outdoors for an environmental impact noise rating.

Sufficient reference data is provided on the worksheet for examples within this chapter and for some real applications. Reference data from any source can easily be added. This chapter allows for most practical examples of mechanical plant to be assessed quickly and without having to deal with the equations themselves. Data is provided for frequencies from 125 Hz to 4000 Hz as this range is likely to cover the important noise levels for comfort. The range of frequencies can be added to should the need arise. The reader may wish to study the principles of acoustics in the appropriate text books and the references made as it is not the intention of this chapter to teach the subject in its entirety. However, it is the purpose of this chapter to provide an easily understandable method of analysing practical noise applications. Consequently, the reader should not find it difficult to enter correct data and acquire suitable results for educational reasons and in practical design office cases. The worksheet DBDUCT.WKS is used to calculate the noise rating in the target room that is produced by noise being transmitted from the air-conditioning fan through the ductwork system. Further examples of spreadsheet applications and explanation of spreadsheet use are provided in *Building Services Engineering Spreadsheets* (Chadderton, 1997).

Acoustic principles

The building services design engineer is primarily concerned with controlling the sound produced by items of plant such as boilers, supply air and exhaust air fans, refrigeration compressors, water pumps, diesel or gas engine-driven electrical generating sets and air compressors. An excess of sound that is produced by the plant, above that which is acceptable to the recipient, is termed noise. All the mechanical service equipment and distribution systems to be installed within an occupied building are capable of generating noise.

Sound travels through an elastic, compressible medium, such as air, in the form of waves of sound energy. These waves of energy are in the form of variations in the pressure of the air above and below the atmospheric air pressure. The human ear receives these air pressure fluctuations and converts the vibration generated at the eardrum into electric impulses to the brain. What we understand to be recognizable language, music and noise is the result of human brain activity. Animals and the mythical person from another planet do, or may process what we determine as normal sounds and come to a different conclusion from those of us who are conditioned to life on Earth. These variations in the pressure of the atmospheric air are very small when measured in the Pascal or millibar values that engineers use. A scale of measurement that relates to the subjective response of the human ear is used. Although absolute units of measurement are taken and normal calculation procedures are adopted, it is important to remember that the smallest unit of sound is that which can be detected by the human ear. The waves of air pressure which pass through the atmosphere are measured in relative pressure units. The acoustic energy of the source which caused the air pressure waves has an acoustic power, or rate of producing energy, in the same way that all thermodynamic devices have a power output. There are two ways of assessing the output and transmission of acoustic energy:

1.   source sound power: Watts;
2.   sound wave atmospheric pressure variation: Pascals.

Sound waves are generated at different frequencies measured in cycles per second, Hertz (Hz). The plant which produces the noise has components that rotate, move and vibrate at a range of different speeds, or frequencies of rotation. The flowing fluid is vibrated by the passage of fan or pump blades and it transfers the plant vibration through to downstream parts of the connected services systems. The fluid is water, oil, air, gas, refrigerant or steam, and can either simply transmit the plant vibrations and noise or add to them by means of its own pulsations due to its turbulent flow. Turbulence means that a fluid flow contains recirculating parcels of fluid in the form of eddy currents. These parcels of swirling eddy currents move in all directions, i.e. along with the general direction of the main flow, but also in the reverse direction and transversely across the main flow. Viewing wave action on a beach or a fast river flow from a bridge or at a bend reveals the nature of turbulent flow. The turbulent eddy currents occur at a range of frequencies, parcels of recirculating fluid per second, depending upon the overall diameter of the eddy current. The physical movement of the swirling fluid can vibrate the containing water pipe or air duct, causing vibration and noise. Obstructions in the air or waterway occasioned by sharp edges, dampers, grilles, temperature sensors, and changes in duct cross-section, can cause the turbulent fluid to shear into additional swirling eddy currents and produce more vibration and noise. Turbulent fluid can vibrate air ducts, pipes and terminal heat exchange units. The structure of the building transmits noise by the vibration of its solid material particles and continuous steel frame and reinforcing bars within concrete framework. Acoustic energy is transferred between pressure waves in the air and vibration through solid materials in either direction. The vibration of fans, compressors, engines and pumps is controlled by mounting them on coiled steel springs, rubber feet and rubber matt. Fluid pipes and air ducts are separated from fans, air-handling units and pumps with flexible connections. These minimize, or stop, plant vibration being transferred to the reticulation system. Fluid-borne noise is reduced by selective absorption with a porous lining to the air duct. Sound waves are absorbed into the thickness of the lining material through a perforated surface material which protects the absorber from fluid damage and erosion. Sound energy is dissipated within the absorber by multiple reflections among the fibrous material.

Sound power and pressure levels

Sound power and pressure levels are measured over a range of frequencies that are representative of the response of the human ear to sounds with a hand-held microphone sound pressure measuring instrument. The unit of measurement of sound is the Bel (B). The smallest increment of sound that the human ear can detect is one-tenth of a Bel, one decibel (dB). This means that the smallest change in sound level that is perceptible by the human ear is 1 dB, so any decimal places that are produced from calculations using sound power or pressure level are not relevant. A calculated sound level of 84.86 can only be 84 dB as the 0.86 decimal portion is not detectable by the human ear. The 'A' scale of measurement gives a weighting to each frequency in the range 20 Hz to 20 kHz in the same ratio as can be heard. For example, the human ear is more sensitive to sounds at 1000 Hz than at higher frequencies.

The acoustic output power of a machine is termed its sound power level, *SWL* dB. Think of *SWL* as the sound watts level of the acoustic output power of the machine. The value of acoustic power in watts from building services plant is very small, much less than one watt of power. The word level is used because it is not the actual value of the number of watts that is normally used; it is the sound level produced in acoustic units of measurement, dB, that are taken for practical use. The manufacturer of the plant provides the sound power levels produced by a particular machine from test results and predictions for known ranges of similar equipment. The sound power level of a machine at the range of frequencies from 125 Hz to 8000 Hz is required by the building services design engineer in order to assess the acoustic affects upon the occupied spaces of the building. The overall sound power level for a range of frequencies is also quoted by the manufacturer of a machine.

Sound pressure level

A sound field is created by the sound power output from a machine within a plant room. It is made up of a direct sound field, i.e. directly radiated sound, and a reverberant sound field, i.e. general sound that reflects uniformly from the hard surfaces around the room. The direct sound field reduces with the inverse square of the distance from the sound source and is not normally of importance as it only applies to very short distances from the sound source. The reverberant sound field results from the average value of the sound pressure waves passing around the room. These waves try to escape from the plant room and find their way into the occupied spaces where the air-conditioning engineer is attempting to create a quiet and comfortable environment. The sound pressure level, SPL dB, of the total sound field, direct plus reverberant, that is generated within a room from a sound source of sound power level SWL dB, is found from

$$SPL = SWL + 10 \times \log \left(\frac{Q}{4 \times \pi \times r^2} + \frac{4}{R} \right) dB$$

(CIBSE Guide B, and Sound Research Laboratories Limited), where,

SPL = sound pressure level produced in room   dB
SWL = sound power level of acoustic source   dB
log = logarithm to base 10   dimensionless
Q = geometric directivity factor   dimensionless
r = distance from sound source to the receiver   m
R = room sound absorption constant   m^2

Logarithms to base 10, $\log_{10}$, are used throughout the calculation of acoustic values. A sound source that radiates sound waves uniformly in all directions through unobstructed space will create an expanding spherical sound field and have a dimensionless geometric directionality factor Q of 1. A sound source that is on a plane surface radiates all its sound energy into a hemispherical sound field moving away from the surface. This has a directionality factor Q of 2, that is, twice the sound energy passes through a hemisphere. Similarly, if the sound source occurs at the junction of two adjacent surfaces that are at right angles to each other, such as the junction of a wall and ceiling, Q is 4. When there are three adjacent surfaces at the sound source, such as two walls and a ceiling, Q is 8. Distance r is that from the sound source to the receiving person, surface or measurement location, such as an air outlet duct from the plant room or outdoor air grille.

Absorption of sound

The room sound absorption constant, $R\,m^2$, is found from the total surface area of the enclosing room, $S\,m^2$, and the mean sound absorption coefficient of the room surfaces, α, at each of the relevant frequencies:

$$R = \frac{S \times \bar{\alpha}}{1 - \bar{\alpha}}$$

$\bar{\alpha}$ = mean absorption coefficient of room surfaces

S = total room surface area m^2

Mean absorption coefficient is found from the area and absorption coefficient for each surface of the enclosing space. All the absorbing surfaces within the space, such as seats and people in a theatre, are included in the overall sound absorbing ability of the room:

$$\bar{\alpha} = \frac{A_1 \times \alpha_1 + A_2 \times \alpha_2 + A_3 \times \alpha_3}{A_1 + A_2 + A_3}$$

A_1 = surface area of surface number 1 m^2

α_1 = absorption coefficient of surface number 1

Materials absorb different amounts of sound energy at each frequency due to the frequency of natural vibration of their fibres and the method of their construction. Stiff, dense materials, such as brickwork walls, absorb sound by molecular vibration. Highly porous materials, such as glass wool, have large air passageways that allow the sound waves to penetrate the whole of the material thickness quickly. The strands of glass wool are vibrated by the sound waves and the sound energy is dissipated as heat. Dense materials are very efficient at absorbing acoustic energy. The reduction in sound level between the surfaces of a sound barrier is proportional to the mass of the barrier. Sample of absorption coefficients of some common surface materials are given in Table 15.1 for use within this book only. This data is repeated on the worksheet from line 201.

Reverberation time

Reverberation time is the time in seconds taken for a sound to decrease in value by 60 dB. This effectively means the time taken for the sound source to decay to an imperceptible level, as a sound pressure of 30 dB is very quiet to the human ear. An echo is produced by sound waves bouncing, or reverberating, from one or more hard surfaces and this may last for several seconds. A room that has a long reverberation time sounds noisy, lively and it allows echoes. A room having a short reverberation time, less than 1 s, sounds dull and there is no echo. The ultimate in short reverberation time is found in the anechoic chamber that is used for the acoustic testing of equipment. The walls and ceiling of the chamber are lined with thick acoustic absorbent wedges.

Table 15.1 Absorption coefficients of common materials.

Material	Absorption coefficient at					
	125 Hz	250 Hz	500 Hz	1000 Hz	2000 Hz	4000 Hz
25 mm plaster, 18 mm plasterboard, 75 mm cavity	0.3	0.3	0.6	0.8	0.75	0.75
18 mm board floor on timber joists	0.15	0.2	0.1	0.1	0.1	0.1
Brickwork	0.05	0.04	0.04	0.03	0.03	0.02
Concrete	0.02	0.02	0.02	0.04	0.05	0.05
12 mm fibreboard, 25 mm cavity	0.35	−.35	0.2	0.20	0.25	0.3
Plastered wall	0.01	0/01	0.02	0.03	0.04	0.05
Pile carpet on thick underfelt	0.07	0.25	0.5	0.5	0.6	0.65
Fabric curtain folds	0.05	0/15	0.35	0.55	0.65	0.65
15 mm acoustic ceiling tile, suspended 50 mm mineral fibre wool or glass fibre	0.5	0.6	0.65	0.75	0.8	0.75
50 mm polyester acoustic blanket metalized film	0.25	0.55	0.75	1.05	0.8	0.7
50 mm glass fibre blanket, perforated surface finish	0.15	0.4	0.75	0.85	0.8	0.85

The floor is a suspended wire mesh, and beneath the floor more absorbent wedges complete the coverage of all the room surfaces. The sound source radiates outward and upon reaching the surfaces is instantly absorbed, allowing no reverberation or echo. This is as close to a free field test method as can be achieved because there is no reverberant field caused by reflected sound waves.

An interesting example of a large semi-anechoic chamber is a car testing facility. The four walls and the ceiling are covered with acoustic wedges, while the floor is a plain concrete surface. This simulates an open road, a hemispherical acoustic field under laboratory repeatable conditions.

According to CIBSE (2005b) and Sound Research Laboratories Limited, reverberation time of a room is estimated from:

$$\text{reverberation time } T = \frac{0.161 \times V}{S \times \overline{\alpha}}$$

EXAMPLE 15.1

A plant room for an air-conditioning fan is $4\,m \times 3\,m$ in plan and $2.5\,m$ high. It has four brickwork walls, a concrete floor and a pitched sheet steel deck roof having 50 mm thickness of glass fibre and an aluminium foil finish to the underside. Ignore the effects of the metal plant, air ductwork and the door into the plant room. Calculate the room constant and the reverberation time for the plant room.

The surface absorption coefficients are selected from Table 15.1. It can be seen that there will be a different room constant and reverberation time for each frequency. The solution is presented in Table 15.2.

$$\text{Room volume } V = 4 \times 3 \times 2.5\,m^2$$

$$= 30\,m^3$$

$$\text{floor area} = 12\,m^2$$

$$\text{ceiling area} = 35\,m^2$$

$$\text{wall area} = 35\,m^2$$

Table 15.2 Solution to Example 15.1.

	Absorption data at frequency					
Surface	125 Hz	250 Hz	500 Hz	1 kHz	2 kHz	4 kHz
Floor α	0.02	0.02	0.02	0.04	0.05	0.05
Ceiling α	0.15	0.4	0.75	0.85	0.8	0.85
Walls α	0.05	0.04	0.04	0.03	0.03	0.02
Floor ($S \times \alpha$)	0.24	0.24	0.24	0.48	0.6	0.6
Ceiling ($S \times \alpha$)	1.8	4.8	9.0	10.2	9.6	10.2
Walls ($S \times \alpha$)	1.75	1.4	1.4	1.05	1.05	0.7
Mean α	0.064	0.109	0.18	0.199	0.191	0.195
Room constant Rm^2	4.03	7.21	12.95	14.66	13.93	14.29
Reverberation Ts	1.28	0.75	0.45	0.41	0.43	0.42

Room surface area $A = (2 \times 4 \times 3) + (4 + 4 + 3 + 3) \times 2.5 \, \text{m}^2$

$$= 59 \, \text{m}^2$$

For 125 Hz, the mean absorption coefficient is,

$$\bar{\alpha} = \frac{12 \times 0.02 + 12 \times 0.15 + 35 \times 0.05}{12 + 12 + 35}$$

$$= 0.064$$

room constant $R = \dfrac{S \times \bar{\alpha}}{1 - \bar{\alpha}} \, \text{m}^2$

$$= \frac{59 \times 0.064}{1 - 0.064} \, \text{m}^2$$

$$= 4.03 \, \text{m}^2$$

reverberation time $T = \dfrac{0.161 \times V}{S \times \bar{\alpha}}$

$$= \frac{0.161 \times 30}{59 \times 0.064} \, \text{s}$$

$$= 1.28 \, \text{s}$$

EXAMPLE 15.2

An air-conditioning centrifugal fan has an overall acoustic output power level SWL of 87 dB on the 'A' scale. The fan is to be installed centrally within the air-handling plant room described in Example 15.1. Calculate the sound pressure level that will be produced in the plant room at 1000 Hz when the fan is operating, close to the fan and also generally within the room.

Room absorption constant from Example 15.1 at 1000 Hz,

$$R = 14.66 \, \text{m}^2$$

The fan is in the centre of a concrete floor in the plant room. Sound pressure waves leaving the fan will radiate into a hemispherical field above floor level. The sound waves are concentrated into half of a completely free field. The directivity, Q, of the sound field is 2. A person within the plant room can stand in the range of 100 mm to 2 m away from the fan. A typical distance between the fan and the recipient is 1 m. The room sound pressure level is calculated for 100 mm and 1 m distances from the sound source. When,

$r = 100 \, \text{mm}$

$$SPL = SWL + 10 \times \log_{10} \left(\frac{Q}{4 \times \pi \times r^2} + \frac{4}{R} \right) \text{db}$$

$$= 87 + 10 \times \log_{10} \left(\frac{2}{4 \times \pi \times 0.1^2} + \frac{4}{14.66} \right) \text{db}$$

$$= 87 + 10 \times \log_{10} (16.188) \text{ dB}$$

$$= 87 + 10 \times 1.2092 \text{ dB}$$

$$= 99 \text{ dB}$$

The smallest change in sound level that is perceptible by the human ear is 1 dB, so the decimal places are not relevant. The plant room sound pressure level at 100 mm radius from the fan is 99 dB. At 1 m from the fan, the recipient experiences a sound pressure level of

$$r = 1 \text{ m}$$

$$SPL = 87 + 10 \times \log_{10} \left(\frac{2}{4 \times \pi \times 1^2} + \frac{4}{14.66} \right) \text{ dB}$$

$$= 87 + 10 \times \log_{10} (0.432) \text{ dB}$$

$$= 87 + 10 \times -0.3645 \text{ dB}$$

$$= 83 \text{ dB}$$

The direct sound field diminishes with distance from the source. The reverberant sound field establishes the general room sound pressure level when the recipient is sufficiently far away from the source.

EXAMPLE 15.3

The spectrum of sound power levels produced by the centrifugal fan being installed in the 4 m × 3 m × 2.5 m high plant room in Example 15.1 is 78 dB at 125 Hz, 82 dB at 250 Hz, 86 dB at 500 Hz, 87 dB at 1 kHz, 70 dB at 2 kHz, and 60 dB at 4 kHz. Use the surface absorption data from Example 15.1 and calculate the room sound pressure level at a radius of 1.5 m from the fan for each frequency from 125 Hz to 4 kHz.

At 125 Hz,

$$SWL \text{ is 78 dB.}$$

$$r = 1.5 \text{ m}$$

$$Q = 2$$

$$R = 4.03 \text{ m}^2$$

$$SPL = SWL + 10 \times \log_{10} \left(\frac{Q}{4 \times \pi \times r^2} + \frac{4}{R} \right) \text{ dB}$$

$$= 78 + 10 \times \log_{10} \left(\frac{2}{4 \times \pi \times 1.5^2} + \frac{4}{4.03} \right) \text{ dB}$$

$$= 78 \text{ dB}$$

The results are shown in Table 15.3.

Table 15.3 Fan sound spectrum in Example 15.3.

Item	Data at frequency					
	125 Hz	250 Hz	500 Hz	1 kHz	2 kHz	4 kHz
Room constant Rm^2	4.03	7.21	12.95	14.66	13.93	14.29
Reverberation Ts	1.28	0.75	0.45	0.41	0.43	0.42
Fan SWL dB	78	82	86	87	70	60
Room SPL dB	78	79	81	82	65	55

Plant sound power level

The design engineer requires to know the sound power level of the, potentially, noise-producing items of plant. These plant items will be the supply air fan, extract air fan, exhaust fans from toilets, kitchens and some store rooms, fan coil units in ceiling spaces above occupied rooms, packaged air-handling units incorporating fans, direct refrigerant expansion outdoor condensing units, direct refrigerant expansion packaged air-conditioning roof-mounted units, gas- and oil-fired boilers, packaged air conditioners and heat pumps within rooms, external cooling towers and dry air-cooled heat exchangers, refrigeration compressors and water chilling refrigeration plant. In addition to these major items of plant, supply air grilles, extract air grilles, room terminal air-handling units, dampers, air volume control boxes and fan-powered variable air volume control boxes, can also generate noise. The manufacturer of these items will provide the results of acoustic test data for the building services design engineer. Current acoustic data, rather than catalogue information, is acquired and the manufacturer then becomes responsible for the numbers used. The designer needs the sound power level at each frequency that is to be analysed. These are normally 125 Hz to 4000 Hz. Often the critical frequency for design will be 1000 Hz and this corresponds to a sensitive band in the human ear response.

For the worked examples and questions within this book, sound power levels are provided, either in the form of a discrete value for each frequency, or a single value for all frequencies for the plant item. Figure 15.1 gives an indication of the variations in sound power level from a single value for centrifugal fans, axial fans, and refrigeration compressors, cooling towers, fan coil units and boilers. The reader will find the spectral sound power level by subtracting the variances from the single value quoted in the example or question. This data is not to be used in real design work as is provided for illustration purposes only. The numbers that were used to produce Figure 15.1 are listed in Table 15.4. Figure 15.1 is also provided as a chart on the worksheet file.

Transmission of sound

The sound pressure within a space will cause the flow of acoustic energy to an area that has a lower acoustic pressure. Sound energy converts into structural vibration and passes through solid barriers. A reduced level of sound pressure is established in the adjacent space due to the attenuation of the separating partition, wall, floor or ceiling. Air passageways through the separating partition act as sound channels that have little, or no, sound-reducing property, or attenuation. The reader can validate this effect by partially opening a window when the outdoor sound level is substantial. Compare the open and closed window performance when a train, lorry or high traffic volumes are present. A well-air-sealed single-glazed window imposes a sound reduction of 30 dB on external noise but a poorly sealed or open window has little attenuation.

Table 15.4 Illustrative sound power level variances from Figure 15.1.

Plant item	Sound power level dB variance at frequency								
	31.5 Hz	63 Hz	125 Hz	250 Hz	500 Hz	1 kHz	2 kHz	4 kHz	8 kHz
Centrifugal fan	−2	−5	−8	−10	−14	−18	−23	−30	−40
Axial fan	−2	−3	−4	−6	−8	−10	−13	−16	−20
Refrigeration compressor	−15	−25	−15	−10	−13	−20	−21	−25	−30
Cooling tower	−10	−11	−12	−15	−18	−22	−23	−28	−33
Fan coil unit	−25	−30	−25	−28	−22	−25	−30	−35	−40
Boiler	−8	−3	−1	−5	−10	−12	−15	−20	−25

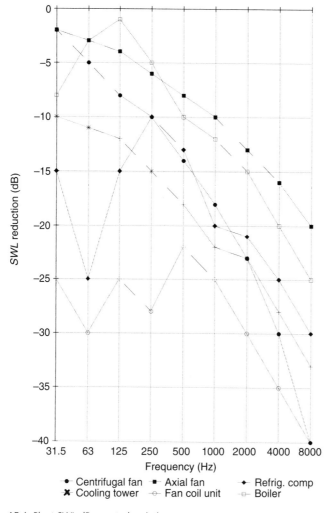

15.1 Plant SWL dB, spectral variation.

Sound reduction by a surface is from the reflection of sound waves striking the surface and by the absorption of sound energy into a porous material. Absorbed acoustic energy is dissipated as heat within the solid components of the absorber. Dense materials are often efficient sound attenuators. The exception is metal. Sound travels easily through metals for great distances due to their molecular vibration. When the imposed sound frequency coincides with a natural frequency of vibration of the metal, resonance occurs and an increased sound level may be generated. This happens in particular when the shape of the metal creates an air space for the sound waves to resonate within, such as in a bell, an empty tank or a pipe. The structural steel within a building, service pipe work, air ducts and railways lines can all transfer noise and vibration over long distances.

The sound pressure level generated within a room by mechanical plant, or sound systems for entertainment, will be passed through sound barrier materials and constructions such as walls and the ceiling, to adjacent spaces, occupied rooms and to the external environment around the building. The sound pressure levels received at each frequency depend upon the barrier attenuation, distance between the sound source and the recipient and the acoustic properties of the receiving space. The low-frequency sound waves, below 1000 Hz, are more difficult to attenuate than those above 1000 Hz. This is because the commonly used building and sound absorbing materials and vibration isolating rubber all have a low natural frequency of vibration. They will resonate at a frequency often as low as 100 Hz. A material loses its attenuation property at the resonant frequency. Worse still, of course, is that when a rotary machine passes through or runs at its natural frequency of vibration, during start-up procedures, additional noise can be generated and the amplitude of its vibration may escalate to the point of physical destruction. It is vital that variable speed controllers run the rotary machine speed through its resonant frequency band as quickly as possible to minimize noise and vibration. Attenuation materials such as brick, concrete, timber and acoustic fabric are good at absorbing sounds at the higher frequencies. The human ear is most sensitive to sounds around 1000 Hz, making this the critical frequency for the acoustic design engineer.

Sound pressure level in a plant room

The sound source space is normally the mechanical services plant room. The reverberant sound pressure level in a plant room can be taken as:

$$SPL_1 = SWL + 10 \times \log(T_1) - 10 \times \log(V_1) + 14\,dB$$

(Source: Sound Research Laboratories Limited; see also, Smith et al. (1985))
where:

SPL_1 = sound pressure level in plant room	dB
SWL = sound power level of source mechanical plant	dB
T_1 = reverberation time of plant room	s
V_1 = volume of plant room	m^3

The reverberant sound pressure level is independent of the measurement location within the room. When a sound pressure level is required at a known location, the earlier equation is used with the radius from the source, r m,

$$SPL = SWL + 10 \times \log_{10}\left(\frac{Q}{4 \times \pi \times r^2} + \frac{4}{R}\right)\,dB$$

EXAMPLE 15.4

A refrigeration compressor has an overall sound power level of 86 dB on the 'A' scale. The plant room has a reverberation time of 2 s and a volume of 70 m³. Calculate the plant room reverberant sound pressure level.

$SWL = 86\,\text{dBA}$

$T_1 = 2\,\text{s}$

$V_1 = 70\,\text{m}^3$

$SPL_1 = SWL + 10 \times \log(T_1) - 10 \times \log(V_1) + 14\,\text{dB}$

$\qquad = 86 + 10 \times \log(2) - 10 \times \log(70) + 14\,\text{dB}$

$\qquad = 83 + 3 - 18 + 14\,\text{dBA ignoring decimal places}$

$\qquad = 85\,\text{dBA}$

Outdoor sound pressure level

The sound pressure level in the outdoor environment immediately external to the plant room can be taken as:

$$SPL_2 = SPL_1 - B + 10 \times \log(S_2) - 20 \times \log(d) + DI - 17\,\text{dB}$$

(Source: Sound Research Laboratories Limited), where:

SPL_2 = outdoor air sound pressure level   dB
SPL_1 = sound pressure level in source room   dB
B = sound reduction index of exterior wall or roof   dB
S_2 = surface area of external wall or roof   m²
d = distance between plant room surface and recipient   m
DI = directivity index   dB

EXAMPLE 15.5

A refrigeration compressor generates an overall sound pressure level of 85 dBA within a plant room. The plant room has an external wall of 12 m² that has an acoustic attenuation of 30 dB. Sound radiates from the plant room wall into a hemispherical field that has a directivity index of 2 dB. Hotel bedroom windows are at a distance of 4 m from the plant room wall. Calculate the external sound pressure level at the hotel windows.

$SPL_1 = 85$ dBA

$B = 30$ dBA

$S_2 = 12$ m^2

$d = 4$ m

$DI = 2$ dB

$SPL_2 = SPL_1 - B + 10 \times \log(S_2) - 20 \times \log(d) + DI - 17$ db

$\qquad = 85 - 30 + 10 \times \log(12) - 20 \times \log(4) + 2 - 17$ dB

$\qquad = 85 - 30 + 10 - 12 + 2 - 17$ dB

$\qquad = 38$ dBA

Sound pressure level in an intermediate space

The sound which is generated within a plant room may be transferred to an intermediate space within a building before being received in the target occupied room. Such intermediate spaces are corridors, store rooms, service ducts or roof voids. While it may not be important what the sound pressure level is within the intermediate space, the acoustic performance of this space affects the overall transfer of sound to the target occupied area. When the intermediate space is very large and has thermally insulated surfaces, for example, in a roof space, a considerable attenuation is possible. The sound pressure level in such an intermediate room or space can be taken as:

$$SPL_3 = SPL_1 - SRI + 10 \times \log(S_4) + 10 \times \log(T_2) - 10\log(0.16 \times V_2)\, dB$$

(Source: Sound Research Laboratories Limited), where:

SPL_3 = sound pressure level in intermediate space   dB
SPL_1 = sound pressure level in plant room   dB
SRI = sound reduction index of common surface   dB
S_4 = area of surface common to both rooms   m^2
T_2 = reverberation time of intermediate space   s
V_2 = volume of intermediate space   m^3

EXAMPLE 15.6

A showroom has floor dimensions of 25 m × 10 m and a height of 3.6 m to a suspended tile ceiling. The average height of the ceiling void is 1.8 m. An air conditioning system has distribution ductwork in the roof void above the suspended acoustic ceiling tiles. The air-handling plant room is adjacent to the roof void and there is a common plant room wall of 5 m × 2.5 m high in the roof void. The sound pressure level in the plant room is expected to be 50 dB. The reverberation time of the roof void is 0.8 s. The plant room wall adjoining the roof void has a sound reduction index of 10 dB. Calculate the sound pressure level that is produced within the roof void as the result of the air-handling plant room noise.

$SPL_3 = 50\,dB$

$SRI = 10\,dB$

$S_4 = 12.5\,m^2$

$T_2 = 0.8\,s$

$V_2 = 25 \times 10 \times 1.8\,m^3$

$\quad = 450\,m^3$

$SPL_3 = SPL_1 - SRI + 10 \times \log(T_2) - 10 \times \log(0.16 \times V_2)\,dB$

$\quad = 50 - 10 + 10 \times \log(12.5) + 10 \times \log(0.8) - 10 \times \log(0.16 \times 450)\,dB$

$\quad = 50 - 10 + 10 + 0 - 11\,dB$

$\quad = 39\,dB$

Sound pressure level in the target room

The sound pressure level in the target occupied room or space can be taken as:

$$SPL_4 = SPL_3 - SRI + 10 \times \log(S_5) + 10 \times \log(T_3) - 10 \times \log(0.16 \times V_3)\,dB$$

(Source: Sound Research Laboratories Limited), where:

SPL_4 = sound pressure level in target room	dB
SPL_3 = sound pressure in adjacent room	dB
SRI = sound reduction index of common surface	dB
S_5 = area of surface common to both rooms	m^2
T_3 = reverberation time of target room	s
V_3 = volume of target room	m^3

The target room may be adjacent to, or close to, the plant room, or it may not be influenced by the plant room other than by the transfer of noise through the interconnected air-ductwork system. Analysis of the ductwork route for noise transfer is calculated separately and is not covered in this book.

EXAMPLE 15.7

The showroom in Example 15.6 has floor dimensions of 25 m × 10 m and a height of 3.6 m to a suspended tile ceiling. The reverberation time of the showroom is 0.5 s. The air-conditioning plant room generates a sound pressure level of 39 dB in the roof space. The acoustic tile ceiling has a sound reduction index of 12 dB. Calculate the sound pressure level that is produced within the showroom by the air-conditioning plant.

$SPL_3 = 39\,dB$

$SRI = 12\,dB$

$S_5 = 250\,m^2$

$T_3 = 0.5\,s$

$V_3 = 25 \times 10 \times 3.6\,m^2$

$\quad = 900\,m^3$

$SPL_4 = SPL_3 - SRI + 10\log(S_5) + 10 \times \log(T_3) - 10 \times \log(0.16 \times V_3)\,dB$

$\quad = 50 - 12 + 10 \times \log(250) + 10 \times \log(0.5) - 10 \times \log(0.16 \times 900)\,dB$

$\quad = 50 - 12 + 23 - 3 - 21\,dB$

$\quad = 37\,dB$

Noise rating

The human ear has a different response to each frequency within the audible range of 20 Hz to 20000 Hz. It has been found that a low-frequency noise can be tolerated at a greater sound pressure level than a high-frequency noise. Noise rating (NR) curves are used to specify the loudness of sounds. Each curve is a representation of the response of the human ear in the range of audible frequencies.

The design engineer makes a comparison between the sound pressure level produced in the room at each frequency and the noise rating curve data at the same frequency. When all the noise levels within the room fall on or below a noise rating curve, that noise rating is attributed to the room. Noise rating curves for NR 25 to NR 50 are shown in Figure 15.2. The values are plotted from:

$SPL = NR_f \times B_f + A_f\,dB$

where:

SPL = sound pressure level at frequency f and noise rating NR dB
NR_f = noise rating at frequency f Hz   dimensionless
B_f and A_f = physical constants   dB
f = frequency   Hz

The values of the physical constants to calculate noise rating are shown in Table 15.5 (Australian Standard AS 1469–1983).

The normal applications of noise rating are shown in Table 15.6.

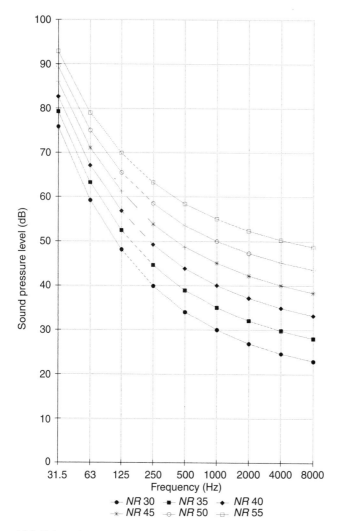

15.2 Noise rating curves.

Table 15.5 Physical constants for noise rating calculation.

Frequency f Hz	A_f dB	B_f dB
31.5	55.4	0.681
63	35.5	0.79
125	22.0	0.87
250	12.0	0.93
500	4.8	0.974
1000	0	1.0
2000	−3.5	1.015
4000	−6.1	1.025
8000	−8.0	1.03

Table 15.6 Noise rating applications.

Application	Noise rating	Comment
Acoustic laboratory	NR 15	Critical acoustics
Radio studio	NR 15	Critical acoustics
Concert hall	NR 15	Critical acoustics
TV studio	NR 20	Excellent listening
Large conference room	NR 25	Very good listening
Hospital, home, hotel	NR 30	Sleeping, relaxing
Library, private office	NR 35	Good listening
Office, restaurant, retail	NR 40	Fair listening
Cafeteria, corridor, workshop	NR 45	Moderate listening
Commercial garage, factory	NR 50	Minimum speech interference
Manufacturing	NR 55	Speech interference
Heavy engineering to industrial	NR 60 to NR 80	Sound levels judged on merits, leading to risk of hearing damage

EXAMPLE 15.8

An air-conditioning fan produces the sound spectrum shown in Table 15.7 in an occupied room. Calculate the sound pressure levels for noise ratings NR 35, NR 40, NR 45, NR 50 and NR 55, and plot the noise rating curves for the frequency range from 31.5 Hz to 8000 Hz. Plot the room sound pressure levels on the same graph and find which noise rating is not exceeded.

Table 15.7 Noise spectrum in Example 15.8.

Frequency f Hz	31.5	63	125	250	500	1 k	2 k	4 k	8 k
Room SPL dB	39	44	48	52	55	49	36	33	28

A manually calculated example for one noise rating curve is shown. The reader should use the spreadsheet graph or chart facilities to plot the whole figure. Calculate the SPL values for NR 55.

$$NR = 55$$

$$SPL = NR_f \times B_f + A_f \text{ dB}$$

Calculate the SPL at each frequency for the values of B_f and A_f from Table 15.5. For 31.5 Hz:

$$SPL = 55 \times 0.681 + 55.4 \text{ dB}$$

$$= 92 \text{ dB}$$

For 63 Hz:

$$SPL = 55 \times 0.79 + 35.5 \text{ dB}$$

$$= 78 \text{ dB}$$

For 125 Hz,

$$SPL = 55 \times 0.87 + 22 \text{ dB}$$
$$= 69 \text{ dB}$$

For 250 Hz:

$$SPL = 55 \times 0.93 + 12 \text{ dB}$$
$$= 63 \text{ dB}$$

For 500 Hz:

$$SPL = 55 \times 0.974 + 4.8 \text{ dB}$$
$$= 58 \text{ dB}$$

For 1000 Hz:

$$SPL = 55 \times 1.0 + 0 \text{ dB}$$
$$= 55 \text{ dB}$$

For 2000 Hz:

$$SPL = 55 \times 1.015 - 3.5 \text{ dB}$$
$$= 52 \text{ dB}$$

For 4000 Hz:

$$SPL = 55 \times 1.025 - 3.5 \text{ dB}$$
$$= 50 \text{ dB}$$

For 8000 Hz:

$$SPL = 55 \times 1.03 - 8.0 \text{ dB}$$
$$= 48 \text{ dB}$$

These sound pressure levels are compared to the room data in Table 15.8.

Table 15.8 Noise spectrum comparison in Example 15.8.

Frequency f Hz	31.5	63	125	250	500	1 k	2 k	4k	8 k
NR 55 SPL dB	92	78	69	63	58	55	52	50	48
Room SPL dB	39	44	48	52	55	49	36	33	28

The closest approach to the SPL limit for NR 55 occurs at 500 Hz. Check that NR 50 is exceeded. For 500 Hz,

$$SPL = 50 \times 0.974 + 4.8 \, dB$$

$$= 53 \, dB$$

It should now be possible to check manually any sound pressure level against noise rating. Once the frequency that produces the greatest sound pressure level from the sound source is identified, other SPL values can be obtained for the peak frequency to check which NR is not exceeded.

The room does not exceed the NR 55 curve data but it does exceed NR 50. Plot the chart with the spreadsheet functions for NR 35 to NR 55 and with the room noise SPL. The spreadsheet will produce curves for six sets of data, so five NR curves and the one room curve can be displayed on one chart. The results are shown in Figure 15.3.

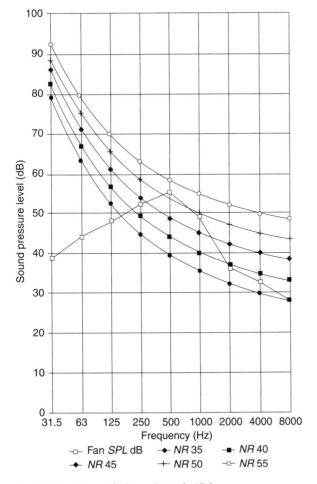

15.3 Noise ratings, solution to Example 15.8.

Questions

Questions 1–31 do not need to be evaluated on the worksheet. The worksheet is to be used for Questions 32 onwards. The solution to Question 32 is shown on the original file DBPLANT.WKS. Solutions to descriptive questions are to be found in the chapter, except where specific answers are provided.

1. List the sources of noise that could be found within an air-conditioned building.
2. What is meant by noise?
3. State which items of mechanical services plant, equipment and systems within an occupied building are not likely to create noise.
4. Explain how sound travels from one location to another.
5. Explain what is meant by the term sound pressure wave.
6. Why is sound important?
7. Explain how we 'hear' sounds.
8. State what is meant by sound power and sound pressure level. State the units of measurement for sound power, sound pressure, sound power level and sound pressure level.
9. Explain why any decimal fraction of a decibel is not used in engineering design.
10. List the ways in which mechanical and electrical services plant, equipment and systems generate sound.
11. Explain, with the aid of sketches and examples, how sound is transferred, or can be, through a normally serviced multi-storey occupied building.
12. Discuss the statement: 'Turbulent flows in building services systems create a noise nuisance.'
13. State how the building structure transfers sound.
14. Explain, with the aid of sketches, ways in which the noise and vibration produced by the mechanical and electrical services of a building can be reduced before they become a nuisance for the building's users.
15. Explain how sound energy is dissipated into the environment.
16. State the range of frequencies that are detectable by the human ear and the frequencies that are used in acoustic design calculations. State the reasons for these two ranges being different, if they are.
17. Define the terms 'sound power level' and 'sound pressure level'.
18. Explain what is meant by direct and reverberant sound fields.
19. A plant room for a refrigeration compressor is $6\,m \times 4\,m$ in plan and 3m high. It has four brickwork walls, a concrete floor and a concrete roof. Select the surface absorption coefficients for the frequency range $125\,Hz$ to $4000\,Hz$. Calculate the room absorption constant and the reverberation time for the plant room at each frequency. Do the calculations manually and then enter the same data onto the worksheet to validate the results.
20. An air-conditioning centrifugal fan has an overall sound power level SWL of 75 dBA. The fan is to be installed centrally within a plant room that has a room absorption constant R of $12\,m^2$. Calculate the sound pressure level that will be produced close to the fan, in the plant room at 1000 Hz when the fan is operating, and also generally within the room.
21. A 900 mm diameter axial fan is to be installed on the concrete floor of an $8\,m \times 4\,m \times 3\,m$ high plant room. The fan sound power level at 1000 Hz is 89 dB. The room absorption constant R at 1000 Hz is $8\,m^2$ and the reverberation time is 0.4 s. Calculate the room sound pressure level at a radius of 300 mm from the fan, and the reverberant room sound pressure level.

22. A reciprocating water chilling refrigeration compressor has an overall sound power level of 92 dBA. It is to be located within a concrete-and-brick plant room that has a reverberation time of 1.8 s and a volume of 250 m³. Calculate the plant room reverberant sound pressure level.

23. An air-handling plant has an overall sound power level of 81 dB. The plant room has an external wall of 10 m² that has an acoustic attenuation of 35 dB and ventilation openings having a free area of 3 m². The windows of residential and office buildings are at a distance of 12 m from the plant room wall. Calculate the external sound pressure level at the windows and recommend what, if any, attenuation is needed at the plant room.

24. A forced draught gas-fired boiler has an overall sound pressure level of 96 dB. The boiler plant room has an external wall of 60 m² that has an acoustic attenuation of 25 dB and two louver doors to admit air for combustion. Calculate the external sound pressure level at a distance of 20 m from the plant room wall. State your recommendations for the attenuation of the boiler and the plant room.

25. A single-storey office building has floor dimensions of 40 m × 30 m and a height of 3 m to a suspended acoustic tile ceiling. The average height of the ceiling void is 1.5 m. A plant room is adjacent to the roof void. There is a common plant room wall of 10 m × 1.5 m high in the roof void. The sound pressure level in the plant room is expected to be 61 dB. The reverberation time of the roof void is 0.6 s. The plant room wall adjoining the roof void has a sound reduction index of 13 dB. Calculate the sound pressure level that is produced within the roof void as the result of the plant room noise. Comment on the resulting sound pressure level.

26. A hospital waiting area has floor dimensions of 8 m × 12 m and a height of 3 m to a plasterboard ceiling. A packaged air conditioning unit is housed in an adjacent room. There is a common wall of 15 m² and sound reduction index of 35 dB to the two rooms. The sound pressure level in the plant room is expected to be 72 dB. The reverberation time of the waiting room is 1.3 s. Calculate the sound pressure level that will be produced in the waiting room.

27. A meeting room has floor dimensions of 8 m × 6 m and a height of 2.7 m to a suspended tile ceiling. The reverberation time of the room is 0.7 s. A fan coil heating and cooling unit creates a sound pressure level of 43 dB in the ceiling space. The acoustic tile ceiling has a sound reduction index of 8 dB. Calculate the sound pressure level in the meeting room.

28. A hotel bedroom is 6 m long, 5 m wide and 2.8 m high and it has a reverberation time of 0.4 s. The air-conditioning plant room generates a sound pressure level of 56 dB in the service space above the ceiling of the bedroom. The plasterboard ceiling has a sound reduction index of 16 dB. Calculate the sound pressure level in the bedroom.

29. Explain how noise rating curves relate to the response of the human ear and are used in the design of mechanical services plant and systems.

30. The centrifugal fan in an air-handling plant produces the noise spectrum shown in Table 15.9 in an office. Calculate the sound pressure levels for noise ratings NR 35, NR 40, NR 45, NR 50 and NR 55 and plot the noise rating curves for the frequency range 31.5 Hz to 8 kHz. Plot the room sound pressure levels on the same graph and find which noise rating is not exceeded.

Table 15.9 Noise spectrum in Question 30.

Frequency f Hz	31.5	63	125	250	500	1 k	2 k	4 k	8 k
Room SPL dB	30	35	32	40	42	31	28	20	10

31. A model XT45 water chiller is to be located within a plant room on the roof of a hotel in a city centre. The plant room is 12 m long, 10 m wide and 3 m high. The room directivity index is 2. The plant operator will normally be 1 m from the noise source. The floor is concrete; the roof is lined internally with a 50 mm polyester acoustic blanket with a metallised film surface. The plant room walls are 115 mm brickwork. There are no windows. The water chiller manufacturer provided the sound power levels as 100 dB overall, 74 dB at 63 Hz, 89 dB at 125 Hz, 95 dB at 250 Hz, 97 dB at 500 Hz, 99 Hz at 1 kHz, 97 dB at 2 kHz and 90 dB at 4 kHz.

   1. Check that the correct data is entered onto the working copy of the original worksheet file DBPLANT.WKS and find the noise rating that is not exceeded within the plant room.
   2. The plant room has three external walls. The nearest openable window in nearby buildings is at a distance of 15 m from a plant room wall. There is no acoustic barrier between a plant room wall and the recipient's window. The directivity index for the outward projection of sound is taken as 3 dB. Find the noise rating at the recipient's window and state what the result means.
   3. A corridor adjoins the plant room. The target sound space, an office, is on the opposite side of the corridor. The corridor is 10 m long, 1 m wide and 3 m high. It has a room directivity index of 2, a carpeted concrete floor, plastered brick walls and a plasterboard ceiling. The common wall between the plant room and the corridor is 10 m long, it is constructed with 115 mm plastered brickwork and it does not have a door. There are no windows. There is no other sound barrier. Find the noise rating which would be found at a distance of 0.5 m from the plant room wall while within the corridor.
   4. The target office is 10 m long, 10 m wide and 3 m high. The room directivity index is 2. The nearest sedentary occupant of the office will be 1 m from the corridor wall. The floor has pile carpet, the walls are plastered brick and there is a suspended ceiling of 15 mm acoustic tile and 50 mm glass fibre-matt. The office has four 2 m × 2 m single-glazed windows on two external walls. The office wall that adjoins the corridor is 115 mm plastered brickwork and it has one 2 m² door into the corridor. Find the noise rating, *NR*, and sound pressure levels, *SPL* dB that are experienced in the target office. State what affect the office and plant room doors will have on the noise rating in the target room. Recommend appropriate action to be taken with these doors.

32. A centrifugal fan is located within the basement plant room of an office building. The plant room is 8 m long, 6 m wide and 3 m high. The room directivity index is 2 and the plant operator will normally be 1 m from the noise source. The floor and ceiling are concrete, there are four 230 mm brick walls and one acoustically treated door. There are no windows in the plant room. The sound power levels of the fan are: 86 dB overall, 64 dB at 63 Hz, 66 dB at 125 Hz, 72 dB at 250 Hz, 80 dB at 500 Hz, 86 Hz at 1 kHz, 82 dB at 2 kHz and 77 dB at 4 kHz.

   1. Find the noise rating that is not exceeded within the plant room.
   2. A corridor and staircase connect the plant room to the Reception area of the building. The corridor is 6 m long, 1 m wide and 3 m high. It has a room directivity index of 2. The corridor has a concrete floor, plastered brick walls and a plasterboard ceiling. The common wall between the plant room and corridor is 2 m long. The sound reduction index of the plant room door is 20 dB at each frequency from 125 Hz to 4 kHz. There is no other sound barrier. Find the noise rating that would be found at a distance of 1 m from the plant room in the corridor.
   3. The Reception area is 12 m long, 8 m wide and 3 m high. The room directivity index is 2. There are 10 m² of single-glazed windows in Reception. There is a door at the top of the staircase down to the plant room. The stairs door is 1 m wide, 2 m high and it has a sound reduction index of 20 dB at each frequency from 125 Hz to 4 kHz. The nearest

occupant will be 1 m from the stairs door. The floor has thermoplastic tiles on concrete, the walls are plastered brick and there is a plasterboard ceiling. Find the noise rating which is not exceeded in Reception.

33. Oil-fired hot water boilers are located in a plant room in the basement of an exhibition and trade centre building in a city centre. The plant room is 10 m long, 10 m wide and 5 m high. The room directivity index is 2. The floor, walls and ceiling are concrete. There are no windows. The reference sound power level of the boiler plant is 88 dBA.

1. Find the anticipated spectral variation in the sound power level for the frequency range from 63 Hz to 4 kHz from Table 15.4 and Figure 15.1, enter the data onto the worksheet and find the noise rating that is not exceeded within the boiler plant room.
2. The plant room has three 100 mm concrete external walls. The nearest recipient can be 1 m from the external surface of a boiler plant room wall. There is no acoustic barrier between a plant room wall and a recipient. The directivity index for the outward projection of sound is taken as 3 dB. Find the noise rating at the nearest recipient's position and state what the result means.
3. A hot-water pipe and electrical cable service duct connects the boiler plant room to other parts of the building. The concrete-lined service duct is 30 m long, 2 m wide and 1 m high. Both ends of the service duct have a 100 mm concrete wall. Calculate the noise rating within the service duct at its opposite end from the boiler plant room.
4. A conference room 115 mm brick wall adjoins the service duct at the furthest end from the boiler plant room. The conference room is 12 m long, 10 m wide and 4 m high. The room directivity index is 2. The nearest sedentary occupant will be 0.5 m from the service duct wall. The floor has pile carpet, the walls are plastered brick and there is a suspended ceiling of 15 mm acoustic tile and 50 mm glass fibre matt. There are no windows. Find the noise rating that is produced in the conference room by the boiler plant.

34. A four-pipe chilled- and hot-water fan coil unit is located within the false ceiling space above an office in an air-conditioned building. Conditioned outdoor air is passed to the fan coil unit through a duct system. The office is 5 m long, 4 m wide and 3 m high. The room directivity index is 2. The office has a concrete floor with thermoplastic tiles and 115 mm plastered brick walls. The 700 mm deep suspended ceiling has 12 mm fibreboard acoustic tiles, recessed fluorescent luminaires, ducted supply and return air with a supply air diffuser, a return air grille and a concrete ribbed slab for the floor above. The office has a double-glazed window of 2 m x 2 m. The reference sound power level of the fan coil unit is 85 dBA. Enter the ceiling space as the plant room and bypass the intermediate space data as directed.

1. Find the anticipated spectral variation in the sound power level of the fan coil unit for the frequency range from 63 Hz to 4 kHz from Table 15.4 and Fig. 15.1. Enter the data onto the worksheet and find the noise rating that is not exceeded within the ceiling space above the office.
2. Find the noise rating that is not expected to be exceeded within the office at head height. Assume that the sound reduction of the acoustic tile ceiling is maintained across the whole ceiling area.
3. Sketch a cross-section of the fan coil unit installation and identify all the possible noise paths into the office.
4. List the ways in which the potential noise paths into the office can be, or may need to be, attenuated.

35. Presbycusis is:

1. Hearing loss due to long-term exposure to noise above 90 dBA.

2.   Hearing loss due to ear disease.
3.   Normal deterioration in hearing due to ageing.
4.   A church presbytery committee.
5.   Temporary shift in hearing ability from exposure to high industrial noise levels above 95 dBC.

36. Hearing range is:

1.   20 Hz to 20 kHz.
2.   2 Hz to 20 MHz.
3.   200 Hz to 200 MHz.
4.   Infinitely wide.
5.   2 kHz to 20 MHz.

37. What is noise?

1.   Sound.
2.   Acoustic power.
3.   Unwanted sound.
4.   Age-related sound.
5.   Traffic, aeroplanes, pneumatic drills, fans, refrigeration compressors.

38. How do we judge sound?

1.   With absolute measurement.
2.   Comparing a sound with absolute zero sound level.
3.   Relatively.
4.   Subjectively.
5.   Qualitative judgement.

39. What is sound?

1.   Electromagnetic radiation.
2.   Molecular vibration of solid materials.
3.   Radio frequency waves.
4.   Anything that causes an ear response.
5.   Pressure waves.

40. Sound travels through air because it is:

1.   Incompressible.
2.   Supporting molecular vibration.
3.   Compressible.
4.   Inelastic.
5.   Plastic.

41. Reference point for sound level measurement is:

1.   Absolute zero sound.
2.   Lowest audible level by a domestic animal.
3.   Smallest sound detectable by human ear.
4.   Zero atmospheric pressure as found in space.
5.   Inaudible level created in a test laboratory.

42. Sound waves repeat at a frequency due to:

1.   Absorption by porous surfaces.
2.   Wind forces.

3. Multiple sources of sound.
4. Passage of blades in a rotary machine such as a compressor, pump or turbine.
5. Variations in air pressure.

43. An eight cylinder Formula 1 car engine peaks at 20000 RPM. One of the sound frequencies it produces is:

1. 8 Hz.
2. 20 kHz.
3. 400 Hz.
4. 2000 Hz.
5. 2667 Hz.

44. A gas turbine rotates at 60000 RPM and has 50 blades on its largest diameter. One of the sound frequencies it produces is around:

1. 50 kHz.
2. 50 Hz.
3. 5000 Hz.
4. 60 kHz.
5. 20 kHz.

45. How can the structure of a building transmit noise?

1. Concrete framed structures cannot as noise is dampened.
2. Steel and concrete structures absorb all acoustic energy.
3. Structures always absorb acoustic energy and dissipate it as heat.
4. Molecular vibration.
5. Physical movement.

46. How is noise transmission from plant reduced?

1. Cannot be reduced, only contained within plant room.
2. Select quieter plant.
3. Seal plant room doors.
4. Locate plant room away from occupied rooms.
5. Flexible rubber and spring mountings.

47. Which is the smallest increment of sound pressure level detectable by the human ear?

1. 1 W/m^2.
2. 1 Bel.
3. 60 Bel.
4. 100 N/m^2.
5. 1 decibel.

48. Explain the meaning of SWL:

1. Selective wind loading.
2. Sound wind level.
3. Sound watts level, meaning power.
4. Sound pressure level, meaning energy.
5. Sound watts loudness, meaning loudness power.

49. How much acoustic power is experienced within buildings?

1. 10% of electric motor power becomes acoustic energy.

2. Around 10 W/m^2 floor area.
3. Above 1 kW.
4. Always below 500 W.
5. Less than 1 Watt.

50. Frequency range used for assessment of sound power level, SWL, from machines:

    1. 0 to 200 MHz.
    2. 1 kHz to 2 MHz.
    3. 125 Hz to 8 kHz.
    4. 63 Hz to 20000 Hz.
    5. 125 kHz to 8 MHz.

51. By what mechanism do ears respond to sound power level, SWL?

    1. Ears have no mechanism.
    2. Sound power radiates to vibrate the eardrum.
    3. Acoustic vibration energy vibrates the body, which transfers through the body muscle and bone structure, to vibrate eardrums.
    4. Acoustic power raises air pressure on eardrums.
    5. Acoustic output power pulsates and vibrates air, raising air pressure waves; eardrum vibrates from air pressure waves.

52. What is a reverberant sound field?

    1. Sound transmitted over a large distance.
    2. Sound passing through a structure.
    3. What remains within an enclosure after source energy is absorbed by the building structure.
    4. Reflected sound.
    5. Sound pressure level measured in an anechoic laboratory chamber.

53. Direct sound field:

    1. Increases in intensity further from the source.
    2. Remains at a constant noise at any distance from the source while hearer remains in the source plant room.
    3. Decreases linearly with distance from source.
    4. Falls with the inverse square of the distance from the sound source.
    5. Doubles the value of the reverberant sound field.

54. Which of these is not correct about absorbing sound energy?

    1. Dense materials absorb acoustic energy efficiently.
    2. Highly porous materials are good sound absorbers.
    3. The denser the material mass, the greater the sound absorption.
    4. A 75 mm air cavity behind a sheet of plasterboard is a good sound absorber.
    5. A plastered brick wall has a low sound absorption coefficient.

55. Which is not correct about reverberation time?

    1. When short, below a second, room seems lively.
    2. Long reverberation time causes room to sound noisy and echoes.
    3. A lecture theatre needs a short reverberation time.
    4. A large volume car manufacturing building has a long reverberation time.
    5. When short, below a second, room seems dull.

56. Which is not correct about an anechoic chamber?

    1. Used to measure sound power level from acoustic sources such as fans and compressors.
    2. Must have no reverberant sound field.
    3. Lined with fully absorbent foam wedges.
    4. Sounds perfectly dull.
    5. Used to measure reverberant field sound pressure level from acoustic sources such as fans and compressors.

57. What does natural frequency of vibration mean?

    1. Damped vibration.
    2. Strike a guitar string and it vibrates at up to four times its natural frequency depending on volume of sound box.
    3. Bounce a coil spring and it vibrates at its natural frequency of vibration.
    4. A frequency mechanically forced upon an item, such as by a motor.
    5. A material never vibrates at this frequency.

58. Which does NR stand for?

    1. Noise resonance.
    2. Normal rating.
    3. No resonance.
    4. Noise ratification.
    5. Noise rating.

59. How are noises related to human ear response?

    1. Humans respond to sound power level within a range of audible frequencies.
    2. Humans respond to loudness produced over a range of audible frequencies.
    3. Sound pressure levels are added to create an overall relationship to ear response.
    4. Sound power levels are added to create an overall relationship to ear response.
    5. Loudest sound at any frequency is taken as ear response.

60. Should we be concerned with any linkage between the *HM Government Carbon Plan, 2011* and acoustics?

    1. No, there is no connection.
    2. Acoustic energy dissipates in porous materials and raises its temperature causing a cooling load for the refrigeration system; yes we are concerned.
    3. Only people create unwanted sound.
    4. Noise reconverts back into useful energy.
    5. Yes, noise means energy is used somewhere.

16 Mechanical transportation

Learning objectives

Study of this chapter will enable the reader to:

1. understand the principles of passenger and goods transportation within and between buildings;
2. discuss the applications of passenger lift systems;
3. know the speeds and carrying capacities of lifts;
4. know how lift systems are controlled and used during the outbreak of fire in a building;
5. know the principles of electric motor and hydraulic lifts;
6. know the principles and carrying capacities of escalators and passenger conveyors;
7. understand the importance of lift shaft and motor room ventilation;
8. recognize the builder's work required for a lift installation.

Key terms and concepts

car speed 342; carrying capacity 341; collecting mode 341; driving motor 344; driving pulley 343; escalator 343; fireman's lift 342; goods lift 342; hydraulic lift 346; lift motor room 344; lift shaft 347; noise 344; passenger conveyor 344; paternoster 343; passenger lift 341; service lift 342; ventilation 347.

Introduction

The mechanical transportation of people and goods is an energy-using service which needs the designer's attention at the earliest stages of building design. Standards of service rise with expectations of quality by the final user and with the provision of access for disabled people. Principles of transportation systems are outlined and reference is made to movement between buildings. Their energy consumption may be low, but the electrical power requirement is significant for short periods. Integration with other services, fire protection, means of escape and correct maintenance are of the highest importance.

Transportation systems

Mechanical transportation of people and equipment around and between buildings is of considerable importance in relation to the degree of satisfactory service provided. Increasing usage of computer networks, the internet, and digital communications will gradually reduce travel requirements. How many indoor workers might be able to work from home through the wired or wireless internet with visual conferencing and only attend a commercial building for meetings? Very many. Cost-effective and energy-efficient transportation will always be in demand. Walking and cycling are the supreme of personal low cost mobility for the majority of the population. City express cycle lanes for aerodynamically enclosed human-powered vehicles can be provided with present-day technology for cruising speeds of 45.0 km/h. Tunnels and covered above-ground routes could be used for conventional bicycle traffic on large building developments or around towns. When the total concept of global sustainability and the *HM Government Carbon Plan, 2011* (DECC, 2011a) is envisaged, alternatives to consuming highly refined forms primary energy, as electricity, must be considered.

Lifts

Lifts or elevators, use up to 5% of the electrical energy consumed in commercial buildings, 2000–15000 kWh/yr per lift (http://en.wikipedia.org/wiki/Elevator). At 10 p/kWh a single lift might cost £200–£1500 per year while a high rise city centre major building may have 40 lifts on each floor, serving 100 storeys, split vertically into three levels of sky lobby, making 120 elevators in total, maybe costing £30,000 annually plus maintenance, repair and refurbishments. Such energy use is largely unpredictable in terms of time of use, although regular patterns will be established. Once installed, it is peak hour energy use that cannot be significantly reduced. It is unregulated energy; that is, outside the building regulation controls. It represents the energy cost of our style of working and living as compared to ground-level activities. Elevators also serve our below-ground structures, train travel, vehicle parking, storage spaces, high security bunkers, recreational activities as well as in industrial mining. Excessive reliance on elevators might be said to be a contributory cause of human obesity and lack of physical fitness. Those who can walk up and down a few floors may feel better about themselves when opting for the stairways. Ask your doctor about it.

Passenger lifts are provided for buildings of over three storeys, or less when wheelchair movement is needed. The minimum standard of service is one lift for each four storeys, with a maximum walking distance of 45.0 m between workstation and lobby. Higher standards of service are provided in direct proportion to rent-earning potential of the building and prestige requirements. The peak demand for lift service is assessed from the building size, shape, height and population. Up to 25% of the population will require transportation during a 5-minute peak period. Congestion at peak travel times is minimized by arranging the lift lobbies in a cul-de-sac of, say, two lift doors on either side of a walkway, rather than in a line of four doors along one wall. Computer-controlled installations can be programmed to maximize their performance in a particular direction at different times of the day. Each lift car can be parked at an appropriate level to minimize waiting time. Two lifts of 680 kg-carrying capacity, maximum 10 people, provide a better service than one 1360 kg, 20-person lift. The large single lift would run only partly loaded during the major part of the day, with a resulting decrease in efficiency and increased running cost. One of the smaller lifts could be parked for long periods to reduce costs. The advantages of using two smaller lifts may be considered partly to outweigh the additional capital and maintenance costs.

Car speed is determined by travel distance and standard of service. Buildings of more than 15 storeys may have some high speed lifts, not stopping for the lower 10 storeys. A 49-storey tower in the City of London has double deck cars serving two reception floors levels simultaneously. Sky lobbies are halfway up and near the top of the tower; non-stop service is provided in both lower and upper sectors between the main lobbies, at a speed of 7 m/s. Intermediate floors are served with lower speed stopping lifts. Travel from basement car parking to the main street lobby is usually a dedicated low speed service. Car speeds for various travel distances are around 7 m/s for 15 floors, 3 m/s for less floors and 1 m/s for 4 floors. Car speed is chosen so that the driving motor can be run at full speed for much of the running time to maximize the efficiency of power consumption. Starting and accelerating power is greater than steady speed energy use as it is with a road motor vehicle. Deceleration during braking dissipates the momentum gained by the car and counterweight in friction-generated heat, lost to the atmosphere and into the lift motor plant room at the top of the shaft. The overall speed of operation is determined by the acceleration time, braking time, and maximum car speed, speed of door opening, degree of advance door opening, floor levelling accuracy required, switch timing and variation of car performance with car load. An automatic control system should function in an upward collecting and downward collecting mode. Requests for service made sufficiently early at a lobby cause the car in that shaft to break its original journey instruction and stop. Computer controls are used to optimize the overall performance of the installation, by causing the nearest car to stop and to minimize electricity consumption.

In the event of a fire within the building, the central fire detector and alarm system signal causes all the lifts to run to the ground floor. Where the building extends out of the reach of conventional fire-fighting turntable ladders on vehicles, at least one of the lifts is designated as the fireman's lift. These have a minimum carrying capacity, must reach the top of the building in 1 minute, doors remain open on any floor and have overriding fire brigade control.

Goods lifts travel at a maximum of 1.0 m/s and have full width doors, sometimes at each end of the car. Accurate floor levelling, to within 5.0 mm, may be provided to facilitate smooth passage of trolleys carrying fragile goods, fluids or patients in a hospital. Passengers can use goods lifts but service is slow. A variable-voltage electrical or hydraulic power supply is used. Hydraulically operated lifts have the advantage of very quiet operation and low running costs. The only power-consuming plant item is a small hydraulic pump immersed in an oil tank. Goods lifts complete each journey instruction before accepting another. Door operation can be manual or automatic. Additional structural supports are needed for lifts with high carrying capacity and well-designed brakes. Non-metallic serrated inserts may be fitted into the grooves of the driving pulley (sheave) to reduce wear and increase traction.

Service lifts are small goods lifts with car floor areas of up to 1.20 m^2 and heights of less than 1.40 m. The serving level may be 0.850 m above floor level to coincide with the working plane. Documents, goods or food are carried at up to 0.50 m/s with a maximum contract load of 260 kg. Control can be manual or semi-automatic. Prefabricated service lifts can be installed in a day and require minimal builders' work. Controls and the electric driving motor can be at the base of the shaft and a chain drive used.

Service conveyors for document transportation combine horizontal and vertical movement. An installation in the City of London transfers documents throughout a bank complex using briefcase-sized carriages on a continuously moving railway track with sidings for loading and unloading. A tunnel connects buildings on either side of a street. The Post Office unmanned underground electric mail trains are another example.

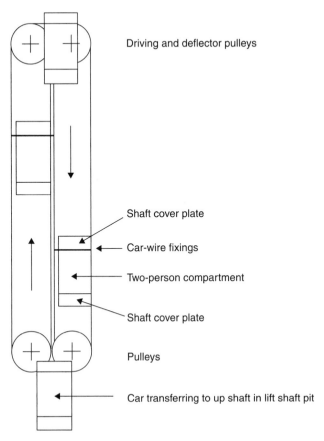

Driving and deflector pulleys

Shaft cover plate

Car-wire fixings

Two-person compartment

Shaft cover plate

Pulleys

Car transferring to up shaft in lift shaft pit

16.1 Paternoster lift.

Paternoster

A number of doorless two-person lift cars travel continuously in a clockwise direction within a common vertical shaft. The principle of operation is shown in Figure 16.1. This system is used in an office or an adult education building which requires regular inter-floor travel over short distances for normally agile people. A conventional passenger lift is installed alongside the paternoster for the carriage of more passengers, goods, children, the elderly or the infirm. Paternoster performance over more than about six storeys is limited by its low speed, 0.40 m/s. Its carrying capacity is 720 passengers per hour. A degree of acclimatization is needed for its use. Car spacing is 4.0 m. A chain drive from electric motors is employed. The driving motor, brake, gearing and control equipment are fitted in a machine room at the top of the shaft. Emergency stop buttons are at each floor level and are linked to an audible alarm.

Escalator

Escalators and passenger conveyors are primarily used where large numbers of passengers form surges at discharge times from offices, railway underground stations and airport terminals. Crowd flow, in plan, is similar to two-dimensional turbulent fluid behaviour and design for passenger routes can be regarded in a similar manner. Escalators provide suitable transport for all ages, laden or unladen. Their operating direction is reversible to correspond to peak travel times. Tread widths are from 0.60 m to 1.050 m. For a given quality of service, they require less horizontal floor space than a lift. The angle of inclination is normally 30°, but 35° can be used for a vertical rise of less than 6.0 m and a speed of less than 0.5 m/s. Speeds of up to 0.75 m/s are permissible as this is the maximum safe entry and exit velocity.

Passenger conveyors

Passenger conveyors are moving pavements which can carry wheeled vehicles such as shopping or luggage trolleys, prams or wheelchairs. Distances of up to 300 m can be travelled at speeds of up to 0.90 m/s with an 8° slope. Combinations of down-grade, horizontal and up-grade can be included, as in a road underpass. An S-shaped track overcame the limitation of entry and exit speed restriction on journey time by having its floor constructed from metal plates which slide relative to each other. The plates bunch together at the entry and exit curves of the track but spread out along the central straight which may be up to 1000 m length. Travel speed along the straight can be five times the entry speed. An electric motor of 19 kW drives the conveyor through a reduction gearbox and chain and 7200 passengers per hour can be carried and. A concrete ramp forms the structural base for the entire conveyor. Emergency stop buttons are provided.

Driving machinery

The lift car and its load are partly balanced with a counterweight and this reduces motor power consumption. Motor power is used to overcome friction, acceleration, inertia and the unbalanced load during lifting. Power is transmitted to the traction sheaves through a gearbox, two speed or variable speed motor driving sets, sometimes using direct current motors. The motor and driving sheaves are mounted on a load-bearing concrete base at the top of the lift shaft. Considerable heat output is created and lift motor room natural or mechanical ventilation is essential, plus in some cases, mechanical cooling from an air handling unit, in hot climates. The wire rope configuration for higher speed lifts is shown in Figure 16.2.

The lift motor room must be maintained at between 10°C and 40°C by natural or mechanical heating and ventilation if necessary. This ensures a condensation-free atmosphere in winter and adequate motor cooling in summer. A smoke extract ventilation grille of 0.10 m^2 free area must be provided at the top of each lift shaft. Some noise is produced in the motor room and its escape from this area is limited by a vibration-isolating concrete machine base and the concrete construction of the lift motor room. The external noise level produced is considered in conjunction with nearby room usage to assess whether additional attenuation is required. Air bypass holes, as large and as frequent as possible, cross-connect adjacent lift shafts to allow car-induced draughts to circulate with minimum restriction. An emergency electrical power generator is often installed so that, in the event of mains failure, one lift at a time can be run to the ground floor and the doors opened. One lift can then be made available for the emergency services, and emergency lighting is provided.

Hydraulic drive is often used for lift speeds up to 1.75 m/s for passenger travel and goods lifts of up to six storeys. Alternative bore hole and rope drum drive operation principles are shown in Figures 16.3 and 16.4.

Service lifts can use an electric-motor-driven winding drum with a deflector pulley at the top of the shaft, a pulley on top of the car and the motor at the bottom of the shaft. This is less efficient than counterweighted designs but saves space and complexity.

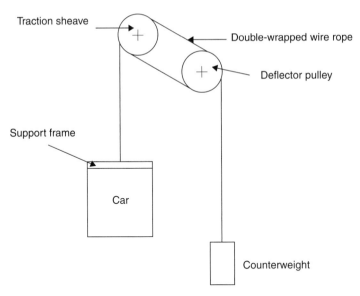

Traction sheave

Double-wrapped wire rope

Deflector pulley

Support frame

Car

Counterweight

16.2 Traction arrangement for a high-speed passenger lift.

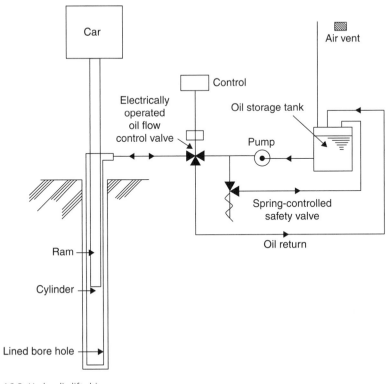

Car

Air vent

Control

Electrically
operated
oil flow
control valve

Oil storage tank

Pump

Spring-controlled
safety valve

Oil return

Ram

Cylinder

Lined bore hole

16.3 Hydraulic lift drive.

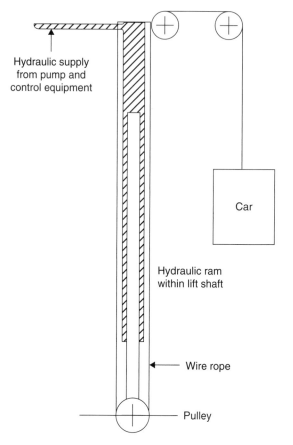

16.4 In-shaft hydraulic lift drive.

Each passenger lift car has a service hatch in its ceiling. When this is opened, an electrical interlock switch opens to prevent the lift from being operated while escapees or a maintenance engineer may be in the shaft. A mechanical extract fan is often fitted in the roof of the car.

A lift motor room has the following features:

1. a concrete machine base incorporating a vibration-isolating cork slab to separate its upper and lower parts;
2. motor and brake equipment bolted to the upper, vibration-isolated, concrete slab;
3. flexible armoured electrical cable connections to the motor;
4. lift motor main isolating switch close to the plant room door;
5. access hatch into the lift shaft;
6. electrical control panel, switches or digital;
7. lifting beam built into the structure;
8. adequate artificial illumination;
9. natural or mechanical ventilation;
10. 13 A power point;
11. locked door;
12. light coloured walls and ceiling;
13. emergency telephone.

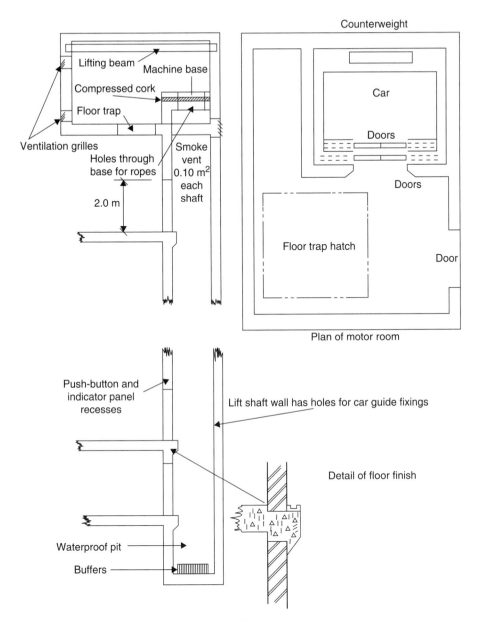

16.5 Construction features of a passenger lift shaft.

The main constructional features of a lift shaft and motor room are shown in Figure 16.5.

Questions

1. Explain how transportation systems are employed within and between buildings to assist the movement of people and goods. Include details of the main characteristics of the systems, their performance and costs.

2. Explain with the aid of sketches, drawings or illustrations from the internet, how high speed electrically driven lifts operate and where they are used.
3. Give examples of mechanical transportation systems used in buildings. Take photographs of installations available to you if permission is granted. Use the internet and manufacturers resources to illustrate current practices.
4. Relate the *HM Government Carbon Plan, 2011* to the use of mechanical transportation within and between buildings. Can you identify any conflicts of intention? How could the use of primary energy for mechanical transportation be reduced? Do you consider that commercial buildings have become too high and wasteful of energy for transport? What could be done about it? How could it be justifiable to travel less in our work to reduce carbon emissions?
5. Comment on improvements which might be made in the current designs of mechanical transportation systems used in buildings.
6. Explain with the aid of sketches, drawings or illustrations from the internet, how goods and hospital lifts operate and where they are suitable.
7. Suggest some probable future developments that may occur for mechanical transportation systems used in buildings, giving reasons.
8. Discuss the statement, 'Having machines to move people through buildings contributes significantly to obesity and lack of physical fitness.'
9. Compare the use of primary energy to operate mechanical transportation systems used in buildings with alternatives and comment on the differences.
10. Discuss the statement, 'Having tall buildings where vertical travel can only be accomplished by consuming primary energy, is not sustainable.'
11. Explain with the aid of sketches, drawings or illustrations from the internet, how hydraulic lifts operate and where they are suitable.
12. Visit a lift motor room in a large building, with the permission of the building manager. Photograph, report and draw a detailed description of the installation, room space dimensions, control cabinets, motor and drive arrangement, ventilation and temperature control systems.
13. Use a lift system in a building where you have permission to enter, and measure the typical round trip time for a journey from the entrance foyer to the top of the building and back to the foyer. Record the number of stops, passengers carried, type of lift system and time taken by the door systems. Comment on the quality of service provided.
14. Visit an escalator system, with the permission of the building manager if it is not a public facility. Photograph (if possible), report and draw a detailed description of the installation, space dimensions and suitability for the application. Comment on the quality of service provided.
15. Sketch, draw or illustration from the internet and publications, how drive systems for mechanical transportation systems work. Clearly show their principles of operation, how they use primary energy and where they impose loads on the building structure.
16. List the builders' work associated with lift installations. Show by means of sketches or drawings where these items will be located and how they will be performed by the construction team. At what stage of the building will the lift installation be constructed and how will it be achieved?
17. Which are correct about escalators?

    1. Run continuously.
    2. Reversible in flow direction.
    3. Stop and start frequently to save energy.

4. Require electrical drive.
5. Provide sloping and horizontal travel.

18. Which are correct about lifts?

1. Lift capacity is designed to move the whole population of the building in or out within a 10-minute period.
2. Number and speed of lifts in an office building are designed to match the likely inflow of people into the building.
3. Need negligible maintenance.
4. Lift cars have to be air-tight.
5. Lift cars have an access hatch.

19. Quality of a lift service is determined from:

1. Number of passengers carried in each car.
2. Highest car speed attainable.
3. Round trip time for one lift divided by the number of lifts available.
4. Passenger comfort.
5. Quietness of the lift system.

20. Which are correct about lift motor rooms?

1. There aren't any, all components are within the lift shafts.
2. The motor sits on top of each car and has a pulley at the top of the shaft.
3. Electrical lift motor room is situated at the top of each lift shaft.
4. Hydraulic lift motor room is at the base of the lift shaft.
5. Lift motor rooms are always halfway up the lift shaft.

21. Which is correct about lift and escalator installations?

1. Lift systems can be added to the building design after the important design decisions have been agreed.
2. Escalators are just motorized stairways and do not need much design work.
3. Lift systems are an afterthought in the overall concept of the building.
4. Transportation systems do not cost much.
5. Mechanical movement systems for the building's occupants are critical to the success of the project.

22. Which is correct about transportation systems?

1. Allow people to become unfit, we should walk.
2. Are not essential for several categories of buildings and building user.
3. Allow rapid movement around large sites.
4. Do not need much energy to operate.
5. Need no maintenance.

23. Where could a double deck lift system apply?

1. Combined goods and passenger lift cars.
2. Commercial buildings of around 20 storeys.
3. High speed long travel lifts.
4. Low speed short travel lifts.
5. Hospitals.

17 Question bank

Learning objectives

Study of this chapter will enable the reader to:

1. practise answering short questions from a multiple choice of answers;
2. prepare for tests, assignments and written examinations where this form of questioning is provided;
3. evaluate why some answers are not entirely correct for the question;
4. have discussions with peers and instructors over the meaning of the incorrect answers;
5. lead to further study and investigation;
6. answer a wide range of questions from the sixth edition;
7. stimulate the enjoyment of learning as much as the author had in preparing the sixth edition.

Key terms and concepts

acronyms 351; benchmark 353; *Carbon Plan* 351; CCS 352; CO_2 351; data transmission 352; design 352; emissions 352; energy 351; ETS 351; green 354; greenhouse 354; heat transfer 352; humidity 353; hydrocarbons 353; insulation 352; inverter 351; Kelvin 354; low energy building 354; obesity 351; odours 353; PPD 354; refrigeration 354; sustainability 355; tax 351; temperature 354; thermal comfort 354; zero carbon 352.

Introduction

This a general knowledge section of questions covering topics within the sixth edition. Some questions may require the reader to look up answers in additional resources or search the internet. There is only one correct answer to each question unless specified as having more. Incorrect answers may be partially true but not considered by the author to be the entirely correct response for the purpose of this book; these may stimulate additional study, discussion, questioning with peers or the instructor.

Question bank

1. Which is correct about International Energy Agency data?

    1. Shows that all countries are reducing emissions.
    2. Advertises good countries.
    3. Rates countries according to their energy consumption.
    4. Shows that global CO_2 emissions are on target for compliance with the Kyoto Protocol.
    5. Demonstrates what is really happening with world emissions.

2. What does inverter mean?

    1. Alternating current phases are reversed.
    2. It is an electronic soft starter for a three phase motor.
    3. Incoming 50 Hz alternating current is digitally reformed into an output frequency to a motor in the range from 0 to 20000 Hz.
    4. Incoming 50 Hz alternating current is digitally reformed into an output frequency to a motor in the range from 0 to 50 Hz.
    5. Alternating current is electronically converted into direct current to drive a motor.

3. Does taxing hydrocarbons reduce global CO_2 emissions?

    1. Of course it does, proven during recent 100 years of industrialization.
    2. Forces the design of modest energy-using buildings.
    3. Reduces personal transportation emissions.
    4. Not done so yet.
    5. Taxation is a punishment for emitters.

4. Which of these acronyms is correct?

    1. ASHRAE means Australian Society for Heating, Refrigerating and Air Engineering.
    2. AIRAH means American Institute for Refrigeration and Air Heating.
    3. CIBSE stands for The Chartered Institution of Building Services Engineers.
    4. BSRIA stands for British Services Refrigeration Institute for Air Conditioning.
    5. CIC is the Council for Industry and Construction.

5. Which does the *HM Government Carbon Plan, 2011* do?

    1. Has little effect on gas and petroleum consumption.
    2. Diminishes energy use in buildings.
    3. Prescribes how buildings are to be insulated and operated.
    4. Is an alternative to the EU ETS.
    5. Encourages reducing CO_2 emissions.

6. Do air conditioning and lift systems lead to obesity?

    1. Yes, absolutely.
    2. Obesity is due to overeating only.
    3. Obesity is due to working at computers too long each day.
    4. Metabolism slows and we eat more in air-conditioned environments.
    5. Modern facilities tend to reduce our physical activity and we eat less natural food.

7. Which is correct for the level of CO_2 emissions from the UK?

    1. 500 mega tonnes per year.
    2. 500 tonnes per year.
    3. 500 giga tonnes per year.
    4. There aren't any.
    5. 500 million kilograms per year.

8. Which of these is not a common standard for data transmission?

    1.   Ethernet.
    2.   RS484.
    3.   RS232.
    4.   RS124.
    5.   C-bus.

9. Which describes a zero carbon building?

    1.   Supplied from solar power systems.
    2.   Forested timber, no glass, naturally heated and ventilated.
    3.   Consumes a minimum amount of energy for all uses.
    4.   Net exporter of electricity.
    5.   Probably no such thing.

10. Which correctly describes heat transfer?

    1.   Sensible heat transfer comprises all types.
    2.   Latent heat transfer raises temperature.
    3.   Sensible heat transfer is logged by a thermocouple and thermistor.
    4.   Latent heat transfer is hidden from view.
    5.   Sensible heat transfer only takes place through conduction and convection.

11. Which is correct about CCS?

    1.   Centralized carbon separation.
    2.   Too expensive to contemplate.
    3.   Potential to return most of carbon from flue gas back to the earth.
    4.   Good idea from the oil and gas industry.
    5.   A political nightmare.

12. Economic thickness of thermal insulation is which of these?

    1.   Found by calculation of minimum total cost.
    2.   When graph of capital cost of additional thermal insulation become horizontal.
    3.   When a graph of energy savings value reaches a peak.
    4.   Always occurs when payback from energy cost savings reaches two and a half years.
    5.   Thickest amount the building designer can accommodate.

13. AR, BER, SER, DEC and iSBEM all relate to what?

    1.   Electrical engineering services and systems.
    2.   They are meaningless acronyms.
    3.   Types of zero carbon buildings.
    4.   Emission rating.
    5.   Types of energy audit.

14. Which is correct?

    1.   Building designers accurately calculate future energy use.
    2.   Architects do feedback studies of all their buildings.
    3.   New buildings always work perfectly as design predictions.
    4.   PROBE stands for probable recycling of built environment.
    5.   Only careful analytical review establishes how a new building achieves its objectives.

15. What does sustainability mean for low energy buildings?

    1.   No such thing as a modern sustainable building.
    2.   Everything used in the buildings service life comes from globally sustainable resources.

3. This building is an example of good modern design practice.
4. All waste output from this building is recycled.
5. The building has been constructed from organically grown materials.

16. How is air leakage by a building measured?

   1. Anemometer readings at every opening to outdoors.
   2. Fill with smoke and time taken to fully disperse.
   3. Cannot be done at all.
   4. Large fan sucks air out of whole building at −50 Pa and flow measured.
   5. Large fan pressurizes building at 50 Pa and flow measured.

17. Which are the energy benchmarks used?

   1. Best practice and normal for that building type.
   2. Zero energy and worst example.
   3. BREEAM zero score and LEED maximum allowed.
   4. Asset Rating scale A–G.
   5. A 1950 heated and naturally ventilated office and a fully air-conditioned tower office.

18. When designing the shape of a building:

   1. Maximize exposure to solar warming;
   2. Ignore location, make it impressive;
   3. Minimize solar overheating;
   4. Square in plan is always better for energy saving;
   5. Minimize external surface area.

19. What does BREEAM stand for?

   1. Building Rehabilitation Electrical Energy Alternative Methodology.
   2. Building Research Establishment Energy Audit Methodology.
   3. Building Recycling Energy Effectiveness Association Member.
   4. Brick Recycling Energy and Environment Assessment Method.
   5. Building Research Establishment Environmental Assessment Method.

20. We sense odours by:

   1. Identifying smells;
   2. Breathing onto others;
   3. A measuring instrument;
   4. Tasting them in our mouth;
   5. Olfactory response.

21. What are fluorinated hydrocarbons used for?

   1. Swimming pool water treatment.
   2. Biocide decontamination of cooling towers.
   3. Ozone-depleting refrigerants.
   4. Non-CFC foam insulation and furnishings.
   5. Combustible gaseous fuel.

22. Which is correct about the density of humid air?

   1. Decreases with increasing pressure.
   2. Increases with increasing air temperature.
   3. Varies with air temperature and pressure.
   4. Not affected by humidity.
   5. Increases as air velocity increases.

23. Satisfactory air quality may be deemed when:

    1. 100% of the full-time occupants are satisfied.
    2. 85% of the full-time occupants are satisfied.
    3. 50% of the full-time occupants are satisfied.
    4. Complaints cease.
    5. Odours have been eliminated.

24. What does Greenhouse Rating of a building stand for?

    1. The higher the greenhouse gas production due to the building, the higher the greenhouse rating.
    2. A 10-star building produces no greenhouse gases.
    3. Assessed greenhouse gas emission standard of a building.
    4. Any new building even if not painted green.
    5. An emission standard applied to all types of buildings.

25. Which is correct about Kelvin?

    1. Name of engineer who designed the first steam engine.
    2. Unit of heat.
    3. Measured in kJ/kg s.
    4. Temperature scale.
    5. Absolute temperature.

26. Which is correct about low energy buildings?

    1. A low energy building is one that requires the minimum amount of primary resource energy to build it.
    2. A low energy building may consume more energy to construct.
    3. A low energy building consumes less energy during its 100+ years of use that an equivalent building.
    4. We have no idea what an equivalent building is for a specific site.
    5. All building consume uncontrolled amounts of energy.

27. Thermal comfort PPD means?

    1. Personal preferences determined.
    2. Personal preferences determination.
    3. Has no meaning.
    4. Percentile people dissatisfied.
    5. Predicted percentage of dissatisfied people.

28. Which might be a means of reducing the refrigeration system energy usage in a small retail premises where food refrigeration, deep freezers and reverse cycle air conditioning are all needed?

    1. Install smallest capacity compressors possible.
    2. Carry out frequent maintenance checks and parts replacement.
    3. Switch reciprocating compressors off for as long as possible and maintain wide temperature differentials.
    4. Use same outdoor air cooled condenser for all three systems.
    5. Variable refrigerant volume scroll compressor with software-controlled digital operation programmable for all variations in year-round duties.

29. Which is correct about low energy building designs?

    1. Are always modern and look impressive.

2. Are always found to be ideally comfortable by users.
3. Must have large windows and glazed walling.
4. Must have small windows and high levels of thermal insulation.
5. Should consume a minimum of primary energy when compared with similar types and sizes of buildings.

30. What does sustainability mean for low energy buildings?

1. The mechanical and electrical services within this building all have a low maintenance requirement.
2. All the water, sewerage, paper and plastic waste output from this building go to recycling.
3. All the light bulbs and tubes from this building are recyclable.
4. Somebody has found a good argument why this design of building is less harmful to the global environment than competitive designs.
5. This building has consumed, and will continue to consume, more of the Earth's physical resources than it can ever put back.

31. And finally, what have I learnt from this study?

1. Nothing, it is all a fog to me!
2. Mechanical and electrical services within a building are not very important to the overall concept of the design and construction.
3. I can design or construct buildings; someone else must worry about the fiddly bits.
4. The building will work without the mechanical and electrical services anyway.
5. I now appreciate the importance and main features of the essential and desirable building services!

18 Understanding units

Learning objectives

Study of this chapter will enable the reader to:

1. practise answering short questions from a multiple choice of answers;
2. test own understanding of using units of measurement;
3. prepare for tests, assignments and written examinations where this form of questioning is provided;
4. evaluate why some answers are not entirely correct for the question;
5. have discussions with peers and instructors over the meaning of the incorrect answers;
6. lead into further study and investigation;
7. answer a range of questions relevant to the fifth edition and building services engineering work generally.

Key terms and concepts

acceleration due to gravity 357; atmospheric pressure 361; Celsius 362; density of water 358; electrical units 360; exponential 359; frequency 361; humid air 358; joule 360; Kelvin 362; Newton 360; pressure 361; pressure drop rate 363; specific heat capacity 358; standard atmosphere 357; Stefan Boltzmann 358; time 359; volume 363; watt 360.

Introduction

This a general knowledge section of questions covering measurement units. All questions should be understood by students of this topic area. There is only one absolutely correct answer to each question. Tackling these may stimulate additional study, discussion, questioning with peers or the instructor.

Questions

1. Which of these equals one standard atmosphere at sea level?

    1. 1.013 tonne/m^2.
    2. 1 bar.
    3. 10000 N/m^2.
    4. 1013.25 mb.
    5. 10^6 N/m^2.

2. Which of these equals one standard atmosphere at sea level?

    1. 1013 tonne/m^2.
    2. 10^5 bar.
    3. 10^9 N/m^2.
    4. 14.7 lb/in^2.
    5. 10^6 N/m^2.

3. Which of these equals one standard atmosphere at sea level?

    1. 1×10^5 Pascals, Pa.
    2. 1.01325×10^5 N/m^2.
    3. 1×10^4 N/m^2.
    4. 30 m H_2O.
    5. 1013.25 mm Hg.

4. Which of these equals one standard atmosphere at sea level?

    1. 9.807 m H_2O.
    2. 29.35 m H_2O.
    3. 10.3 m H_2O.
    4. 101325 kJ/m^2.
    5. 1.205 kg/m^2.

5. Which of these is the acceleration due to gravity, g?

    1. 10 m/s^2.
    2. 30 ft/s^2.
    3. 186000 miles per hour.
    4. gravity is static.
    5. 9.807 m/s^2.

6. Which of these describes the acceleration due to gravity, g?

    1. Calculated from a 1 kg weight freefalling from a height.
    2. Relative to distance from the Moon.
    3. Constantly 9.807 m/s^2.
    4. Varies with height above sea level.
    5. Inversely proportional to depth below sea level.

7. Which is the Specific Heat Capacity of air?

    1. Sensible heat content kJ/kg.
    2. Total heat content kJ/kg.
    3. 1.205 kJ/kg.
    4. 1.012 kJ/kg K.
    5. 4.186 kJ/kg K.

8. Which is the Specific Heat Capacity of air?

    1. Ratio of $\frac{Cp}{Cv}$.
    2. Cannot be defined.
    3. Varies with atmospheric pressure.
    4. 1.012 kg K/W.
    5. 1.012 kJ/kg K.

9. Which is the Specific Heat Capacity of water?

    1. 1.013 kW/kg K.
    2. 1.012 MJ/kg K.
    3. 4.186 kg K/kW.
    4. 4.186 kJ/kg K.
    5. 4.2 kg K/Kj.

10. Which is correct about the Specific Heat Capacity of water?

    1. Varies with water pressure.
    2. 4.19 kW s/kg K.
    3. A ratio.
    4. 1.102 kJ/kg K.
    5. Used to calculate the flow rate of heating and cooling system water.

11. Which is correct about the Stefan Boltzmann constant?

    1. Used to calculate convective heat transfer.
    2. 4.186 kJ/kg K.
    3. 1.012 kJ/kg K.
    4. 5.67×10^{-8} W/m^2 K^4.
    5. Combines convective and radiant heat transfer.

12. Which is correct about the density of humid air?

    1. Decreases with increasing pressure.
    2. Increases with increasing air temperature.
    3. Varies with air temperature and pressure.
    4. Not affected by humidity.
    5. Increases as air velocity increases.

13. Which is correct about the density of humid air?

    1. 4.186 kg/m^3 at 21°C, 60% relative humidity.
    2. 1.013 kg/m^3 at 20°C, sea level.
    3. 0.802 m^3/kg.
    4. 1.205 kg/m^3 at 20°C, 1013.25 mb.
    5. 5.67 kg/m^3.

14. Which is correct about the density of water?

    1. Always 10^3 kg/m^3.
    2. 1013.25 m^3/kg.
    3. 101325 kg/m^3.
    4. 1.205 MJ/m^3 K.
    5. 1000 kg/m^3 at 4°C.

15. Which is correct about the density of water?

    1. Cannot be measured.
    2. Cannot be measured accurately.
    3. Always relative to the specific gravity number.
    4. 1000 times that of air.
    5. Specific gravity is 1.0.

16. Which is correct about the density of water?

    1. 1.205×10^5 kg/m^3.
    2. 1 tonne/m^3 at 4°C.
    3. 1.012×10^3 kJ/m^3.
    4. 1.27 kJ/kg K.
    5. 100 g/cm^3.

17. Which is correct about the density of water?

    1. 1 g/cm^3.
    2. 1.2×10^3 kg/m^3.
    3. Specific gravity is 4.186.
    4. 1000 tonne/m^3 at 10°C.
    5. 1 kg per litre.

18. What does the exponential, e, mean?

    1. A logarithm.
    2. A variable number.
    3. Always 10^x, ten to the power x.
    4. 2.718.
    5. Has no meaning.

19. What does the exponential, e, mean?

    1. Something which is raised to a power.
    2. $e = 10^x$.
    3. The sum of an infinite series.
    4. $e = \sqrt{-1}$.
    5. $e^1 = 2.718$.

20. Which of these has the correct units?

    1. Mass is measured in kilos.
    2. 1 tonne $= 10^6$ kg.
    3. 1 kg $= 10^9$ mg.
    4. 10^3 kg/m^3 $= 1$ kg/l.
    5. 10^6 m $= 1$ km.

21. Which of these are not the correct units?

    1. 1 hour $= 3600$ seconds.
    2. 60 hours $= 3600$ minutes.
    3. 3.6×10^3 seconds $= 1$ hour.
    4. 1 year $= 8760$ hours.
    5. 1 hour $= 360$ seconds.

22. Which of these has the correct units?

    1. 1 Newton $= 1$ kg $\times$ 1 m/s^2.
    2. 1 Joule $= 1$ kg $\times$ 1 m.
    3. 1 Watt $= 1$ kg $\times$ g m/s^2.
    4. 10^3 Joules $= 3600$ kN/m^2.
    5. 1 Joule $= 1$ N/m^2.

23. Which of these has the correct units?

    1. 1 Joule $= 1$ Newton $\times$ 1 metre.
    2. 1 Joule $= 1$ Watt/s.
    3. 10^3 J $= 1$ kW/s.
    4. 1 Watt $= 10^3$ Joule/s.
    5. 1 MJ $= 10^3$ kW/s.

24. Which of these has the correct units?

    1. 1 W $= 1$ Nms.
    2. 1 W $= 1$ Js.
    3. 1 W/s $= 10^3$ J.
    4. 1 W $= 1$ Nm/s.
    5. 1 kW/h $= 10^3$ J/h.

25. Which of these has the correct electrical units?

    1. 1 MW $= 10^3$ W.
    2. 10^3 kJ $= 10^3$ kW/s.
    3. 1 Watt $= 1$ volt $\times$ 1 ampere.
    4. Electrical energy meters accumulate kW/h.
    5. 10^3 W $= 10^3$ V $\times$ 10^3 A.

26. Which of these has the correct electrical units?

    1. 10^3 kW/h $= 10^3 \times 3600$ W/s.
    2. kWh $=$ energy.
    3. 1 GJ $= 10^6$ V $\times$ 1 A.
    4. 1 MJ $= 10^6$ V $\times$ 1 A.
    5. 1 kJ/s $= 10^3$ V $\times$ 1 A.

27. Which of these has the correct electrical units?

    1. 10^3 W $= 10^3$ V $\times$ 1 A.
    2. 1 MWh $= 10^3$ Ws.
    3. 1 kWh $= 10^3$ Ws.
    4. 1 kWh $= 1000$ W/s.
    5. 1 kWh $= 1000$ W/h.

28. Which of these has the correct units?

    1. 1 atmosphere $= 10^3$ b.
    2. 1 Pascal $= 1$ N/m^2.
    3. Pascal is a unit of radiation measurement.
    4. 1 kN/m^2 $= 1$ b.
    5. 1 mb $= 10^3$ N/m^2.

29. Which of these has the correct pressure units?

    1.   1.01325 mb = 1 atmosphere.
    2.   1 MN = 10^3 kN/m^2.
    3.   1 b = 1 kN/m^2.
    4.   13.6 mb = 13.6 N/m^2.
    5.   1 b = 10^5 N/m^2.

30. Which of these has the correct pressure units?

    1.   1 mb = 1 N/m^2.
    2.   1 b = 10^3 mb.
    3.   1 mb = 10^3 N/m^2.
    4. 10^3 kN/m^2 = 1 b.
    5.   1 mb = 10^6 b.

31. Which of these has the correct pressure units?

    1.   1 Nm = 1 Pa.
    2.   1000 Pa = 1 atmosphere.
    3.   1 kPa = 1 kN/m^2.
    4.   1 Pa = 1 mb.
    5.   1 Pa = 1 N/m^2.

32. Which is the correct meaning for frequency?

    1.   Number of times an event is repeated.
    2.   Cyclic repetition of an event.
    3.   Number of complete rotations per unit time.
    4.   Statistical correlation.
    5.   Occasional reoccurrence.

33. Which is the correct meaning for frequency?

    1.   Alternating current rate of increase.
    2.   Electrical single or three phase.
    3.   Torque of a motor.
    4.   Air changes per hour.
    5.   Revolutions per minute.

34. Which is not correct in relation to frequency?

    1.   3000 RPM = 50 Hz.
    2.   1 Hz = 1 Nm/s.
    3.   High frequency fluorescent lamps work at 20000 Hz.
    4.   VFD means variable frequency drive.
    5.   60 Hz = 3600 RPM.

35. Which is correct about Kelvin?

    1.   Name of engineer who designed the first steam engine.
    2.   Unit of heat.
    3.   Measured in kJ/kg s.
    4.   Temperature scale.
    5.   Absolute temperature.

36. Which is correct about Kelvin?

    1.   Where absolute zero gravity starts.
    2.   Something to do with temperature.
    3.   First name of Dr. K. Celsius.
    4.   °C + 273.
    5.   Engineered the first closed circuit piped heating system.

37. Which is correct about Kelvin degrees?

    1.   Celsius scale plus 180.
    2.   Are always negative values of Celsius degrees.
    3.   Symbol K.
    4.   Awarded by Kelvin University, Peebles, Scotland.
    5. $K = C \times \dfrac{9}{5} + 32$.

38. Which is correct about Kelvin?

    1.   Name of a famous Scottish scientist.
    2.   Invented first bicycle in Scotland.
    3. $K = C + 273$.
    4.   Kelvin McAdam invented tarmacadam for road surfacing.
    5.   Degrees measured above absolute zero at $-180°F$.

39. Which is correct about Kelvin degrees?

    1.   Measurement of room air temperature.
    2.   Always used in heat transfer units.
    3.   Used to specify absolute temperature and temperature difference.
    4.   Fahrenheit plus 180.
    5.   Zero scale commences at $-40°C$.

40. Which is correct about Celsius?

    1.   Latin name of inventor of Roman hypocaust under floor heating system in 200 BC.
    2.   Fahrenheit minus 32.
    3. $C = F \times \dfrac{5}{9} + 32$.
    4. $C = 32 - F \times \dfrac{5}{9}$.
    5. $C = (F - 32) \times \dfrac{5}{9}$.

41. Which is correct about Celsius?

    1.   Called °C units.
    2.   Kelvin degrees plus 273.13.
    3. $C = (F + 32) \times \dfrac{5}{9}$.
    4.   Commonly used for cryogenic applications.
    5. $F = (C - 32) \times \dfrac{9}{5}$.

42. Which is correct about Celsius?

    1.   Temperature scale in the centimetre, grammes, second, CGS, metric system.
    2.   Name of the Roman Senator in 35 AD who stabbed Caesar.

3. $C = (F - 180) \times \dfrac{5}{9}$.

4.   Defines normal human body temperature of 98.4 degrees.

5.   Temperature scale in the metre, kilogramme, second, MKS, metric system.

43. Which is correct about volume?

1.   1 cubic centimetre water occupies 1 litre.
2.   1 tonne water occupies 1000 m^3.
3.   1 m^3 = 1000 litre.
4.   1 litre water weighs 100 kg.
5.   1 litre water weighs 10 kg.

44. Which is correct about volume?

1.   1 m^3 air weighs around 100 kg.
2.   1 m^3 air weighs around 10 kg.
3.   1 m^3 air weighs around 1 kg.
4.   1 litre occupies 1 m^2 area and 100 mm height.
5.   1 litre occupies 1 m^2 area and 10 mm height.

45. Which is correct about volume?

1.   1 litre water is contained in a cube of 100 mm sides.
2.   1 litre air is contained in a cube of 1000 mm sides.
3.   There is such a thing as a volume sensor for a control system.
4.   100 concrete blocks of 300 mm × 200 mm × 100 mm occupy a volume of 6 m^3.
5.   1 tonne water occupies 10 m^3.

46. A room 12 m long, 8 m wide and having an average height of 4 m, has a volume of?

1.   400 m^3.
2.   62 m^3.
3.   462 m^3.
4.   384 m^3.
5.   192 m^3.

47. Which is the correct length of a 1200 m^3 sports hall of average height 4 m and width 12 m?

1.   25 m.
2.   10 m.
3.   250 m.
4.   120 m.
5.   12.5 m.

48. What are the units for pressure drop rate in a pipeline?

1.   m head H_2O/m run.
2. N/m^2.
3.   mb/m.
4. N/m^3.
5. kN/m^3.

49. What does N/m^3 stand for?

1.   Nanometres per m^2 pressure drop per metre run of pipe.
2.   Neurons per cubic metre of room volume.

3.   Newtons per square metre pressure drop per metre run of pipe or duct.
4.   Newton per cubic metre is a density.
5.   Nano-particles of Radon gas per cubic metre of air in a building.

50. What does N/m^3 stand for?

1.   Normalized volumetric air change rate for a room.
2.   Number of people in a building divided by building volume.
3.   Volumetric coefficient.
4.   Noise rating divided by room volume.
5.   Pressure drop rate in a pipe or duct.

19 Answers to questions

1 Climate change

1. Regulated sites, power stations and industrial, do not know how much their surplus allowances will be worth when available to sell.
2. Because we design, build and maintain the plant and buildings that are final consumers of energy. Power station and industrial buildings are within our work.
3. 5
4. 4
5. 3
6. 3
7. 1, 3
9. 3
10. 4
11. 5
12. 2
13. By digging up hydrocarbons from the earth to power our inventions. The most powerful and influential nations have dug up the most, while those less well developed with air-conditioned buildings, transportation and homes, have yet to catch up by doing the same for themselves. The good thing about our mastery of the planet is that whatever amount of carbon we remove from the ground, combust and discharge what have been considered to be waste exhaust gases into the atmosphere, cannot escape from planet Earth; it remains in the atmosphere unless nature brings it back down to ground level. So, there is no change in the total quantity of carbon on planet Earth; it will always remain a constant amount here; that is, not counting those components we send out into space with rocket engines, never to return.

    Another good thing to note is that we have only scratched the surface of planet Earth by digging only a few kilometres or so down. Unknown reserves of valuable resources lie beneath our present attempts at gardening in the near surface. So what is the problem? Centuries of habitation have resulted in us depositing our waste gas straight into the atmosphere at ground level and out of high chimneys when it was considered unacceptable to release products of combustion from large fossil-burning plant too near to people. Dispersal of combustion products relied upon prevailing winds and nobody would be affected, would they? Well, yes, they are. Traffic exhausts and chimneys do have noticeable effects.

What else might have caused the observed growth of greenhouses gases in the atmosphere? Perhaps solar flares, world wars, rainforest clearance, changes in agriculture, burning off stubble, wild forest fires, volcano eruptions, melting polar ice caps, changes in rainfall and evaporation of seawater, plus any number of other suggestions? Who really knows? There are arguments in partial explanation of numerous theories.

14. Post-combustion capture of CO_2 occurs after the burning of fossil fuels. CO_2 is separated from the flue gas through a process called scrubbing. Flue gas is passed through a liquid which causes a chemical reaction and separates the CO_2 ready for transportation and storage. Post-combustion capture technology can be retrofitted to existing fossil fuel power stations.

15. Once pure CO_2 is captured through CCS technologies, it is compressed into a liquid state to a similar density to crude oil at 70 atmospheres and is transported via pipeline and stored safely offshore in depleted oil and gas fields.

16. 1

17. Specific heat load is 16.2 W/m^2 against an allowed 10 W/m^2. Incoming air from the heat exchanger is allowed to be electrically heated to 19°C to comply.

18. Regulated demands for heating and cooling 47.36 kW each, annual regulated load 102,717 kWh, unregulated electrical load 151,008 kWh, summer cooling load 110 kW, winter cooling load 15 kW. Internal heat gains exceed regulated heat loss.

19. Emission 4.7MtCO$_2$/yr, payable carbon tax on CO_2 emissions from the power station adds 5 cents/kWh to the cost of production, and that will be passed to distributors and final customers. A household that consumes 10,000 kWh per year will pay an additional €500. Power station operator could have €141M/yr to invest in CCS technology.

20. Capped allowance 2 MtCO$_2$, sale of allowances €28.8M but uncertain, sale of CO_2 uncertain value, might be viable.

21. No relationship as abatement is an engineering cost while allowances are only valued by their supply and demand to traders.

22. 5

23. Answer 1, 105% (14,618 in 1971 to 29,939 in 2008, it doubled.)

24. 5

25. 2, 4

26. 4

27. 1. An ancient standard of living; 2. Green building; 3. BREEAM; 4. Green building stars; 5. LEED.

28. 1. NABERS; 2. Zero carbon building; 3. Zero carbon building; 4. Zero carbon building; 5. Zero carbon building.

29. 1. Hyperthermia tower; 2. Sun-blinded office workers; 3. Net zero energy cost building; 4. Allowable technology minimizes energy consumption; 5. *Passivhaus*.

30. 1. Sunstroke problem; 2. Zero net source energy use building; 3. Net zero emission building; 4. Large-scale solar, wind and wave power providers; 5. Zero grid supply to the building.

2 Post occupancy

1. Designers' passion for the appearance of their building, the big picture, contrasts with the microscopic view of the users on how it functions to meet their own needs. Users' attention focuses on any dysfunction, technical fault and discomfort to themselves. They have no say in what the building looks like but they can report that the lift doors are slow, a tap leaks,

rain comes through the skylights because the motor controller and rain sensor are too slow, there are holes in the flooring, or they sit in a draught. Architects and engineers create impressive-looking buildings having the latest technology but the user lives with it for 25 years.

2. 3
3. 5
4. 2

3 Built environment

4. 17.9°C, 17.2°C, 19.3°C, 20.3°C
5. 13°C, 11.3°C, 15.7°C, 20°C
6. 16.8°C, 15.4°C, 19.2°C, 21.4°C
8. All
9. All
10. 2, 3, 4.

4 Energy economics

9. 3
13. 2
14. 1

5 Ventilation and air conditioning

1. From

$$Q = \frac{SH \text{ kW}}{t_r - t_s} \times \frac{(273 + t_s)}{357} m^3/s$$

$$357Q(t_r - t_s) = SH(273 + t_3)$$

$$357Qt_r - 357Qt_s = 273SH + SH \times t_s$$

$$357Qt_r - 273SH = SH \times t_s + 357Qt_s$$

$$357Qt_r - 273SH = t_s(SH + 357Q)$$

and

$$t_s = \frac{357Qt_r - 273SH}{SH + 357Q}$$

$$= \frac{357 \times 5 \times 23 - 273 \times 50}{50 + 357 \times 5} = 14.94°C$$

2. 0.793 m^3/s
4. 0.007469 kg H_2O/kg air
5. 1. no; 2. 21.2°C w.b., 0.877 m^3/kg; 3. 6.186 kW
6. 1. 4.25 m^3/s; 2. 4.86 m^3/s; 3. 87.45%; 4. 4.13 m^3/s; 5. 0.61 m^3/s
7. 20 air changes/h
14. 1680 mm × 930 mm

16. 2.68 m^3/s, 10.72 air changes/h, 0.0076 kgH$_2$O/kg air
17. t_s 28.6°C d.b., reduce supply air quantity to 1.7 m^3/s and use t_s, 30°C d.b. if the room air change rate will not be less than 4 changes/h
21. 15 air changes/h, 710 mm × 710 mm, 2 m^3/s fresh air, 2 m^3/s recirculated air, 3.6 m^3/s extract air, 4 m^3/s supply air-duct 0.4 m^3/s natural exfiltration

6 Heat demand

4. 2330.5 W.
5. 20.112 kW.
6. Allowed heat loss per degree Celsius difference inside to outside is 3746.8 W/K; thus the proposal complies. Proposed heat loss 3407 W/K.
7. 83.14 kW.
8. 43%.
12. R_{si} 0.1 m^{2}K/W, Q 50 W, U 2.78 W/m^{2}K, R_n 6.67 m^{2}K/W, 221 mm.
13. R_{si} 0.12 m^{2}K/W, Q 19.2 W, 114 mm, 120 mm used, U_n 0.29 W/m^{2}K, 17.4°C.
14. R_{si} 0.1 m^{2}K/W, Q 20 W, U 1.82 W/m^{2}K, extra R_a 0.18 m^{2}K/W; 81.75 mm, 90 mm used, U_n 0.23 W/m^{2}K, new Q 4.83 W, 15.5°C.
15. 4
16. 4
17. 5
18. 5
19. 5
20. 5
21. 5
22. 2
23. 2
24. 2
25. 3
26. 5
27. 4

7 Heating

11. 0.95 litre/s.
12. 2.4 m long × 700 mm high.
13. X 42 mm, Y 35 mm, Z 28 mm, radiator 1 22 mm, radiator 2 28 mm.
14. Expected internal temperature 26.5°C, system performance is satisfactory.
17. 4
18. 2
19. 3
20. 5
21. 2
22. 2
23. 3
24. 5
25. 3
26. 3

8 Water services

12. 13.44 kW
13. 0.56 h
14. 0.05 kg/s. 3.15 m head, pump C.
20. 5
21. 5
22. 3
23. 4
24. 2
25. 5
26. 1
27. 2
28. 4
29. 1
30. 2
31. 3
32. 1.28 h
39. 5
40. 3
41. 1
45. 3
46. 50 l/s, at least 3
48. Storage volume 2.4 m^3, one pit diameter 1.25 m

9 Electrical installations

 5. 0.00172 ohm.
 6. 0.2867 ohm.
 7. 10.7%.
10. 12.6 kVA.
11. 19.2 ohm.
12. 0.2857 mA.
14. 9716.8 kW.
15. 28.6 m.
16. 1. 18.25 kVA, 25.4 A; 2. £691.95.
26. Three earth rods give a total system resistance of 8.937 ohm.
27. 2, 4, 5
28. 5
29. 3
30. 1, 5
31. 4
32. 2, 4
33. 4, 5
34. 5
35. 1
36. 3
37. 5

38. 4
39. 1
40. 5
41. 2
42. 4
43. 1
44. 5
45. 2
46. 5
47. 3
48. 5
49. 5
50. 4
51. 4
52. 2
53. 2
54. 4
55. 2

10 Lighting

9. 59%.
11. Room index 3, *UF* 0.73, *MF* 0.9, 36 luminaires in 3 rows of 12 along the 20 m dimension, 16.8 W/m^2, 21 A.
22. Lighting 3750 h/yr, 3.81×10^6 lm, tungsten, 1814 lamps, replace 3401 per year, total annual cost £80255 per year, fluorescent 569 lamps, replace 178 per year, total annual cost £14403 per year, sodium 139 lamps, replace 22 per year, total annual cost £10,917 per year.
23. Lighting 1200 h/yr, 350685 lm, tungsten 352 lamps, replace 211 per year, total annual cost £3559 per year, fluorescent 95 lamps, replace 16 per year, total annual cost £772 per year, halogen 37 lamps, replace 2 per year, total annual cost £423 per year.
24. 1
25. 4
26. 1
27. 1, 3, 5
28. 3
29. 4
30. 2
31. 3
32. 4
33. 3, 5
34. 4
35. 3
36. 1
37. 2
38. 3
39. 5
40. 4
41. 2

42. 3
43. 3
44. 4
45. 2
46. 2
47. 3
48. 4
49. 5

11 Condensation in buildings

15. 1. 6.79°C, 6.11°C; 2. 13.89°C, 13.55°C, 2.89°C, 2.55°C; 3. 18.47°C, 17.83°C, 6.71°C, 2.91°C, 0.27°C; 4. 19.45°C, 19.06°C, −0.38°C.
16. 2.72°C.
17. −7.46°C.
18. 28.1 mm.
19. U value 0.46 W/m^2 K, heat flow 9.26 W/m^2. Thermal temperature gradient is 22°C, 21.07°C, 20.38°C, 18.71°C, 11.76°C, 3.08°C, 2.9°C, 2.7°C, 2°C. Indoor dew-point 11.3°C, vapour pressure 1300 Pa, outdoor air −0.8°C and 568 Pa. Vapour resistance R_v 6.265 GN s/kg, mass flow of vapour G 1.168×10^{-7} kg/m^2 s. Dew-points at the same interfaces as the thermal temperatures are 11.3°C, 11.3°C, 10.5°C, 10.5°C, 8.8°C, −0.7°C, −0.8°C, −0.8°C, −0.8°C. Condensation does not occur.
20. U value 0.6 W/m^2 K, heat flow 7.74 W/m^2. Thermal temperature gradient is 14°C, 13.07°C, 12.84°C, 11.32°C, 3.58°C, 2.19°C, 1.22°C, 1°C. Indoor dew-point 6.5°C, vapour pressure 936 Pa, outdoor air −1.8°C and 531 Pa. Concrete block work resistivity taken as 200 GN s/kg m. Vapour resistance R_v 25.35 GN s/kg, mass flow of vapour G 1.6×10^{-8} kg/m^{2}s. Dew-points at the same interfaces as the thermal temperatures are 6.5°C, 6.5°C, 6.3°C, 0.09°C, −0.06°C, −0.06°C, −1.8°C, −1.8°C. Condensation does not occur.
21. 3
22. 4
23. 5
24. 3
25. 4
26. 1
27. 3
28. 4
29. 3
30. 1
31. 3
32. 4

12 Gas

1. 1.026 litre/s, 125 mmH$_2$O.
2. 5.394 mb, 3.5 mb, 0.75 mb, 15 mb, 1050 mb.
3. 3.333 Pa/m.
4. 27.17 m.
5. 1.47 litre/s, 2.609 Pa/m, 32 mm.

6. 72 Pa.
11. 4, 5
12. 3
13. 1

13 Plant and service areas

10. 5
11. 3
12. 1
13. 4
14. 2
15. 3
16. 2
17. 4
18. 2
19. 5
20. 2

14 Fire protection

10. 2, 3, 5
11. 4
12. 2, 3
13. All
14. 3
15. All
16. 5
17. 4
18. 3
19. 5
20. 1
21. 5

15 Room acoustics

19. Reverberation time T 2.901 s at 125 Hz, 3.462 s at 250 Hz, 3.462 s at 500 Hz, 3.157 s at 1 kHz, 2.752 s at 2 kHz and 3.253 s at 4 kHz.
20. r 100 mm SPL 87 dB; r 1 m SPL 71 dB.
21. Directivity Q 2, r 0.5 m SPL 92 dB, reverberant SPL 79 dB.
22. 84 dBA.
23. Through the wall SPL_2 19 dB; through air vent 49 dB; open air vent causes noise to bypass the attenuation of the wall and may need acoustic louvers or an acoustic barrier.
24. Through the wall SPL_2 47 dB; through air vents in doors 59 dB; open air vent causes noise to bypass the attenuation of the wall; burner needs an acoustic enclosure.
25. SPL in roof is 32 dB; the large volume and short reverberation time assist in attenuating the plant room noise.
26. 33 dB.

27. 37 dB.
28. 39 dB.
29. See chapter explanation.
30. *NR* 40 is not exceeded in the room.
31. 1. *NR* 80; 2. *NR* 25, no intrusive noise from the chiller; 3. *NR* 45; 4. *NR* 20 when doors have equal sound reduction to the walls, have air-tight seals and are closed.
32. 1. *NR* 80; 2. 65 dB due to sound escape through door; 3. *NR* 35.
33. 1. *NR* 75; 2. *NR* 45, equivalent to the background noise level in a corridor; 3. *NR* 35; 4. *NR* 20, there is no intrusive noise.
34. 1. *NR* 60; 2. *NR* 40; 3. through the supply and return air ducts, noise radiation from the outer case of the fan coil unit, from the ceiling space through ceiling tiles, light fittings, noise break-in from the ceiling space into the supply and return air ducts and then into the office, structurally transmitted vibration from the fans, main air-handling plant noise through the outside air duct to the fan coil unit; 4. acoustic lining in the outdoor air, supply air and return air ducts, anti-vibration rubber mounts for the fan coil unit and the fan within it, acoustic lining within the fan coil unit, acoustic blanket above the recessed luminaires and above the ceiling tiles.
35. 3
36. 1
37. 3
38. 3
39. 5
40. 3
41. 3
42. 4
43. 5
44. 1
45. 4
46. 5
47. 5
48. 3
49. 5
50. 3
51. 5
52. 4
53. 4
54. 2
55. 1
56. 5
57. 3
58. 5
59. 2
60. 5

16 Mechanical transportation

17. 1, 2, 4, 5
18. 2, 5

19. 3
20. 3, 4
21. 5
22. 3
23. 2

17 Question bank

1. 5
2. 3
3. 4
4. 3
5. 5
6. 5
7. 1
8. 5
9. 5
10. 3
11. 3
12. 1
13. 4
14. 5
15. 3
16. 5
17. 1
18. 5
19. 5
20. 5
21. 3
22. 3
23. 4
24. 3
25. 5
26. 3
27. 5
28. 3
29. 5
30. 4
31. 5

18 Understanding units

1. 4
2. 4
3. 2
4. 3
5. 5
6. 3

7. 4
8. 5
9. 4
10. 5
11. 4
12. 3
13. 4
14. 5
15. 5
16. 2
17. 5
18. 4
19. 5
20. 4
21. 5
22. 1
23. 1
24. 4
25. 3
26. 2
27. 1
28. 2
29. 5
30. 2
31. 3
32. 3
33. 5
34. 2
35. 5
36. 4
37. 3
38. 3
39. 3
40. 5
41. 1
42. 5
43. 3
44. 3
45. 1
46. 4
47. 1
48. 4
49. 3
50. 5

References

Bowyer, A. (1979) Space allowances for building services: outline design stage, *Technical Note TN4/79*, Bracknell: Building Services Research and Information Association, available at: http://www.bsria.co.uk/.

BRE Digest (1992) *Interstitial Condensation and Fabric Degradation*, No. 369. Garston, Watford: Building Research Establishment, available at: http://www.bre.co.uk/.

British Gas (1980) *The Effect of Thermal Insulation in Houses on Gas Consumption*, London: Watson House.

Brundtland Commission of the United Nations General Assembly (1987), March 20, available at: http://www.un-documents.net/ocf-02.htm.

Building Regulations, Part L, Fuel and Power, available at: http://www.communities.gov.uk/planningandbuilding/buildingregulations/publications/.

Butler, H. (1979b) Space requirement for building services distribution systems: detail design stage, *Technical Note TN3/79*, Bracknell: Building Services Research and Information Association.

Chadderton, D.V. (1997a) *Air Conditioning: A Practical Introduction*, 2nd edn, London: E & FN Spon.

Chadderton, D.V. (1997b) *Building Services Engineering Spreadsheets*, London: E & FN Spon.

CIBSE (n.d.) *CIBSE Guide*, London: Chartered Institution of Building Services Engineers, available at: http://www.cibse.org/.

CIBSE (2004) *Energy Efficiency in Buildings Guide F*, 2nd edn, London: The Chartered Institution of Building Services Engineers.

CIBSE (2005a) *CIBSE Guide K Electricity in Buildings, 2005*, London: The Chartered Institution of Building Services Engineers.

CIBSE (2005b) *Guide B Heating, Ventilating Air Conditioning and Refrigeration, 2005*, London: The Chartered Institution of Building Services Engineers, and Sound Research Laboratories Limited.

CIBSE (2007) *CIBSE Guide C Reference Data, 2007*, London: The Chartered Institution of Building Services Engineers.

CIBSE (2010a) *CIBSE Guide E Fire Safety Engineering, 2010*, London: The Chartered Institution of Building Services Engineers.

CIBSE (2010b) *CIBSE Lighting Design Guide 5: Lighting for Education, 2011*, London: The Chartered Institution of Building Services Engineers.

CIBSE Journal, http://www.cibsejournal.com/.

CIBSE PROBE reports available for members to download from: https://www.cibse.org/membersservices/downloads/listings.asp?pid=373.

DECC (2011a) *HM Government Carbon Plan, 2011*, London: The Department of Energy & Climate Change, available at: http://www.decc.gov.uk.

DECC (2011b) *Digest of UK Energy Statistics* (DUKES) 2011, available at: http://www.decc.gov.uk/en/content/cms/statistics/publications/ecuk/ecuk.aspx.

Hulme, M. (2010) The Hartwell Paper, CIBSE Annual Lecture, 23 November, University of East Anglia.

Inventory of Carbon & Energy, Sustainable Energy Research Team, University of Bath, Bath, available at: http://www.bath.ac.uk/mech-eng/sert/

Jones, W.P. (1985) *Air Conditioning Engineering*, 3rd edn, London: Edward Arnold.

Productivity Commission (2011) *Productivity Commission 2011, Carbon Emission Policies in Key Economies*, Research Report, Canberra, Australia. Media and Publications, Productivity Commission, available at: maps@pc.gov.au.

Smith, B.J., Peters, R.J. and Owen, S. (1985) *Acoustics and Noise Control*, Harlow: Longman.

US Green Building Council, *Leadership in Energy and Environmental Design, LEED*, available at: http://www.usgbc.org/.

Other resources

A User Guide to iSBEM, version 4.1c March 2011, http://www.ncm.bre.co.uk/.

BRE Standard Assessment Procedure (SAP 2009), http://www.bre.co.uk/sap2009/page.jsp?id=1642.

BREEAM, UK Building Research Establishment Environmental Assessment Method, http://www.breeam.org/.

Carbon Dioxide Information Analysis Center, U.S. Department of Energy, Oak Ridge National Laboratory, Bethel Valley Road, P.O. Box 2008, Building 2040, Oak Ridge, Tennessee 37831-6290, USA, http://cdiac.ornl.gov/.

CIBSE Heritage Group, *The Quest for Comfort, Centenary 1897–1997, Centenary of the CIBSE Heritage Group*.

European Union Emissions Trading System, http://www.decc.gov.uk/en/content/cms/emissions/eu_ets/eu_ets.aspx.

Green Building Council, http://www.ukgbc.org/site/home.

Heating and Ventilating Contractors Association, HVCA Green Building of the Year award in 1995, http://www.hvca.org.uk/index.php.

International Energy Agency, 9, rue de la Fédération, 75739 Paris Cedex 15, France, CO_2 emission from fuel combustion and Kyoto Protocol targets, http://www.iea.org.

Kyoto Protocol 1997, http://unfccc.int/kyoto_protocol/items/2830.php.

Met Office, FitzRoy Road, Exeter, Devon, EX1 3PB, United Kingdom, http://www.metoffice.gov.uk/.

National Australian Built Environment Rating System, NABERS, NSW Office of Environment and Heritage, PO Box A290, Sydney South NSW 1232, Australia. http://www.nabers.com.au/.

Queens Building, De Monfort University, Leicester, DUALL Project, Probe 4 BSJ April 1996. http://duall.iesd.dmu.ac.uk/1010buildings/.

The Carbon Trust Ltd, 6th Floor, 5 New Street Square, London EC4A 3BF, United Kingdom, http://www.carbontrust.co.uk.

Trading Carbon, Thomson Reuters Point Carbon, Second floor, 102-108 Clerkenwell Road, London, EC1M 5SA, United Kingdom, http://www.pointcarbon.com/.

Index